D1268229

K. Bucher · M. Frey

Petrogenesis of Metamorphic Rocks

6th Edition
Complete Revision of Winkler's Textbook

With 93 Figures and 28 Tables

Springer-Verlag
Berlin Heidelberg New York
London Paris Tokyo
Hong Kong Barcelona
Budapest

Professor Dr. Kurt Bucher
Mineralogisch-Petrographisches Institut
Albertstr. 23 b
D-79104 Freiburg im Breisgau

Professor Dr. Martin Frey
Mineralogisch-Petrographisches Institut
Bernoullistr. 30
CH-4056 Basel

ISBN 3-540-57567-7 Springer-Verlag Berlin Heidelberg New York
ISBN 0-387-57567-7 Springer-Verlag New York Berlin Heidelberg

CIP data applied for

© Springer-Verlag Berlin Heidelberg 1994
Printed in Germany

Typesetting: Data conversion by Elsner & Behrens GmbH, Oftersheim
SPIN: 10010049   32/3130 – 5 4 3 2 1 0 – Printed on acid-free paper

Preface

Metamorphic rocks are one of the three classes of rocks. Seen on a global scale they constitute the dominant material of the Earth. The understanding of the petrogenesis and significance of metamorphic rocks is, therefore, a fundamental topic of geological education. There are, of course, many different possible ways to lecture on this theme. This book addresses rock metamorphism from a relatively pragmatic view point. It has been written for the senior undergraduate or graduate student who needs practical knowledge of how to interpret various groups of minerals found in metamorphic rocks. The book is also of interest for the non-specialist and non-petrologist professional who is interested in learning more about the geological messages that metamorphic mineral assemblages are sending, as well as pressure and temperature conditions of formation.

The book is organized into two parts. The first part introduces the different types of metamorphism, defines some names, terms and graphs used to describe metamorphic rocks, and discusses principal aspects of metamorphic processes. Part I introduces the causes of metamorphism on various scales in time and space, and some principles of chemical reactions in rocks that accompany metamorphism, but without treating these principles in detail, and presenting the thermodynamic basis for quantitative analysis of reactions and their equilibria in metamorphism. Part I also presents concepts of metamorphic grade or intensity of metamorphism, such as the metamorphic-facies concept. Also, a brief presentation of techniques and methods for estimating pressure and temperature conditions at which metamorphic rocks once were formed a long time ago can be found in Part I (geothermobarometry).

Part II deals systematically with prograde metamorphism of six different classes of bulk rock compositions. The categories of rock compositions treated cover most of the sedimentary, igneous and metamorphic precursor material that is commonly encountered. Some more exotic rock compositions, for example ironstones, evaporites, laterites and manganese-rich rocks are not covered by the book. Part II can, in principle, be used alone, provided that the reader is familiar with common types of petrologic phase diagrams. Part II makes extensive use of computed phase diagrams. The routine computation of phase diagrams is one of the major developments in petrology of the last few years. It was made possible by a continuous and tremendous effort by scientists in experimental laboratories, who have vastly expanded the database of material of geological interest through phase equilibria and thermochemical studies. Advanced mathematical and thermodynamic analysis of the experimental data resulted in the compilation of thermodynamic data of phase components of geologic substances and in the development of solution models for solid mineral solutions and gas mixtures at high pressures and temperatures. Finally, several computer software packages developed during recent years make the computation of phase diagrams convenient and easy. Computation of phase diagrams is becoming increasingly widespread among geologists, and as the reader will see, this book in its present form could not have been written 5 years ago.

In general, the book focusses on the interpretation and significance of mineral assemblages and chemical reactions in metamorphic rocks. Several important topics concerning metamorphism and metamorphic rocks are not covered by this book. These topics include metamorphic micro-structures and their interpretation, deformation and metamorphism, geodynamic aspects of metamorphism and the thermodynamic basis for the quantitative treatment of chemical reactions in rocks.

The references are cited at the end of each chapter, the lists of references include a selection of important papers and books on the subject covered by the respective chapter that are highly recommended for further reading. The bibliography includes classical papers in the field as well as more recent contributions that may help to provide access to research in specialized fields via the references cited in the listed research (e.g. granulite-facies metamorphism of pelitic rocks etc.).

Many of the phase diagram figures show numbered reaction equilibria that are formulated and listed in tables. Reaction numbers are also frequently used in the text. We recommend copying the tables with the reaction stoichiometries and keeping them on hand, which will help eliminate the need to flip back and forth between text and reaction tables.

We also strongly recommend regular reading of current issues of scientific journals that arrive at your library. In the field of metamorphic petrology, the Journal of Metamorphic Geology is essential reading and a few of the other particularly relevant journals are Journal of Petrology, Contributions to Mineralogy and Petrology, American Mineralogist, European Journal of Mineralogy, Mineralogical Magazine, Lithos, Chemical Geology and Earth and Planetary Science Letters.

Freiburg and Basel   Kurt Bucher and
(March 1994)   Martin Frey

Acknowledgements

KB started the work on this book during his Oslo period and it was completed at the University of Freiburg. During the process of writing this book, parts of the text were taught to student courses at the University of Oslo and Freiburg. The patience of the students in following the lectures at various stages and their helpful comments are gratefully acknowledged.

KB thanks the Institute für Mineralogie und Petrographie, ETH Zürich, where Chapter 5 was written. Chapter 9 was written during another sabbatical that was partly spent at Johns Hopkins University. MF thanks the Department of Geology, Stanford University, where Chapter 4 was written during a sabbatical.

KB also wishes to thank John Schumacher for help and useful comments and suggestions. KB is indebted to Ernie Perkins and James Connolly, who provided various versions of their phase diagram software and showed great interest in the book project. MF is grateful to Steven Bohlen for reviewing Chapter 4, to Christian de Capitani for help in computing phase diagrams of Chapter 8 and for reviewing Chapter 8, and to Meinert Rahn for drafting of the figures of Chapters 1, 4, 8 and 10.

One of the authors, Kurt Bucher, has also published under his extended name Bucher-Nurminen but found it more convenient to simplify his name for the textbook and he will continue this practice in future publications. We hope this will not be the cause of much confusion.

Acknowledgements

Contents

Part I Basic Principles

1 Definition, Limits and Types of Metamorphism

The definition given below follows that given by the IUGS subcommission on the systematics of metamorphic rocks.

Metamorphism is the process leading to changes in the mineralogy and/or structure and/or chemical composition in a rock. These changes are due to physical and/or chemical conditions that differ from those normally occurring in the zone of weathering, cementation and diagenesis. They may include partial melting as long as the bulk of the rock remains in a solid state. If change in bulk composition is the dominant metamorphic process, the term metasomatism is applied.

1.1 Limits of Metamorphism

The limits of metamorphism are discussed with respect to two important physical variables, i.e. temperature and pressure.

Low-Temperature Limit of Metamorphism

Temperatures at which transformations set in are strongly dependent on the material under investigation. Important transformations of evaporites, of vitreous material and of organic material, for example, begin to take place at considerably lower temperatures than such transformations of most silicate and carbonate rocks. This book is not concerned with the transformation of organic material, i.e. coalification (maturation). For a recent review on this subject the reader is referred to Teichmüller (1987).

Correlations between the rank of coalification and mineralogical changes in slightly metamorphosed sediments and volcanic rocks are reviewed by Kisch (1987).

In many rocks, phase transformations begin shortly after sedimentation and continue to take place with increasing burial. Whether such transformations are called diagenetic or metamorphic is largely arbitrary. However, many authors would agree that in silicate rocks the low-temperature limit of metamorphism is around $150 \pm 50\,°C$. The first appearance of the following minerals is taken to indicate the beginning of metamorphism: Fe-Mg-carpholite, glaucophane, lawsonite, paragonite, prehnite, pumpellyite or stilpnomelane. Note, however, that these minerals may also be found as detrital grains in unmetamorphosed sediments. In this case, textural evidence from thin section will distinguish between a neoformation or a detrital origin. For a more detailed discussion of problems dealing with the low-temperature limit of metamorphism and its delimitation from diagenesis the reader is referred to Frey and Kisch (1987).

High-Temperature Limit of Metamorphism

According to the definition given above, the beginning of melting is included in the field of metamorphism as long as rocks remain predominantly in their solid state. The experimental determination of the melting temperature of rocks yields, therefore, an estimate for the high-temperature limit of metamorphism. These melting temperatures are strongly dependent on pressure, rock composition and the amount of water present. As an example, assume a pressure of

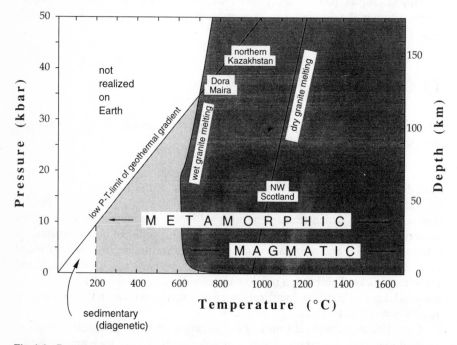

Fig. 1.1. Pressure (p)-temperature (T) range of metamorphic processes. The boundary between diagenesis and metamorphism is gradational, but a T value of 200 °C is shown for convenience. Note that the metamorphic field has no upper p and T limit on this diagram, and that there is a large overlap for metamorphic and magmatic conditions. Extreme pT conditions are shown for the following areas: (1) Scourian granulites from NW Scotland (Lamb et al. 1986, Fig. 3); (2) Pyrope-coesite rocks from Dora Maira, western Alps (Schertl et al. 1991); (3) Diamond-bearing metamorphic rocks from the Kokchetav massif, northern Kazakhstan. The *wet granite melting curve* is after Stern and Wyllie (1981), and the *dry granite melting curve* is after Newton (1987, Fig. 2)

5 kb. In the presence of an aqueous fluid, granitic rocks begin to melt at a temperature of ca. 660°C, while basaltic rocks need a much higher temperature of ca. 800°C (Fig. 1.1). Under dry conditions, these temperature values are raised considerably, that is ca. 1000°C for granite and ca. 1120°C for basalt. The highest temperatures recorded from metamorphic rocks of crustal origin by geothermometric methods are 1000–1100°C (e.g. Lamb et al. 1986, Fig. 3). However, metamorphism is not restricted to the Earth's crust. A given volume of rock in the convecting mantle continuously undergoes metamorphic processes such as recrystallization and various phase transformations in the solid state at temperatures in excess of 1500°C.

Summing up, the high-temperature limit of crustal metamorphism may be estimated at about 650–1100°C depending on rock composition, whilst this limit is at much higher temperatures for the Earth's mantle.

Low-Pressure Limit of Metamorphism

If a magma rises towards the surface, metamorphism in contact aureoles may occur at near-surface pressures of a few bars.

High-Pressure Limit of Metamorphism

For a long time it was believed that maximum pressures in metamorphic crustal rocks did not exceed 10 kb, which corresponds to the lithostatic pressure at the base of a normal continental crust with a thickness of 30–40 km. As better calibrations became available, it was found that mineral assemblages in eclogites often recorded pressures of 15–20 kb. Recent discoveries of pyrope-coesite rocks indicate that the pressure range has to be extended to at least 30 kb (Chopin 1984). However, as mentioned above, metamorphism should not be considered as being confined to the Earth's crust. As an example, many garnet-peridotites (or garnet-olivine-pyroxene-granofelses) from ophiolite complexes or from xenoliths in kimberlites record pressures of > 30–40 kbar.

1.2 Types of Metamorphism

On the basis of geological setting, we distinguish between metamorphism of local and regional extent:

1. Regional extent
   Orogenic metamorphism
   Ocean-floor metamorphism
   Burial metamorphism

2. Local extent
   Contact metamorphism
   Cataclastic metamorphism
   Impact metamorphism
   Hydrothermal metamorphism

Such a subdivision is certainly desirable, but it should be kept in mind that transitional forms often exist between these genetic categories of metamorphism.

1.2.1 Orogenic Metamorphism

Following a suggestion by Miyashiro (1973, p. 24), this name is preferred over the commonly used term regional metamorphism or a synonymous designation like dynamothermal metamorphism (Holmes 1920; Winkler 1974, 1976, 1979). **Regional metamorphism** is used here as a general term for metamorphism occurring over a wide area, as opposed to local metamorphism, and includes orogenic, burial and ocean-floor metamorphism.

Orogenic metamorphism is characteristic of orogenic belts where deformation accompanies recrystallization. Such metamorphic rocks in general exhibit a penetrative fabric with preferred orientation of mineral grains, including phyllites, schists and gneisses. Orogenic metamorphism appears to be a long-lasting process of millions or tens of millions of years' duration, including a number of phases of crystallization and deformation. Individual deformational phases appear to have definite characteristics, like attitude and direction of schistosities, folds and lineations. Therefore, several deformational phases may possibly be put into a time sequence by field work. Microscopic observations may unravel the relationships between structural features and mineral growth, and establish the time relations of movement and metamorphism (e.g. Vernon 1976, p. 224).

Rocks subjected to orogenic metamorphism usually extend over large belts, hundreds or thousands of kilometres long and tens or hundreds of kilometres wide.

Some features of orogenic metamorphism are summarized in Table 1.1. Rocks produced by regional orogenic and local contact metamorphism differ significantly in their fabric, being schistose as opposed to non-schistose, respectively. Furthermore, orogenic metamorphism takes place at higher pressures, while the temperature regime is often the same in both cases.

In many higher temperature parts of orogenic metamorphic belts, syn- or late-tectonic granites are abundant. The relationship between orogenic metamorphism and these granitic masses is in dispute. Did these granites carry heat and thereby contribute to the rise of temperature, or were these granitic rocks generated by incipient melting as a consequence of the relevant metamorphism?

Table 1.1 Comparison of orogenic, ocean-floor and contact metamorphism

Type of metamorphism	Orogenic	Ocean-floor	Contact
Geologic setting	In orogenic belts, extending for several 1000 km^2	In oceanic crust and upper mantle, extending for several 1000 km^2	Proximity to contacts to shallow level igneous intrusions; contact aureole of a few m up to some km width
Static/dynamic regime	Dynamic; generally associated with polyphase deformation	+/− static, fracturing, but no penetrative foliation	Static, no foliation
Temperature Lithostatic pressure	150–1100°C 2–30 kbar for crustal rocks	150–500°C (or higher) <3 kbar	150–750°C From a few hundred bars to 3 kbar
Temperature gradients	5–60°C/km (vertical)	50–500°C/km (vertical or horizontal)	100°C/km or higher (horizontal)
Processes	Lithospheric thickening, compression and heating associated with subduction, followed by thermal relaxation	Heat supply by ascending asthenosphere at mid-ocean ridges, combined with circulatation of sea water through fractured hot rocks	Heat supply by igneous intrusions
Typical metamorphic rocks	Slate, phyllite, schist, gneiss, migmatite, marble, quartzite, greenschist, amphibolite, blueschist, eclogite, granulite	Metabasalt, greenstone, metagabbro, serpentinite; original structure often well preserved	Hornfels, marble, calcsilicate, granofels, skarn

1.2.2 Ocean-Floor Metamorphism

This type of regional metamorphism was introduced by Miyashiro et al. (1971) for transformations in the oceanic crust in the vicinity of mid-ocean ridges. The metamorphic rocks thus produced are moved laterally by ocean-floor spreading, covering large areas of the oceanic crust. The ocean-floor metamorphic rocks are mostly of basic and ultrabasic compositions. As most of these rocks are non-schistose, ocean-floor metamorphism resembles continental burial metamorphism (see below). A further characteristic of ocean-floor metamorphic rocks is their extensive veining, produced by the convective circulation of large amounts of heated sea water. This process leads to chemical exchanges between rocks and sea water; in this respect, ocean-floor metamorphism resembles hydrothermal metamorphism (see below).

Some features of ocean-floor metamorphism are stated briefly in Table 1.1.

1.2.3 Other Types of Regional Metamorphism

Burial metamorphism was introduced by Coombs (1961) for low temperature regional metamorphism affecting sediments and interlayered volcanic rocks in a geosyncline without any influence of orogenesis or magmatic intrusion. Lack of schistosity is an essential characteristic of resultant rocks. This means that original rock fabrics are largely preserved. Mineralogical changes commonly are incomplete, so that the newly generated mineral assemblage is intimately associated with relict mineral grains inherited from the original rock. Burial metamorphism in fact merges into and cannot be sharply distinguished from deep-seated diagenesis. Metamorphic changes are often not distinguishable in hand specimens; only in thin sections can they be clearly recognized.

Well-known examples of burial metamorphism include those described from southern New Zealand (e.g. Kawachi 1975; Boles and Coombs 1977), eastern Australia (e.g. Smith et al. 1982), Japan (e.g. Seki 1973), and Chile (e.g. Levi et al. 1989).

Diastathermal metamorphism is a new term recently proposed by Robinson (1987) for burial metamorphism in extensional tectonic settings showing enhanced heat flow. A possible example of this type of regional metamorphism was described from the Welsh Basin (Bevins and Robinson 1988).

1.2.4 Contact Metamorphism

This type of metamorphism takes place in heated rocks in the vicinity of contacts with intrusive or extrusive igneous bodies. Metamorphic changes are effected by the heat and materials emanating from the magma and sometimes by deformation connected with the emplacement of the igneous mass. The zone of contact metamorphism is termed a **contact aureole**. The width of contact metamorphic aureoles varies, but is in most cases in the range of several metres to a few kilometres.

In detail, the width of aureoles depends upon the volume, nature and intrusion depth of a magmatic body as well as upon the properties of the country rocks, especially their fluid content and permeability. A larger volume of magma carries more heat with it than a smaller one, and the temperature rise in the bordering country rock will last long enough to cause mineral reactions. The rocks adjacent to small dikes, sills or lava flows are barely metamorphosed, whereas larger bodies of igneous rocks give rise to a contact aureole of metamorphic rocks.

The nature of a magmatic body determines its magma temperature. As an example, basaltic magmas form at temperatures well over 1000°C. In contrast, granitic melts may form at temperatures several hundred degrees lower.

The intrusion depth of a magmatic body determines the thermal gradient and heat flow between the hot intrusive contact and the country rock. Marked thermal gradients are generally confined to the upper 10 km of the Earth's crust because, at deeper levels, the country rocks are already rather hot and hence marked thermal aureoles are not produced.

The effects of contact metamorphism are most obvious where sedimentary rocks, especially shales and limestones, are in contact with large magmatic bodies. On the other hand, rocks previously subjected to medium- or high-temperature metamorphism commonly will not show signs of contact metamorphic overprinting because the mineral assemblages are persistent at contact metamorphic conditions.

Contact metamorphic rocks are generally fine-grained and lack schistosity. The most typical example is called hornfels (see Chap. 2 for nomenclature); but foliated rocks such as spotted slates and schists are occasionally present.

Pyrometamorphism is a special kind of contact metamorphism. It shows the effects of particularly high temperatures at the contact of a rock with magma under volcanic or quasi-volcanic conditions, e.g. in xenoliths. Partial melting is common, and in this respect pyrometamorphism may be regarded as being intermediate between metamorphism and igneous processes.

1.2.5 Other Types of Small-Scale Metamorphism

Cataclastic metamorphism is confined to the vicinity of faults and overthrusts, and involves purely mechanical forces causing crushing and granulation of the rock fabric. Experiments show that cataclastic metamorphism is favoured by high strain rates under high shear stress at relatively low temperatures. The resulting cataclastic rocks are non-foliated and are known as fault breccia, fault gauge or pseudotachylite. A pseudotachylite consists of an aphinitic groundmass that looks like black basaltic glass (tachylite). Note that mylonites are no longer regarded as cataclastic rocks, because some grain growth by syntectonic recrystallization and neoblastesis is involved (see, e.g. Wise et al. 1984).

Dislocation or dynamic metamorphism is sometimes used as synonym for cataclastic metamorphism; but these terms were initially coined to represent what is now called regional metamorphism. In order to avoid misunderstanding, the name cataclastic metamorphism will be used here.

Impact metamorphism is a type of shock metamorphism (Dietz 1961) in which the shock waves and the observed changes in rocks and minerals result from the hypervelocity impact of a meteorite. The duration is very short, i.e. a few microseconds. Mineralogical characteristics involve the presence of shocked quartz and the neoformation of coesite and stishovite as well as minute diamonds. A recent review on impact metamorphism is provided by Grieve (1987) and Bischoff and Stöffler (1992) and is not considered further in this book.

Hydrothermal metamorphism was a term coined by Coombs (1961). Here, hot solutions or gases have percolated through fractures, causing mineralogical and chemical changes in the neighbouring rocks. This process is particularly relevant to problems of ore genesis, rock alteration and geothermal energy. In addition, hydrothermal metamorphism is also of academic interest, because active geothermal fields make possible the study of present-day metamorphism. Here, temperatures, pressures and fluid compositions can be directly measured in bore holes. This contrasts to the usual study of metamorphic rocks, where we are looking at the cool and dry final product of rock transformation, and where temperature, pressure and fluid composition must be derived indirectly through the study of mineral assemblages (as will be detailed in later chapters).

Examples of hydrothermal metamorphism in active geothermal areas are known from California, Iceland, Japan and New Zealand.

References

Bevins RE, Robinson D (1988) Low grade metamorphism of the Welsh Basin Lower Paleozoic succession: an example of diastathermal metamorphism? J Geol Soc London 145:363–366

Bischoff A, Stöffler D (1992) Shock metamorphism as a fundamental process in the evolution of planetary bodies: Information from meteorites. Eur J Mineral 4:707–755

Boles JR, Coombs DS (1977) Zeolite facies alteration of sandstones in the Southland Syncline, New Zealand. Amer J Sci 277:982–1012

Chopin C (1984) Coesite and pure pyrope in high-grade blueschists of the western Alps: a first record and some consequences. Contrib Mineral Petrol 86:107–118

Coombs DS (1961) Some recent work on the lower grades of metamorphism. Austr J Sci 24:203–215

Dietz RS (1961) Astroblemes. Scientific American 205:50–58

Frey M, Kisch HJ (1987) Scope of subject. In: Frey M (ed) Low Temperature Metamorphism. Blackie, Glasgow, pp 1–8

Grieve RAF (1987) Terrestrial impact structures. Annual Rev Earth Plan Sci 15:245–270

Holmes A (1920) The nomenclature of petrology. 1st ed. Thomas Murby, London. 284 pp

Kawachi Y (1975) Pumpellyite-actinolite and contiguous facies metamorphism in part of Upper Wakatipu district, South Island, New Zealand. N Z J Geol Geophys 18:401–441

Kisch HJ (1987) Correlation between indicators of very-low-grade metamorphism. In Frey M (ed) Low Temperature Metamorphism. Blackie, Glasgow, pp 227–300

Lamb RC, Smalley PC, Field D (1986) P-T conditions for the Arendal granulites, southern Norway: implications for the roles of P, T and CO_2 in deep crustal LILE-depletion. J metamorphic Geol 4:143–160

Levi B, Aguirre L, Nyström JO, Padilla H, Vergara M (1989) Low-grade regional metamorphism in the Mesozoic-Cenozoic volcanic sequences of the Central Andes. J metamorphic Geol 7:487–495

Miyashiro A (1973) Metamorphism and Metamorphic Belts. Allen and Unwin, London, 492 pp

Miyashiro A, Shido F, Ewing M (1971) Metamorphism in the mid-Atlantic ridge near 24° and 30°N. Phil Trans Roy Soc London A268:589–603

Newton RC (1987) Petrologic aspects of Precambrian granulite facies terrains bearing on their origins. In: Kröner A (ed) Proterozoic Lithospheric Evolution. Am Geophys Union, Geodynamics Series 17:11–26

Robinson D (1987) Transition from diagenesis to metamorphism in extensional and collision settings. Geology 15:866–869

Schertl HP, Schreyer W, Chopin C (1991) The pyrope-coesite rocks and their country rocks at Parigi, Dora Maira Massif, Western Alps: detailed petrography, mineral chemistry and PT-path. Contrib Mineral Petrol 108:1–21

Seki Y (1973) Metamorphic facies of propylitic alteration. J geol Soc Japan 79:771–780

Smith RE, Perdrix JL, Parks TC (1982) Burial metamorphism in the Hamersley Basin, Western Australia. J Petrol 23:75–102

Stern CR, Wyllie PJ (1981) Phase relationships of I-type granite with H_2O to 35 kilobars: the Dinkey Lakes biotite-granite from the Sierra Nevada batholith. J Geophys Res 86(B11):10412–10422

Teichmüller M (1987) Organic material and very low-grade metamorphism. In: Frey M (ed) Low Temperature Metamorphism, Blackie, Glasgow, pp 114–161

Vernon RH (1976) Metamorphic Processes. Allen and Unwin, London, 247 pp

Winkler HGF (1974, 1976, 1979) Petrogenesis of Metamorphic Rocks. 3rd, 4th, and 5th edn., Springer Verlag, New York

Wise DU, Dunn DE, Engelder JT, Geiser PA, Hatcher RD, Kish SA, Odom AL, Schamel S (1984) Fault-related rocks: Suggestions for terminology. Geology 12:391–394

2 Metamorphic Rocks

This chapter deals with the descriptive characterization of metamorphic rocks. Metamorphic rocks are derived from other rocks of igneous, sedimentary or metamorphic origin. The chemical composition of this primary material (= protolith) largely controls the chemical and mineralogical composition of metamorphic rocks. The compositional variation found in the primary material of metamorphic rocks is reviewed in Section 2.1.

The structure of metamorphic rocks can be controlled by the structure of the precursor material. In low-grade metasedimentary rocks, for example, the sedimentary bedding and typical structures of sedimentary rocks such as cross-bedding and graded bedding may be preserved. Ophitic structure characteristic for basaltic lava may be found in mafic metamorphic rocks. Very coarse-grained structure of igneous origin can occasionally be found even in high-grade metamorphic rocks. Most metamorphic rocks, however, exhibit structures that are of distinct metamorphic origin. The structures typically result from the deformation of the rocks which is normally associated with large-scale tectonic processes causing metamorphism. Some descriptive terms for the illustration of metamorphic structures are defined in Section 2.2. Classification principles and nomenclature for metamorphic rocks are explained in Section 2.3.

The large scale tectono-thermal processes that generate metamorphic rocks move each volume element of rock along a unique path in the pressure-temperature-time (pTt) space. Rocks may undergo continuous recrystallization and new minerals may replace old ones in a complex succession. Earlier-formed minerals and groups of minerals often experience metastable survival because of unfavorable reaction kinetics (usually because absence of an aqueous fluid phase). In such rocks metamorphism may proceed episodically. It is of fundamental importance to the study of metamorphic rocks to correctly identify the group of minerals (the mineral assemblage) that may have coexisted in chemical equilibrium at one stage during the evolutionary history of the metamorphic rock. The total succession of mineral assemblages preserved in a metastable state in the structure of a metamorphic rock is designated mineral paragenesis. Some aspects regarding the mineral assemblage and the mineral paragenesis are discussed in Section 2.4.

Discussion and analysis of phase relationships in metamorphic rocks is greatly facilitated by the use of composition phase diagrams. Their construction is explained in the last section of this chapter (Sec. 2.5) about graphic representation of mineral assemblages. The quantitative computation of equilibrium composition phase diagrams is not discussed in this book, however.

2.1 Primary Material of Metamorphic Rocks

All metamorphic rock-forming processes make rocks from other rocks. The precursor rocks of the newly formed rocks during a given rock-forming process may be designated protoliths.

Metamorphism results to a large degree from the addition or removal of heat and matter to discrete volumes of the crust or mantle by tectonic or magmatic processes. Metamorphism, therefore, may affect all possible types of rock present in the Earth's crust or mantle. The chemical composition of the protoliths of metamorphic rocks produced during a specific metamorphic event of interest comprises all realizable chemical compositions of rocks. Precursor rocks of a given metamorphic cycle include the full spectrum of sedimentary, magmatic and metamorphic rocks.

Metamorphic processes also most often change the original composition of the protolith. Addition of heat to rocks typically results in the release of volatiles (H_2O, CO_2 etc.) that are stored in hydrates (e.g. micas, amphiboles), carbonates and other minerals containing volatile components. Therefore, the product rocks of prograde metamorphism are depleted in volatiles relative to the protoliths. Metamorphism that releases only volatiles to its surroundings is, somewhat illogically, termed isochemical. On a volatile-free basis, the chemical composition of protolith and product rock is identical in isochemical metamorphism. In truly isochemical metamorphism, protolith and product rocks are of identical composition including the volatile content. Isochemical metamorphism in such a strict sense is extremely rare.

Many, if not most, metamorphic processes also change the cation composition of the protolith. This type of metamorphism is termed allochemical metamorphism or metasomatism. The dehydration H_2O released during metamorphism may contain dissolved cations from the protolith rocks which are then carried away and lost by the rock system. It has been found, for example, that many granulite facies gneisses are systematically depleted in alkalis (Na and K) relative to their amphibolite facies precursor rocks. This can be explained by loss of alkalis during dehydration. Silica saturation is a general feature of nearly all metamorphic fluids. Pervasive or channelled fluid flow on a regional scale of silica-saturated dehydration fluids may strongly alter silica-deficient rocks (ultramafic rocks, dolomite marbles) with which this fluid comes in contact. Unique metamorphic rock compositions may result from metasomatism on a local or regional scale. Efficient diffusion and infiltration metasomatism require the presence of a fluid phase. Metasomatic processes are examples of fluid rock interaction processes that also are important in sedimentary environments. Interaction of rocks with externally derived fluids having a volatile composition that is not in equilibrium with the rocks' mineralogy is also very common and may also be referred to as allochemical metamorphism. Flushing of rocks with pure H_2O at high-pT conditions may initiate, for example, partial melting, formation of micas and amphibole from pyroxene-bearing rocks, production of wollastonite or periclase in marbles. CO_2 metasomatism is particularly common in very high-grade rocks. Metasomatism can create rocks of extreme composition which, in turn, may serve as protolith in subsequent metamorphic processes. Metasomatic rocks of extreme composition occur very widespread in regional metamorphic terrains and contact aureoles. However, the total volume of such rocks is subordinate. Although interesting petrologically, they do not need to be considered in a systematic discussion of prograde metamorphism of different bulk rock compositions (Part II).

2.1.1 Chemical Composition of Protoliths of Metamorphic Rocks

The average composition of crust and mantle is listed in Table 2.1. The mantle constitutes the largest volume of rocks on planet Earth. From geophysical and petrophysical evidence and from mantle fragments exposed on the surface we know that the mantle is dominated by ultramafic rocks of the peridotite family. The bulk of the mantle is in a solid state and undergoes continuous recrystallization as a result of large-scale convection and tectonic processes. Therefore, nearly all mantle rocks also represent metamorphic rocks. Consequently, the composition of the mantle (Table 2.1) is representative for the most prominent type of metamorphic rock on this planet. However, mantle rocks can only be transported through the cover of crust to the surface of the Earth by active tectonic or magmatic processes. Ultramafic rocks are widespread, though in minor amounts, particularly in orogenic belts.

Crustal rocks may be divided into rocks from oceanic and continental environments. Characteristic compositions of continental and oceanic crust are

Table 2.1 Composition of crust and mantle in wt % (Carmichael 1989)

	Peridotite mantle	Continental crust	Oceanic crust	Basalt	Tonalite
SiO_2	45.3	60.2	48.6	47.1	61.52
TiO_2	0.2	0.7	1.4	2.3	0.73
Al_2O_3	3.6	15.2	16.5	14.2	16.48
FeO	7.3	6.3	8.5	11.0	5.6
MgO	41.3	3.1	6.8	12.7	2.8
CaO	1.9	5.5	12.3	9.9	5.42
Na_2O	0.2	3.0	2.6	2.2	3.63
K_2O	0.1	2.8	0.4	0.4	2.1
H_2O	<0.1	1.4	1.1	<1.0	1.2
CO_2	<0.1	1.4	1.4	<1.0	0.1

Table 2.2 Abundance of rocks (vol%) in the crust. (Carmichael 1989)

Magmatic rocks	64.7
Sedimentary rocks	7.9
Metamorphic rocks	27.4
Magmatic rocks (64.7)	
Granites	16.0
Granodiorites/diorites	17.0
Syenites	0.6
Basalts/gabbros	66.0
Peridotites/dunites	0.3
Sedimentary rocks (7.9)	
Shales	82
Sandstones, arkoses	12
Limestones	6

listed in Table 2.1. Typical compositions of basalt (the dominant volcanic rock of oceanic environments) and tonalite (quartz-diorite to granodiorite; the dominant igneous rocks of continental environments) are given also in Table 2.1. It is evident that the average composition of the oceanic crust is well represented by an average basalt composition, and the average composition of the continental crust can be described by an average tonalite composition.

The average distribution of rocks in the crust is shown in Table 2.2, which shows clearly what kind of rock we expect to find predominately as protolith rocks for metamorphic rocks. Igneous rocks of mafic composition (basalts, gabbros) are the most important rocks in the oceanic crust (oceanic crust covers much larger areas than continental crust). The mafic rocks constitute, therefore, an exceptionally important chemical group of metamorphic rocks (mafic schists and mafic gneisses, amphibolites, mafic granulites, eclogites). Typical compositions of basaltic protolith of metamorphic rocks are given in Tables 2.1 and 2.3.

The granites and related rocks such as granodiorites and quartz-diorites (typical compositions given in Table 2.3) largely dominate the continental crust. These rock types constitute the family of metamorphic rock composition that is commonly referred to as meta-granitoids (quartzo-feldspathic rocks). Metagranitoids are granofelses, gneisses and schists derived from granites, granodiorites and quartz-diorites that represent 33% of all magmatic rocks (Table 2.2).

On a global basis, sedimentary rocks are dominated by shales and clays of pelagic and platform environments (82% of all sediments, Table 2.2). Compositions of typical shales from continental and oceanic environments are listed in Table 2.3. Pelagic clays (Table 2.3) represent the typical sediments of the deep oceans. The extremely fine-grained, clay-rich sediments are designated pelites and form the most important type of metamorphic rocks of sedimentary origin (metapelites).

Continentally derived shales typically contain carbonate minerals. This is reflected in the chemical composition of the average shale analysis given in Table 2.3. Carbonate-rich shales are often referred to as marls. Metamorphic

Table 2.3 Chemical compositions (wt %) of sedimentary and igneous rocks. (Carmichael 1989)

	Sandstones, grey wackes	Shales (platforms)	Pelites, pelagic clays clays	Carbonates (platforms)	Tonalite	Granite	Basalt MORB
SiO_2	70.0	50.7	54.9	8.2	61.52	70.11	49.2
TiO_2	0.58	0.78	0.78	–	0.73	0.42	2.03
Al_2O_3	8.2	15.1	16.6	2.2	16.48	14.11	16.09
Fe_2O_3	0.5	4.4	7.7	1.0	–	1.14	2.72
FeO	1.5	2.1	2.0	0.68	5.6	2.62	7.77
MgO	0.9	3.3	3.4	7.7	2.8	0.24	6.44
CaO	4.3	7.2	0.72	40.5	5.42	1.66	10.46
Na_2O	0.58	0.8	1.3	–	3.63	3.03	3.01
K_2O	2.1	3.5	2.7	–	2.1	6.02	0.14
H_2O	3.0	5.0	9.2	–	1.2	0.23	0.70
CO_2	3.9	6.1	–	35.5	0.1		
C	0.26	0.67	–	0.23			

equivalents, e.g. calcareous mica-schists, are important types of metasediments in orogenic belts and will be discussed separately in this book.

With reference to Table 2.2, sandstones (greywackes, arkoses) and limestones (carbonate rocks) are the remaining important groups of rocks. Characteristic compositions are listed in Table 2.3.

The composition of metamorphic protolith rocks undergoing a new cycle of metamorphism is controlled by the composition of the most common sedimentary and igneous rocks as given in Table 2.3. The composition of metamorphic rocks is therefore conveniently grouped into seven classes of characteristic bulk rock compositions.

2.1.2 The Seven Chemical Composition Classes of Metamorphic Rocks and Their Protoliths

The seven classes are arranged according to increasing chemical complexity.

1. *Ultramafic Rocks.* Usually mantle-derived, very Mg-rich family of rocks (typical composition; Table 2.1, peridotite). The prograde metamorphism of ultramafic rocks will be discussed in Chapter 5.
2. *Carbonate Rocks.* Sedimentary rocks modally dominated by carbonate minerals (calcite, dolomite). Examples: limestones, marbles (typical composition; Table 2.3, carbonates). See Chapter 6.
3. *Pelites (Shales).* Pelitic rocks and shales are the most common type of sedimentary rock. Pelagic clays (pelites) are poor in calcium compared with shales from platforms (Table 2.3). Pelites constitute a separate composition group. Their metamorphic equivalents are termed metapelites (metapelitic

schists and gneisses). Metapelites are usually rich in diagnostic metamorphic mineral assemblages and are very important in the study of metamorphic terrains and metamorphic processes (Chap. 7).

4. *Marls.* Marls are shales containing a significant proportion of carbonate minerals (usually calcite). Carbonate-rich shales are characteristic sediments of platform environments (composition; Table 2.3). The prograde metamorphism of marly rocks will be discussed in Chapter 8.

5. *Mafic Rocks.* Metamorphic mafic rocks (e.g. mafic schists and gneisses, amphibolites) are derived from mafic igneous rocks, mainly basalts and, of lesser importance, gabbros. Compositions are given in Tables 2.1 and 2.3. Basalts are volumetrically the most important group of igneous rocks and metabasalts occur very widespread in metamorphic terrains. Metamorphic assemblages found in mafic rocks are used to define the intensity of metamorphism in the metamorphic facies concept (Chap. 4). The prograde metamorphism of mafic rocks will be discussed in Chapter 9.

6. *Quartzo Feldspathic Rocks.* Rocks of sedimentary (sandstones, greywackes) or igneous origin (granites, granodiorites, tonalites, monzonites, syenites etc.) which are modally dominated by quartz and feldspar (typical compositions; Table 2.3). Gneisses derived from igneous rocks of the granite family may be designated metagranitoids (Chap. 10).

7. *Other Bulk Compositions.* All remaining compositions of protoliths of metamorphic rocks are of minor and subordinate importance. They include manganese sediments, ironstones, laterites, evaporites and alkaline igneous rocks, just to name a few.

2.2 The Structure of Metamorphic Rocks

Most metamorphic rocks form as a result of large-scale tectonic processes in the crust and associated changes in pressure and temperature in a given volume of rock. Metamorphism involves chemical reactions in rocks that replace minerals and mineral assemblages in the original material by new minerals or groups of minerals. The orientation and geometric arrangement of the new inequant metamorphic minerals is therefore largely controlled by the anisotropic pressure field associated with the tectonic processes.

Therefore, metamorphic rocks display not only a characteristic metamorphic mineral content they are also characterized by distinctive **metamorphic structures**. The structure of metamorphic rocks is used for the classification of the rocks. The structures of metamorphic rocks memorize essential information about the type of tectonic setting in which they formed and about the nature of metamorphism. Structural and chemical (petrological) aspects of metamorphic rocks are of equal importance in the study of metamorphism. The characterization of metamorphic rocks requires a description of their structure. Some important and typical structures of metamorphic rocks are defined by the following descriptive structure terms[1]:

Structure. The arrangement of the parts of a rock mass irrespective of scale, including geometric interrelationships between the parts, the shapes and internal features of the parts. The terms micro-, meso- and mega- can be used as a prefix describing the scale of the feature. Micro- is used on the thin-section scale, meso- on the hand-specimen and outcrop scale, mega- on larger scales.

Fabric. The kind and degree of preferred orientation of parts of a rock mass. The term is used to describe the crystallographic and/or shape orientation of mineral grains or groups of grains, but also can be used on the meso- and mega-scale.

Layer. One of a sequence of near parallel tabular-shaped rock bodies. The sequence is referred to as being layered (equivalent expressions: bands, banded).

Foliation. Any repetitively occurring or penetrative planar structural feature in a rock body. Some examples
* regular layering on a centimeter or smaller scale;
* preferred planar orientation of inequant mineral grains;
* preferred planar orientation of lenticular (elongate) grain aggregates.
More than one kind of foliation with more than one orientation may be present in a rock. Foliations may become curved (folded) or distorted. The surfaces to which they are parallel are designated **s-surfaces**.

Schistosity. A type of foliation produced by deformation and/or recrystallization resulting in a preferred orientation of inequant mineral grains. It is common practice in phyllosilicate-rich rocks to use the term slaty cleavage instead of schistosity when individual grains are too small to be seen by the unaided eye.

Cleavage. A type of foliation consisting of a regular set of parallel or sub-parallel closely spaced surfaces produced by deformation along which a rock body will usually preferentially split. More than one cleavage may be present in a rock.

Slaty Cleavage. Perfectly developed foliation independent of bedding resulting from the parallel arrangement of very fine-grained phyllosilicates.

Spaced Cleavage. Regularly spaced cleavage lamellae separated by slices of rocks which are structurally distinct from the cleavage lamellae, this structure is mesoscopic.

Fracture Cleavage. A type of cleavage defined by a regular set of closely spaced fractures.

---

[1] The definition of terms given in this chapter partly follows preliminary unpublished recommendations of the International Commission on the Nomenclature of Metamorphic Rocks.

Crenulation Cleavage. A type of cleavage related to microfolding (crenulation) of a pre-existing foliation. It is commonly associated with varying degrees of metamorphic segregation.

Gneissose Structure. A type of foliation on hand-specimen scale, produced by deformation and recrystallization, defined by:
- irregular or poorly defined layering;
- augen and/or lenticular aggregates of mineral grains (augen structure, flaser structure);
- inequant mineral grains which are present, however, only in small amounts or which display only a weak preferred orientation, thus defining only a poorly developed schistosity.

Lineation. Any repetitively occurring or penetrative visible linear feature in a rock body. This may be defined by, for example
- alignment of the long axes of elongate mineral grains (= mineral lineation);
- alignment of elongate mineral aggregates;
- alignment of elongate objects, bodies (e.g. strongly deformed pebbles in a meta-conglomerate);
- common axis of intersection of tabular mineral grains (or bodies);
- intersection of two foliations (intersection lineation);
- parallelism of hinge lines of small-scale folds;
- slickenside striations;
- striations due to flexural slip.

More than one kind of lineation, with more than one orientation, may be present in a rock. Lineations may become curved or distorted. The lines to which they are parallel are called **l-lines**. Reference to a lineation is incomplete without indication of the type concerned.

Joint. A single fracture in a rock with or without a small amount (< 1 cm) of either dilatational or shear displacement (joints may be sealed by mineral deposits during or after their formation).

Cataclasis. Rock deformation accomplished by some combination of fracturing, rotation and frictional sliding producing mineral grain and/or rock fragments of various sizes and often of angular shape.

Metamorphic differentiation. Redistribution of mineral grains and/or chemical components in a rock as a result of metamorphic processes. Metamorphic process by which mineral grains or chemical components are redistributed in such a way to increase the modal or chemical anisotropy of a rock (or portion of a rock) without changing the overall chemical composition.

2.3 Classification and Names of Metamorphic Rocks

The names of metamorphic rocks are usually straightforward and self-explanatory. The number of special terms and cryptic expressions is relatively small (in contrast to the occult language of magmatic petrologists and mineralogists). Nevertheless in order to be able to communicate with other geologists working with metamorphic rocks, it is necessary to define commonly used names and expressions and to briefly review currently used classification principles for metamorphic rocks.

There is not one sole classification principle used for the description of metamorphic rocks, which consequently means that all metamorphic rocks may have a series of perfectly correct and accepted names. However, modal mineralogical composition and mesoscopic structure are the main criteria for naming metamorphic rocks. In addition, the composition and the nature of the protolith (original material) is an important classification criterion. Finally, a number of well-established special names are used also in metamorphic geology.

The names of metamorphic rocks consist of a **root** name and a series of **prefixes**. The root of the name may be a special name (e.g. amphibolite) or a name describing the structure of the rock (e.g. gneiss). The root name always embraces some modally dominant metamorphic minerals (amphibolite is mainly composed of amphibole + plagioclase; gneiss is mainly composed of feldspar ± quartz). The rock may be further characterized by adding prefixes to the root name. The prefixes may specify some typical structural features of the rock or may give some additional mineralogical information (e.g. banded epidote-bearing garnet-amphibolite, folded leucocratic garnet-hornblende gneiss). The prefixes are optional and the name may consist of the root only.

2.3.1 Structurally Defined Rock Names

The characteristic distribution of the parts (minerals, aggregates, layers etc.) of metamorphic rocks (= structure) results from the geometrical arrangement of minerals, mineral aggregates with non-isotropic crystal shapes and other structural features, as discussed above. This structure is largely controlled by mechanical deformation and chemical segregation processes which are almost always associated with metamorphism. Deformation and recrystallization are two equally important aspects of metamorphism. Expressions that mainly characterize the structure of metamorphic rocks are often used as root names. Metamorphic rock classification by **descriptive structural terms** is the primary classification principle for metamorphic rocks. The most important of these terms are:

Gneiss. A metamorphic rock displaying a gneissose structure. The term gneiss may also be applied to rocks displaying a dominant linear fabric rather than a gneissose structure, in which case the term lineated gneiss may be used. This term is almost exclusively used for rocks containing abundant feldspar (± quartz), but may also be used in exceptional cases for other compositions (e.g. feldspar-free cordierite-anthophyllite gneiss). Examples: garnet-biotite gneiss, granitic gneiss, ortho-gneiss, migmatitic gneiss, banded gneiss, garnet-hornblende gneiss, mafic gneiss).

Schist. A metamorphic rock displaying on the hand-specimen scale a pervasive, well-developed schistosity defined by the preferred orientation of abundant inequant mineral grains. For phyllosilicate-rich rocks the term schist is usually reserved for medium- to coarse-grained varieties, whilst finer-grained rocks are termed slates or phyllites. The term schist may also be used for rocks displaying a strong linear fabric rather than a schistose structure. Examples: epidote-bearing actinolite-chlorite schist (= greenschist), garnet-biotite schist, micaschist, calcareous micaschist, antigorite schist (= serpentinite), talc-kyanite schist (= whiteschist).

Slate. A very fine-grained rock of very low metamorphic grade displaying slaty cleavage.

Phyllite. A fine-grained rock of low metamorphic grade displaying a perfect penetrative schistosity resulting from parallel arrangement of phyllosilicates. Foliation surfaces commonly show a lustrous sheen.

Granofels. A metamorphic rock lacking schistosity, gneissose structure, and mineral lineations.

Names for High-Strain Rocks

Metamorphism may locally be associated with an extremely high degree of rock deformation. Localized high strain in metamorphic terrains may produce rocks with distinctive structures. Some widely used special names for high-strain rocks are defined below:

Mylonite. A rock produced by mechanical reduction of grain size as a result of ductile, non-cataclastic deformation in localized zones (shear zones, fault zones), resulting in the development of a penetrative fine-scale foliation, and often with an associated mineral and stretching lineation.

Ultramylonite. A mylonite in which most of the megacrysts or lithic fragments have been eliminated (> 90% fine-grained matrix).

Augen Mylonite (Blastomylonite). A mylonite containing distinctive large crystals or lithic fragments around which the fine-grained banding is wrapped.

Cataclasite. A rock which underwent cataclasis.

Fault Breccia. Cataclasite with breccia-like structure formed in a fault zone.

Pseudotachylite. Ultra-fine-grained vitreous-looking material, flinty in appearance, occurring as thin veins, injection veins, or as a matrix to pseudo-conglomerates or -breccias, which seals dilatancy in host rocks displaying various degrees of fracturing.

2.3.2 Special terms

Some commonly used and approved special terms (names, suffixes, prefixes) include:

Mafic Minerals. Collective expression for ferro-magnesian minerals.

Felsic Minerals. Collective term for quartz, feldspar, feldspathoids and scapolite.

Mafic Rock. Rock mainly consisting of mafic minerals (mainly consisting $\equiv$ modally > 50%).

Felsic Rock. Rock mainly consisting of felsic minerals.

Meta-. If a sedimentary or magmatic origin of a metamorphic rock can identified, the original magmatic or sedimentary rock term preceded by "meta-" may be used (e.g. metagabbro, metapelite, metasediment, metasupracrustal). Also used to generally indicate that the rock in question is metamorphic (e.g. metabasite).

Ortho- and **para-.** A prefix indicating, when placed in front of a metamorphic rock name, that the rock is derived from an igneous or sedimentary rock respectively (e.g. paragneiss).

Acid, Intermediate, Basic, Ultrabasic. Terms defining the SiO_2 content of igneous and metamorphic rocks (> 63, 63-52, 52-45, < 45 wt.% SiO_2).

Greenschist and Greenstone. Schistose (greenschist) or non-schistose (greenstone) metamorphic rock whose green color is due to the presence of minerals such as chlorite, actinolite, epidote and pumpellyite (greenschist for e.g. epidote-bearing actinolite-chlorite schist; greenstone for e.g. chlorite-epidote granofels).

Blueschist. Schistose rock whose bluish color is due to the presence of sodic amphibole (e.g. glaucophane schist). However, the "blue" color of a blueschist will not be recognized by a non-geologist (i.e. it is not really blue). Blueschists are schistose rocks containing amphiboles with significant amounts of the M4 cation position in the amphibole structure occupied by Na (riebekite, crossite, glaucophane).

Amphibolite. Mafic rock predominantly composed of amphibole (> 40%) and plagioclase.

Granulite. Metamorphic rock in or from a granulite facies terrain exhibiting characteristic granulite facies assemblages. Anhydrous mafic minerals are modally more abundant than hydrous mafic minerals. Muscovite is absent in such rocks. Characteristic is the occurrence of orthopyroxene in mafic and felsic rocks. The term is not used for marbles and ultramafic rocks in granulite facies.

Charnockite, Mangerite, Jotunite, Enderbyite. Terms applied to orthopyroxene-bearing rocks with igneous structure and granitic (charnokite), monzonitic (mangerite, jotunite) and tonalitic (enderbyite) composition, respectively, irrespective of whether this rock is igneous or metamorphic.

Eclogite. A plagioclase-free mafic rock mainly composed of omphacite and garnet, both of which are present in a large proportion.

Eclogitic Rock. Rock of any composition containing diagnostic mineral assemblages of the eclogite facies (e.g. jadeite-kyanite-talc granofels).

Marble. A metamorphic rock mainly composed of calcite and/or dolomite (e.g. dolomite marble).

Calc-Silicate Rock. Metamorphic rock which, besides 0–50% carbonates, is mainly composed of Ca-silicates such as epidote, zoisite, vesuvianite, diopside-hedenbergite, Ca-garnet (grossular-andradite), wollastonite, anorthite, scapolite, Ca-amphibole.

Skarn. A metasomatic Ca-Fe-Mg-(Mn)-silicate rock often with sequences of compositional zones and bands, formed by the interaction of a carbonate and a silicate system in mutual contact. Typical skarn mineralogy includes: wollastonite, diopside-salite, grossular, zoisite, anorthite, scapolite, margarite (Ca skarns); hedenbergite, andradite, ilvaite (Ca-Fe skarns); forsterite, humites, spinel, phlogopite, clintonite, fassaite (Mg skarns); rhodonite, tephroite, piemontite (Mn skarns).

Rodingite. Calc-silicate rock, poor in alkalis and generally poor in carbonates, generated by metasomatic alteration of mafic igneous rocks enclosed in serpentinized ultramafic rocks. The process of rodingitization is associated with oceanic metamorphism (serpentinization of peridotite, rodingitization of enclosed basic igneous rocks such as gabbroic/basaltic dykes). Metarodingite is a prograde metamorphic equivalent of rodingite produced by oceanic metamorphism.

Quartzite. A metamorphic rock containing more than 80% quartz.

Serpentinite. An ultramafic rock composed mainly of minerals of the serpentine group (antigorite, chrysotile, lizardite), e.g. diopside-forsterite-antigorite schist.

Hornfels. Is a non-schistose very fine-grained rock mainly composed of silicate ± oxide minerals and it shows substantial recrystallization due to thermal metamorphism. Hornfelses often retain some features inherited from the original rock such as graded bedding and cross-bedding in hornfelses of metasedimentary origin.

Migmatite. Composite silicate rock, pervasively heterogeneous on a meso- to megascopic scale, found in medium- to high-grade metamorphic terrains (characteristic rocks for the middle and lower continental crust). Migmatites are composed of dark (mafic) parts and light (felsic) parts in complex structural association. The felsic parts formed by crystallization of locally derived partial melts or by metamorphic segregation, the mafic parts represent residues of the inferred partial melting process or formed by metamorphic segregation. Parts of the felsic phases may represent intruded magmas from a more distant source.

Restite. Remnant of a rock, chemically depleted in some elements relative to its protolith. The depletion is the result of partial melting of that rock.

2.3.3 The Modal Composition Systematics

The mineral constituents of metamorphic rocks are classified as follows. (1) Major constituent: present in amounts >5 vol%. To account for a major constituent not included in the definition of a rock name, the mineral name is set in front of the rock name (e.g. muscovite gneiss, epidote amphibolite). It follows from the above that an epidote amphibolite is a rock which is mainly composed of epidote, amphibole and plagioclase. A garnet-staurolite gneiss is a metamorphic rock which is mainly composed of feldspar and quartz (included in the name root "gneiss"). In addition, it contains modally more staurolite than garnet as major constituents. (2) Minor constituent: present in amounts < 5 vol%. If one wishes to include a minor constituent mineral in the rock name it is connected with the "-bearing" (e.g. rutile-ilmenite-bearing garnet-staurolite gneiss; contains less rutile than ilmenite). (3) Critical mineral (or mineral assemblage): indicating by its presence or absence distinctive conditions for the formation of a rock (it also may indicate a distinct chemical composition of the rock). The critical mineral(s) may be present as major or/and minor constituent(s).

It is up to the geologist to decide how many and which of the minerals that are not yet included in the definition of the rock name he/she wants to prefix. It is also possible to use abbreviations of mineral names in the rock names (e.g. Bt-Ms gneiss for biotite-muscovite gneiss). The recommended standard abbreviations for mineral names are listed in the Appendix.

2.3.4 Names Related to the Origin of the Protolith

The chemical variation found in metamorphic rocks has been grouped into seven composition classes according to the most frequently occurring rock types in crust and mantle.

Metamorphic rocks may be characterized by referring its names to the nature of the original material. Examples of such names include: metapelite, metabasite, metabasalt, metapsamite, metagranite, metagabbro, semi-pelitic gneiss, metamarl (e.g. calcareous micaschist), metaeclogite.

2.4 Mineral Assemblages and Mineral Parageneses

The petrogenesis of metamorphic rocks is often a long-lasting complicated process in time and space which may last for some 30 Ma or so (see Chap. 3). The geologist collects the end product of such a long-lasting process on the Earth's surface (1 bar and 30°C). The geologist is, of course, interested in finding out something about the evolution, the petrogenesis of the collected rock at this sample locality, to compare it with rocks from other localities and finally draw some sound conclusions on the large-scale processes that caused the observed rock metamorphism.

One of the important methods to decipher the petrogenesis and evolution of metamorphic rocks is the application of chemical thermodynamics to the heterogeneous chemical systems typified by rocks. Rocks are usually composed of a number of different types of minerals. The mineral species represent chemical subspaces of the rock, so-called phases. Most metamorphic rocks are also composed of a number of different chemical constituents that are designated system components. The chemical constituents making up the rocks' total composition are distributed among a group of mineral species or phases, each of them with a distinct composition (heterogeneous system). The number, composition and identity of the minerals is uniquely defined by the laws of thermodynamics and depends exclusively on the prevailing intensive variables such as pressure and temperature. The group of minerals that make up a rock at equilibrium is designated **equilibrium mineral assemblage** or equilibrium phase assemblage. The succession of mineral assemblages that follow and replace one another during the metamorphic evolution of a given terrain are designated **mineral parageneses**.

In the practical work with metamorphic rocks it is impossible to demonstrate that a given mineral assemblage once coexisted in chemical equilibrium. Therefore, one is forced to define the mineral assemblage of a rock as an association of mineral species in mutual grain contact, that occur in a given chemically homogeneous portion of a rock. For practical work with thin sections it is convenient to use a matrix table for marking all observed two-phase grain contacts. An example: in a thin section of a metamorphic rock showing a coarse-grained mosaic structure the following mafic minerals have been identified: staurolite, garnet, biotite and kyanite. All observed grain contacts between two of these minerals are marked with an X in Table 2.4. The thin-section observations summarized in Table 2.4 imply that the four minerals indeed represent a mineral assemblage.

In rocks displaying clear disequilibrium structures (reaction rims, symplectitic structures, replacement structures), it is often difficult to determine mineral assemblages that represent a particular stage in the rocks' evolutionary history.

One has to keep in mind that mineral assemblages are usually identified in thin sections which represent a two-dimensional section through a volume of rock. In such sections, only two and a maximum of three minerals may be in mutual grain contact. In the three-dimensional rock a maximum of four minerals can be in contact at a point in the space, three minerals form contacts along a line, and two minerals contact along surfaces (that show as lines in thin sections). It is therefore strongly recommended to study more than one section from the most significant and interesting samples. In a study of the metamorphic evolution of a single sample of a coarse-grained high-grade rock, in a series of 20 thin-sections we found unique assemblages and unique structures in nearly

Table 2.4 Practical determination of a mineral assemblage

Mineral	Staurolite	Garnet	Biotite	Kyanite
Staurolite	×	×	×	×
Garnet		×	×	×
Biotite			×	×
Kyanite				×

all of the sections. The problem is especially acute in coarse-grained samples, where the scale of chemical homogeneity may considerably exceed the size of a thin section.

In extremely fine-grained samples the mineral assemblage must be determined by X-ray techniques. In this case it is not possible to maintain the requirement of mutual grain contact in the definition of an assemblage. However, this shortcoming may be overcome by the use of a scanning electron microscope working in the backscatter mode.

Rocks should always be examined by X-ray techniques in order to identify minerals which are difficult to distinguish under the microscope (e.g. muscovite, paragonite, talc, pyrophyllite, also quartz and untwinned albite). Staining techniques help the distinction of some important rock-forming minerals with similar optical properties (e.g. calcite, dolomite).

Minerals occurring as inclusions in refractory minerals such as garnet but not in the matrix of the rock do not belong to the main matrix assemblage. In our example of the staurolite-garnet-biotite-kyanite assemblage, garnet may show small composite two-phase inclusions of chlorite and chloritoid. Chlorite-chloritoid-garnet constitute in this case another, earlier assemblage of the rock.

In the course of a metamorphic process some earlier-formed minerals may become metastable and react to form a new, more stable assemblage. However, metastable relics of the early assemblage may partly survive. Great care must be taken in the study of metamorphic micro structures in order to avoid mixing up mineral assemblages. The correct identification of successive series of mineral assemblages, that is the parageneses of a metamorphic rock represents the "great art" of metamorphic petrology. It can be learned only by experience.

Figure 2.1 shows some of the aspects related to the recognition of mineral assemblages. Three fictive rocks all contain the minerals quartz, calcite and wollastonite on the scale of a thin section. The general micro-structure of the

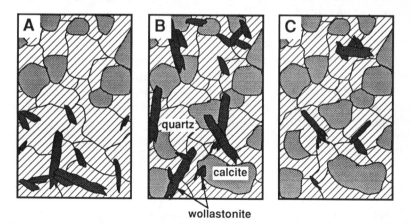

Fig. 2.1A–C. Mineral assemblages in three different rocks all containing wollastonite, calcite and quartz

three sections (Fig. 2.1 A–C) shows the distribution of the three minerals in the respective rocks. Rock A is clearly compositional heterogeneous on the scale of a thin section, the upper part contains the assemblage quartz + calcite, the lower half of the section is free of carbonate and contains the assemblage quartz + wollastonite. The two parts of the section are different in composition. The rock does not contain the assemblage Qtz+Cal+Wo. In rock B obviously all three minerals can be found in mutual grain contact, Qtz+Cal+Wo constitute the characteristic mineral assemblage for this rock. In rock C, which appears to be compositionally homogeneous on thin-section scale, quartz + calcite and quartz + wollastonite form common grain boundaries. However, no wollastonite + calcite grain boundaries can be found. The three phases Qtz+Cal+Wo do not represent a mineral assemblage in the strict sense. Yet many geologists would probably approve the assemblage, despite this.

2.5 Graphical Representation of Metamorphic Mineral Assemblages

Once the mineral assemblage of a rock has been identified, it is convenient or even compulsory to represent the chemical composition of the minerals that constitute the assemblage on a graphical Figure. Such a figure is called a **chemography** and represents a **composition phase diagram**. The geometric arrangement of the phase relationships on such a phase diagram is called the **topology**. Composition phase diagrams can be used just to document the assemblages found in rocks of a given metamorphic terrain or outcrop. However, such diagrams are an indispensable tool in the analysis of metamorphic characteristics and evolution of a terrain. They can be used to deduce sequences of metamorphic mineral reactions that run in a crustal area. Finally, composition phase diagrams can also be calculated theoretically from thermodynamic data of minerals which permit the quantitative calibration of field-derived chemographies.

Composition phase diagrams display the chemical composition of minerals and the topologic relationships of mineral assemblages. The variables on the diagrams are concentrations or amounts of chemical entities. All other variables that control the nature of the stable mineral assemblage such as pressure and temperature must be constant. Composition phase diagrams are isothermal isobaric diagrams. Also, not more than two composition variables can conveniently be displayed on a two-dimensional xy-diagram (sheet of paper, computer screen).

2.5.1 Mole Numbers, Mole Fractions and the Mole Fraction Line

It is useful to change the scale for the compositional variables from wt.% to mol%, mole fractions or mole numbers. Most chemographies use mol% or mole fraction as units for the composition variables. The mineral forsterite

(Fo), for example, is composed of 42.7 wt.% SiO_2 and 57.3 wt.% MgO. Mole numbers and mole fractions for this mineral are calculated as follows:

SiO_2: 42.7/60.1 (molecular weight SiO_2) = **0.71** (number of moles of SiO_2 per 100 gr Fo).
MgO: 57.3/40.3 (molecular weight MgO) = **1.42** (number of moles of MgO per 100 gr Fo).

This mineral has a MgO/SiO_2 ratio of 1.42/0.71 = 2. It has 2 mol MgO per 1 mole SiO_2. The composition of the mineral, forsterite, is reported as Mg_2SiO_4 or (2 MgO SiO_2) or 66.66% MgO + 33.33% SiO_2 or 2/3 MgO + 1/3 SiO_2. This is all equivalent. However, the last version has many advantages $\Rightarrow$ mole fraction basis.
The mole fraction is defined as follows:

$$X_{MgO} = \frac{\text{(number of moles of MgO)}}{\text{(number of moles of MgO)} + \text{(number of moles of } SiO_2)}.$$

For the example above: $X_{MgO} = 2/(2+1) = 1.42/(1.42+0.71) = 0.66$ (dimensionless quantity).

Figure 2.2 shows a graphical representation of the two component system MgO and SiO_2 in a rectangular coordinate system. The two components MgO and SiO_2 define the two-dimensional composition space. The graphic representation of the compositions of forsterite, enstatite, quartz and periclase is a point in the Cartesian coordinate system with the axes: numbers of moles MgO and SiO_2. It is apparent in Fig. 2.2 that the composition of, for example, forsterite can also be viewed as a vector in the composition space with the elements (2,1). Mineral compositions are unequivocally characterized by a vector, the phase vector, in the composition space.

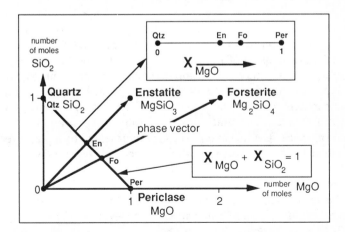

Fig. 2.2. Composition space in the two-component MgO-SiO_2 system

The mineral compositions can also be represented in terms of mole fractions rather than mole numbers. The mole fractions of the two system components MgO and SiO_2 in the four mineral compositions represented in Fig. 2.2 sum to unity. This can be expressed by the general relationship:

$$X_{MgO} + X_{SiO_2} = 1.$$

This equation represents a straight line in Fig. 2.2 connecting $X_{MgO} = 1$ with $X_{SiO2} = 1$. The phase vectors for forsterite and enstatite intersect the mole fraction line at unique positions (En and Fo, respectively; Fig. 2.2). It is, therefore, not necessary to represent phase composition in a two-component system on a two-dimensional diagram. The topologic information is contained on the mole fraction line. The reason for this reduction of the dimension is the simple fact that if the concentration of one component is known in a phase composed of two components, the other concentration is known as well. In an n-component system there are n-1 independent compositional variables, the remaining concentration is given by the equation:

$$\sum_{j=1}^{n} X_j = 1.$$

Note also, for example, that enstatite compositions expressed as $MgSiO_3$, $Mg_2Si_2O_6$, $Mg_4Si_4O_{12}$ all have the same intersection point with the mole fraction line (multiplying the phase vector for enstatite with a scalar preserves its position on the mole fraction line). Phase compositions in a two-component system can be represented on a mole fraction line. The appropriate diagram is shown in Fig. 2.2.

2.5.2 The Mole Fraction Triangle

The mineral talc is composed of three simple oxide components and its composition can be written as: $H_2Mg_3Si_4O_{12}$. Based on a total of 12 oxygens per formula unit, talc consists of 1 mol H_2O, 3 mol MgO and 4 mol SiO_2. A graphic representation of the talc composition in the three-component system is shown in Fig. 2.3. The three components, MgO, SiO_2 and H_2O, define a Cartesian coordinate system with numbers of moles of the components displayed along the coordinate axes. The unit vectors of the system components span the composition space. The talc composition is represented by a phase vector with the elements; $3 \times$ the unit vector of MgO, $4 \times$ the unit vector of SiO_2 and $1 \times$ the unit vector of H_2O.

The total number of moles of system components is 8. The composition of talc can be normalized to a total number of 8 mol of system components. In this case the talc composition will be expressed by $\Rightarrow$ mole fractions: $X_{H_2O} = 1/8$, $X_{M_gO} = 3/8$, $X_{SiO_e} = 4/8$, or: $X_{H_2O} = 0.13$, $X_{MgO} = 0.38$, $X_{SiO_2} = 0.50$; in mole%:

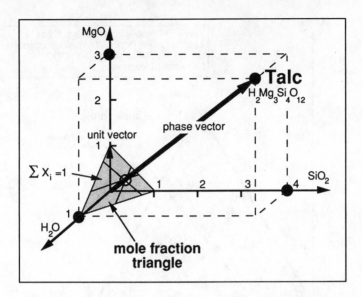

Fig. 2.3. Composition space in the three-component MgO-H_2O-SiO_2 system

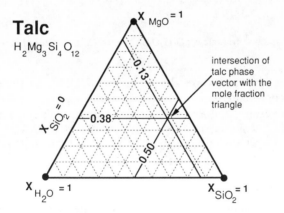

Fig. 2.4. Mole fraction triangle MgO-H_2O-SiO_2 and the intersection coordinates of the talc phase vector with the mole fraction triangle

$H_2O = 13\%$, $MgO = 38\%$, $SiO_2 = 50\%$. The graphic representation of the talc composition on a mole fraction basis is given by the intersection of the talc phase vector in Fig. 2.3 and the plane $\Sigma X_i = 1$. The mole fraction plane is a regular triangle with the corners $X_i = 1$. This triangle is called the mole fraction triangle.

A representation of the mole fraction triangle is shown in Fig. 2.4. The lines of constant X_i are parallel with the base lines of the triangle. This follows from Fig. 2.3, where it can be seen that the planes $X_i = $ constant and $X_{j \neq i} = 0$ intersect the mole fraction plane along lines parallel with the base line $X_i = 0$. Three rulers

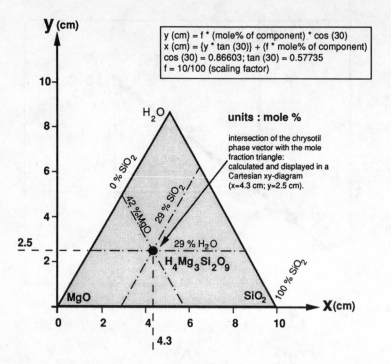

Fig. 2.5. Mole fraction triangle MgO-H$_2$O-SiO$_2$in the Cartesian coordinate system

with X$_i$ = 0.1 increments are shown in Fig. 2.4 for the three components. Triangular coordinate paper is available on the market.

One more example of a mineral composition in a three-component system MgO-SiO$_2$-H$_2$O follows: the chemical formula of chrysotile (one of many serpentine minerals) can be written as: H$_4$Mg$_3$Si$_2$O$_9$. The sum of moles of system components is 7; $\Rightarrow$ mole fractions: X$_{H_2O}$ = 2/7, X$_{MgO}$ = 3/7, X$_{SiO_e}$ = 2/7, or equivalent: X$_{H_2O}$ = 0.29 , X$_{MgO}$ = 0.42 , X$_{SiO_2}$ = 0.29; expressed in mol%: H$_2$O = 29% , MgO = 42% , SiO$_2$ = 29%.

Because triangular coordinate drawing paper is not always at hand for plotting compositional data, the formulas for recalculation into Cartesian coordinates are given in Fig. 2.5 together with the position of the serpentine composition. The value of the scaling factor "f" depends on the desired size of the figure. The scaling factor must be multiplied by 100 when using mole fractions rather than mol%.

2.5.3 Projections

1. Simple Projections. On a mole fraction triangle two compositional variables of a three-component system can be depicted. Most rocks, however, require more than three components to describe and understand the phase relation-

Table 2.5a–c. Phase compositions in the MSH system. **a** Composition matrix in terms of moles; columns are phase vectors, composition space defined by the system components SiO_2, MgO, H_2O. **b** Composition matrix in terms of mole fractions; columns are normalized phase vectors, values are coordinates on the mole fraction triangle. **c** Composition matrix in terms of mole fractions; projected through H_2O, columns are normalized phase vectors, values are coordinates on the SiO_2-MgO binary (mole fraction line)

		Fo	Brc	Tlc	En	Ath	Qtz	Per	Atg	Fl
a	SiO_2	1	0	4	2	8	1	0	2	0
	MgO	2	1	3	2	7	0	1	3	0
	H_2O	0	1	1	0	1	0	0	2	1
	Sum	3	2	8	4	16	1	1	7	1
b	SiO_2	0.33	0.00	0.50	0.50	0.50	1.00	0.00	0.29	0.00
	MgO	0.67	0.50	0.38	0.50	0.44	0.00	1.00	0.43	0.00
	H_2O	0.00	0.50	0.13	0.00	0.06	0.00	0.00	0.29	1.00
	Sum	1	1	1	1	1	1	1	1	1
c	SiO_2	0.33	0.00	0.56	0.50	0.53	1.00	0.00	0.40	0.00
	MgO	0.67	1.00	0.43	0.50	0.47	0.00	1.00	0.60	0.00

ships. Graphic representation of an eight component system requires a seven-dimensional figure. Projection phase diagrams are graphic representations of complex n-component systems that show two composition variables at a time while keeping the other n-3 composition variables constant. The remaining variable is given by the mole fraction equation as outlined above.

In a suite of collected samples of similar bulk composition (e.g. 20 samples of metapelite), one often finds certain mineral species that are present in many of the samples. This circumstance permits projection of the phase compositions from the composition of such a mineral which is present in excess. For instance, in many metapelitic rocks quartz is a modally abundant mineral whereas calcite is present in most marbles. Composition phase diagrams for metapelites can therefore be constructed by projecting through SiO_2 onto an appropriate mole fraction triangle, for marbles by projection the phase compositions from $CaCO_3$.

The system MgO-SiO_2-H_2O (MSH system) will be used to illuminate the basic principle of making projections. Some phase compositions in the MSH system are found in Table 2.5, which represents a **composition matrix** with oxide components as defining unit vectors of the composition space and the mineral compositions as column vectors.

Figure 2.6 shows the chemographical relationships in the ternary system H_2O-MgO-SiO_2. The corners of the triangle represent $X_i = 1$ or 100 mol% of the system components. The lines connecting them are the binary subsystems of the three-component system. The compositions of the black dots occur as stable phases in nature (phase components). The ternary system has three binary sub-

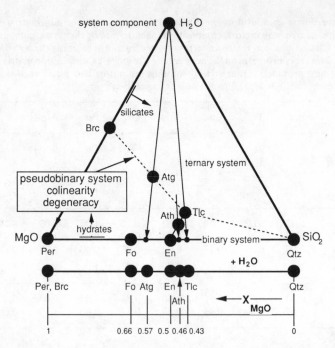

Fig. 2.6. Projection of phase compositions in the $MgO-H_2O-SiO_2$ system through H_2O onto the $MgO-SiO_2$ binary

systems ($MgO-SiO_2$, $MgO-H_2O$, and SiO_2-H_2O). Some phase compositions in the ternary system fall on straight lines such as Tlc-Ath-En and Brc-Atg-Tlc. This means that the phase compositions along the straight line (e.g. Tlc-Ath-En) are linearly dependent; one of these compositions can be expressed by the other two (4 En + Tlc = Ath). Therefore, there are only two components required to describe the compositions of the other phases on the straight line ⇒ **pseudo binary join**. The colinearity is also said to be a **compositional degeneracy** in the system.

Now, one might wish to analyze and discuss phase relationships among the minerals shown in Fig. 2.6 for geologic situations in which water is present in all rocks and in equilibrium with the solid phase assemblage. The presence of excess water in all considered rocks permits projection of the other phase compositions through water onto the $MgO-SiO_2$-binary (in principle on any pseudobinary as well). Imagine that you are standing in the H_2O corner of Fig. 2.6. What you will see from there is shown at the bottom of Fig. 2.6. The chemography on the $MgO-SiO_2$ binary is a projected chemography. The positions of the mineral compositions on this binary is expressed in terms of mole fractions X_{Mg} in Fig. 2.6. They can be calculated from the normalized composition matrix (Table 2.5b) by deleting the row containing the component one wishes to project from (in our case H_2O) and renormalizing the column vectors to unity (Table 2.5c). It follows that projections from quartz, periclase

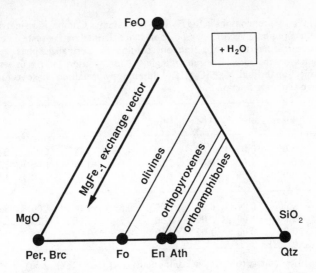

Fig. 2.7. Projection of phase compositions in the FeO-MgO-H_2O-SiO_2 system through H_2O and along the $MgFe_{-1}$ exchange vector onto the MgO-SiO_2 binary

and H_2O can be prepared using the composition matrix of Table 2.5b. It must be stressed again, however, that the projections are meaningful only if the projection component is present as a phase of fixed composition in all rocks of interest. The projected chemography in Fig. 2.6 cannot be used in H_2O-absent situations. Likewise, an analogous projection through MgO onto the SiO_2-H_2O binary requires that periclase is present in all assemblages, and projection through SiO_2 onto the MgO-H_2O-binary requires that quartz is present in all assemblages. It is therefore indispensable to write the projection compositions on any projected composition phase diagram. In simple projections, the projecting phase component is identical to one of the simple oxide system components.

2. Projection of Solid-Solution Phases. The magnesian minerals shown in Fig. 2.6 show in nature a variable substitution of Mg by Fe. If the restriction of water saturation is maintained, the compositions in the FeO-MgO-SiO_2-H_2O system can be projected from H_2O onto the mole fraction triangle FeO-MgO-SiO_2, as shown in Fig. 2.7. In certain applications one does not wish to consider the complexity arising from the Fe-Mg substitution. In such cases, one projects the compositions onto the MgO-SiO_2-binary parallel to the direction of the substitution vector $MgFe_{-1}$. The resulting projection is identical to the binary chemography shown in Fig. 2.6. All solid-solution phases must be projected along exchange vectors. More examples will be given below.

3. Complex Projections. The technique of projecting the phase compositions in the m-dimensional space from one system component onto a subspace with m-1 dimensions is one of the two key steps in producing geologic composition phase

Table 2.6a–c. Phase compositions in the Fo-Brc-Tlc system. **a** Composition matrix in terms of moles; columns are phase vectors, composition space defined by the system components. **b** Composition matrix in terms of mole fractions; columns are normalized phase vectors, values are coordinates on the mole fraction triangle. **c** Composition matrix in terms of mole fractions; projection through Mg_2SiO_4 (Fo), columns are normalized phase vectors, values are coordinates on the mole fraction line

		Fo	Brc	Tlc	En	Ath	Qtz	Per	Atg	Fl
a	Fo	1.00	0.00	0.00	2.00	4.00	−1.00	4.00	0.00	−4.00
	Brc	0.00	1.00	0.00	−1.00	−2.00	−1.00	1.00	3.00	5.00
	Tlc	0.00	0.00	1.00	1.00	5.00	1.00	−1.00	1.00	1.00
b	Fo	1.00	0.00	0.00	1.00	0.57	1.00	1.00	0.00	−2.00
	Brc	0.00	1.00	0.00	−0.50	−0.29	1.00	0.25	0.75	2.50
	Tlc	0.00	0.00	1.00	0.50	0.71	−1.00	−0.25	0.25	0.50

Initial composition matrix (old system components) [A][B_{OC}] (Table 2.5a)

New matrix (new system components) [I][B_{NC}] (Table 2.6a)

Matrix operation: $[A^{-1}][A][B_{OC}] \Rightarrow [I][B_{NC}]$

Old basis [A]			Inverse of old basis [A^{-1}]			Identity matrix [I]		
1	0	4	−0.33	0.67	−0.67	1	0	0
2	1	3	−0.33	0.17	0.83	0	1	0
0	1	1	0.33	−0.17	0.17	0	0	1

		Fo	Brc	Tlc	En	Ath	Qtz	Per	Atg	Fl
c	Brc	0.00	1.00	0.00	−1.00	−0.69	1.00	0.00	0.75	0.83
	Tlc	0.00	0.00	1.00	1.00	1.69	−1.00	0.00	0.25	0.17

diagrams. The projection is made by deleting the row containing the system component from which one wishes to project from and by renormalizing the columns to a constant sum. The columns contain then the coordinates of the phase compositions in the new m-1 dimensional space. In the MSH system, used as an example here, the projection was made from the H_2O component in the three-component space onto the $MgO-SiO_2$ binary. However, often one desires to project from a composition which is not one of the original system components but from some composition in the system which corresponds to the composition of a phase that is present in excess. In order to make this possible, the composition matrix has to be rewritten in terms of a set of new system components. One of these new components has to be the composition one favors to project from.

For example, we would like to prepare a figure representing the phase compositions in the MSH system on a mole fraction triangle with the corners

Mg_2SiO_4 (Fo), $Mg(OH)_2$ (Brc), and $Mg_3Si_4O_{10}(OH)_2$ (Tlc). Secondly, we would like to study phase relationships in rocks which contain excess forsterite and need a projection from Mg_2SiO_4 (forsterite) onto the brucite-talc-binary.

The solution to the problem is shown in Table 2.6. In Table 2.6a the phase compositions are expressed in terms of the new system components Mg_2SiO_4 (Fo), $Mg(OH)_2$ (Brc), and $Mg_3Si_4O_{10}(OH)_2$ (Tlc). As an example, the composition of enstatite can be expressed by 2 Fo − 1 Brc + 1 Tlc, which is equivalent to 1 MgO + 1 SiO_2. Table 2.6b gives the coordinates of the mineral compositions in the mole fraction triangle Fo-Brc-Tlc. The algebraic operation which transforms the composition space expressed in terms of simple oxide components (Table 2.5a) to the composition space expressed in terms of Mg_2SiO_4 (Fo), $Mg(OH)_2$ (Brc) and $Mg_3Si_4O_{10}(OH)_2$ (Tlc) components (Table 2.6a) is also illustrated in Table 2.6. It can be seen that the operation is a pre-multiplication of Table 2.5a by the inverse of the leading 3×3 square matrix in Table 2.5a. The result of the operation is the composition matrix (Table 2.6a) with the mineral compositions expressed by the new set of system components. Today, any standard commercial spreadsheet program running on any PC or MAC will do these algebraic operations for you (e.g. Excel, Wingz). We are now set for the construction of the desired forsterite projection. Just as in simple projections, delete the Fo-row in Table 2.6b and renormalize to constant sum. Table 2.6c shows the coordinates of the phase compositions along the Brc-Tlc-binary. Some of the compositions project to the negative side of talc, periclase cannot be projected onto the Brc-Tlc binary at all. This is apparent also from Fig. 2.6, where it can be recognized that, seen from forsterite, periclase projects away from the Brc-Tlc-binary. It is also clear from the procedure outlined above, that the compositions which one chooses to project from or which one wants to see in the apexes of the mole fraction triangle must be written as column vectors in the original A-matrix.

With the two basic operations, projection and redefinition of system components, one can construct any thermodynamically valid composition phase diagram for any geologic problem (Greenwood, 1975).

4. AFM Projections. A classical example of a composite projection is the AFM projection for metapelitic rocks (see also Chap. 7). Many of the phase relationships in metapelitic rocks can be described in the six-component system K_2O-FeO-MgO-Al_2O_3-SiO_2-H_2O (KFMASH system). A graphical representation of the system requires projection from at least three fixed compositions. Many of the metapelitic rocks contain excess quartz and some aspects of metamorphism can be discussed for water-present conditions. Therefore, a projection from SiO_2 and H_2O can be easily prepared. However, none of the remaining four components is present as a phase in such rocks. Now, in low- and medium-grade metapelites muscovite is often present as an excess phase, in high-grade rocks muscovite is replaced by K-feldspar. A useful projection could therefore be made through muscovite or K-feldspar onto the plane Al_2SiO_5-FeO-MgO. This, in turn, requires that the composition matrix for minerals in metapelitic rocks is rewritten in terms of the new system components

Table 2.7a–d. Composition matrix for minerals in the KFMASH system. **a** Composition matrix in terms of oxide components. **b** Inverse of leading 6×6 square matrix (A-matrix $\Rightarrow$ A^{-1}). **c** Composition matrix in terms of Ky, Qtz, Ms, H$_2$O, FeO and MgO. **d** Renormalized composition matrix, column vectors are coordinates of mineral compositions in the mole fraction triangle Ky (A), FeO (F) and MgO (M) ($\Rightarrow$ AFM diagram), projection through Qtz, Ms and H$_2$O

a

	Ky	Qtz	Ms	Fl	FeO	MgO	Alm	Prp	Ann	Phl	FEs	Es	St	Cld	Crd	OPX	Spl	Chl
SiO$_2$	1	1	6	0	0	0	3	3	6	6	5	5	8	2	5	1.8	0	3
Al$_2$O$_3$	1	0	3	0	0	0	1	1	1	1	2	2	9	2	2	0.2	1	1
FeO	0	0	0	0	1	0	3	0	6	0	5	0	4	2	0	0.5	1	0
MgO	0	0	0	0	0	1	0	3	0	6	0	5	0	0	2	1.3	0	5
K$_2$O	0	0	1	0	0	0	0	0	1	1	1	1	0	0	0	0.0	0	0
H$_2$O	0	0	2	1	0	0	0	0	2	2	2	2	2	2	0	0.0	0	4

b

	Ky	Qtz	Ms	Fl	FeO	MgO
	0.00	1.00	0.00	0.00	-3.00	0.00
	1.00	-1.00	0.00	0.00	-3.00	0.00
	0.00	0.00	0.00	0.00	1.00	0.00
	0.00	0.00	0.00	0.00	-2.00	1.00
	0.00	0.00	1.00	0.00	0.00	0.00
	0.00	0.00	2.00	1.00	0.00	0.00

c

	Ky	Qtz	Ms	Fl	FeO	MgO	Alm	Prp	Ann	Phl	Fe-Es	Es	St	Cld	Crd	OPX	Spl	Chl
Ky	1.00	0.00	0.00	0.00	0.00	0.00	1.00	1.00	-2.00	-2.00	-1.00	-1.00	9.00	2.00	2.00	0.20	1.00	1.00
Qtz	0.00	1.00	0.00	0.00	0.00	0.00	2.00	2.00	2.00	2.00	0.00	0.00	-1.00	0.00	3.00	1.60	-1.00	2.00
Ms	0.00	0.00	1.00	0.00	0.00	0.00	0.00	0.00	1.00	1.00	1.00	1.00	0.00	0.00	0.00	0.00	0.00	0.00
Fl	0.00	0.00	0.00	1.00	0.00	0.00	0.00	0.00	0.00	0.00	0.00	0.00	2.00	2.00	0.00	0.00	0.00	4.00
FeO	0.00	0.00	0.00	0.00	1.00	0.00	3.00	0.00	6.00	0.00	5.00	0.00	4.00	2.00	0.00	0.50	0.00	0.00
MgO	0.00	0.00	0.00	0.00	0.00	1.00	0.00	3.00	0.00	6.00	0.00	5.00	0.00	0.00	2.00	1.30	1.00	5.00

d

	Ky	Qtz	Ms	Fl	FeO	MgO	Alm	Prp	Ann	Phl	Fe-Es	Es	St	Cld	Crd	OPX	Spl	Chl
Ky	1.00	0.00	0.00	0.00	0.00	0.00	0.25	0.25	-0.50	-0.50	-0.25	-0.25	0.69	0.50	0.50	0.10	0.50	0.17
FeO	0.00	0.00	0.00	0.00	1.00	0.00	0.75	0.00	1.50	0.00	1.25	0.00	0.31	0.50	0.00	0.25	0.00	0.00
MgO	0.00	0.00	0.00	0.00	0.00	1.00	0.00	0.75	0.00	1.50	0.00	1.25	0.00	0.00	0.50	0.65	0.50	0.83

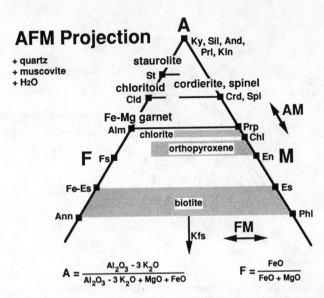

Fig. 2.8. Coordinates of phase compositions in the K_2O-FeO-MgO-Al_2O_3-SiO_2-H_2O system projected through muscovite, quartz and H_2O onto the plane Al_2O_3, FeO and MgO (AFM projection)

$KAl_3Si_3O_{10}(OH)_2$-Al_2SiO_5-FeO-MgO-SiO_2-H_2O as shown in Table 2.7. The coordinates of the phase compositions in the AFM mole fraction triangle can then be represented in an AFM diagram (Fig. 2.8). All Mg-Fe-free aluminium silicates project to the A-apex, pure enstatite and anthophyllite are found in the M apex, ferrosilite and Fe-anthophyllite project to the F apex. Biotites have negative A-coordinates. The iron-magnesium substitution (FeMg$_{-1}$ exchange) in the ferromagnesian minerals is parallel with the AM binary, the Mg-Tschermak substitution (2 Al Si$_{-1}$ Mg$_{-1}$ exchange) is parallel with the AM binary. Minerals such as staurolite, chloritoid, garnet, cordierite and spinel do not show any Tschermak variation, their compositional variation is exclusively parallel to the FeMg$_{-1}$ exchange. Other minerals, such as chlorites, biotites, orthopyroxenes and orthoamphiboles, show both FeMg$_{-1}$ exchange and 2 Al Si$_{-1}$ Mg$_{-1}$ exchange. The composition of these minerals are represented by fields in an AFM diagram. The AFM projection (Fig. 2.8) can be used exclusively for rocks with excess quartz and muscovite and for water-present conditions. Similar projections can readily be prepared from the composition matrix in Table 2.7c if one wants to analyze phase relationships for rocks containing excess muscovite, quartz and alumosilicate but treating H_2O as compositional variable on the diagram. Such a WFM diagram can be constructed by deleting the rows Ky, Qtz and Ms and renormalizing the remaining three rows. A QFM diagram (Qtz-FeO-MgO) projecting through Ms, Ky and H_2O can be constructed from Table 2.7c for discussing rocks with excess alumosilicate and muscovite but not excess quartz. To project from K-feldspar rather than muscovite, the

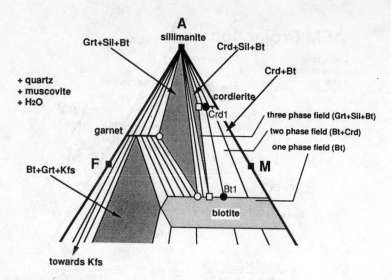

Fig. 2.9. AFM diagram at a specified pressure and temperature showing phase relationships among some typical AFM phases

Ms column in Table 2.7a has to be replaced by the column vector of K-feldspar, the subsequent procedure is as described above. If, for example in low-grade rocks, muscovite contains much Tschermak component (phengite), one may replace the Ms column vector (end member muscovite) in Table 2.7a by the analyzed mica composition.

Figure 2.8 shows the general projection coordinates of AFM phases. However, composition phase diagrams are used to represent phase relationships at a given pressure and temperature. A specific example of an AFM diagram at a distinct P and T is shown in Fig. 2.9. Present in excess are quartz, muscovite and H_2O. The AFM surface is divided into a number of subareas. Three types can be distinguished:

1. One-phase fields: if the total rock composition projects inside, for example, the one phase field for biotite, it will be composed of quartz, muscovite and biotite. The composition of biotite is given by the composition of the rock. The one-phase fields are divariant fields because the composition of the mineral freely changes with the two composition variables of the bulk rock.
2. Two-phase fields: if the total rock composition projects inside the two-phase field biotite + cordierite it will contain the assemblage Ms+Qtz+Crd+Bt. The composition of coexisting biotite and cordierite can be connected with a tie line (isopotential line) passing through the projection point of the bulk rock composition (e.g. Bt1 – Crd1 in Fig. 2.9). Through any bulk rock inside the two-phase field a tie line can be drawn which connects the two coexisting minerals (tie line bundle). It follows that all bulk rock compositions on a specific Crd-Bt tie line (at a given P and T) contain minerals of identical

composition. Two-phase fields are univariant fields. If the total rock projects close to cordierite it will contain much cordierite and little biotite (and vice versa). The mineral compositions are controlled by the Fe-Mg exchange equilibrium between cordierite and biotite. Also, the Tschermak exchange in biotite is controlled by the assemblage. One can also express the same properties of a two-phase field in other words. Consider, for example, a bulk rock which projects in the garnet + biotite field. For a given biotite composition, the composition of the coexisting garnet is fixed by the tie line passing through the given biotite composition and the projection point of the bulk rock.

3. Three-phase fields: if the total rock composition projects inside the three-phase field biotite + cordierite + sillimanite it will contain the assemblage Ms+Qtz+Crd+Bt+Sil. Here, the compositions of all minerals are entirely controlled by the assemblage (open squares in Fig. 2.9). All bulk rock compositions which project in the three-phase field Crd+Bt+Sil will be composed of these three minerals of identical composition in modal proportions depending on the rocks composition. Three-phase fields are invariant fields at constant P and T, because the mineral compositions do not vary with the rock composition. The cordierite coexisting with biotite and sillimanite has the most Fe-rich composition possible at the pT conditions of the AFM diagram. Note that, in general, all three-phase fields on an AFM diagram must be separated from each other by two-phase fields. In the three-phase field Grt+Sil+Bt the compositions of the minerals are controlled by the assemblage as well (open circles in Fig. 2.9). It follows that biotite in this assemblage must be more Fe-rich than biotite in the assemblage Crd+Sil+Bt. The garnet coexisting with Bt+Sil has the most Mg-rich composition possible at the given pressure and temperature.

AFM-type diagrams will be extensively used for discussing metamorphism in metapelitic rocks in Chapter 7.

5. ACF Projections. Phase relationships in marbles, calc-silicate rocks, calcareous metapelites and other rocks with calcic phases can be analyzed, for example, in a simple CFMAS system. Calcic phases may include amphiboles, plagioclase, epidote, diopside and carbonate minerals. A graphic representation of the phase relationships can, for instance, be made by projecting from quartz (if present in excess) and from a CO_2-H_2O fluid phase of constant composition onto the mole fraction triangle Al_2O_3-CaO-FeO (ACF diagram). The ACF diagram is also a projection parallel to the $FeMg_{-1}$ exchange vector. All minerals of the AFM system also can be represented on ACF diagrams provided that one also projects through muscovite or K-feldspar. The coordinates of mineral compositions can be calculated as explained above for the MSH- and KFMASH systems, respectively. Figure 2.10 is a typical ACF diagram, showing phase relationships at a certain pressure and temperature. The Tschermak variation is parallel to the AF binary and affects mainly chlorite and amphibole (and pyroxene which is not present under the pT conditions of the figure). The three-phase fields Pl+Am+Grt and Ky+Grt+Pl are separated

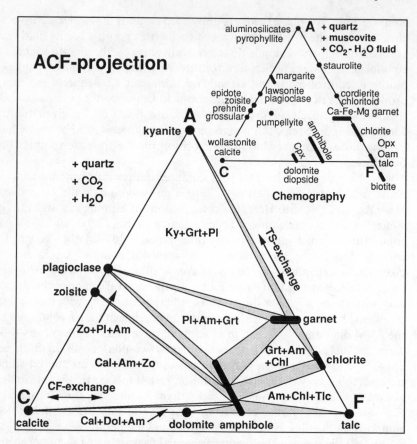

Fig. 2.10. Coordinates of phase compositions in the (K_2O)-CaO-FeO-MgO-Al_2O_3-SiO_2-H_2O-CO_2 system projected through (muscovite), quartz, CO_2 and H_2O and parallel to the $MgFe_{-1}$ exchange vector onto the plane Al_2O_3, FeO and MgO (ACF projection). *Large triangle* Phase relationships among some typical ACF phases at a specified pressure and temperature; *small triangle* cordinates of some ACF phase compositions

by a two-phase field Grt+Pl because of $CaMg_{-1}$ substitution in garnet (grossular component). Both garnet and plagioclase show no compositional variation along the TS vector. Any mineral which can be described by the components K_2O-CaO-FeO-MgO-Al_2O_3-SiO_2-H_2O-CO_2 can be represented on an ACF diagram such as Fig. 2.10. However, the consequences of $FeMg_{-1}$ substitution in minerals cannot be discussed by means of ACF diagrams. Therefore, any discontinuous reaction relationship deduced from an ACF diagram is continuous and dependent on the Fe-Mg variation (if it involves Fe-Mg minerals, of course). For example, the replacement of the garnet-plagioclase tie line by a more stable tie line between kyanite and amphibole can be related to the reaction: $Pl + Grt \Rightarrow Am + Ky \pm Qtz \pm H_2O$. Equilibrium of the reaction, however, depends not only on pressure and temperature but also on the Fe-Mg variation in garnet and amphibole. The projection coordinates of a selection of mineral compositions is shown on the chemography on the upper right of Fig. 2.10.

6. Other Projections. Any other graphic representation of phase relationships on composition phase diagrams can be prepared by the procedure outlined above. The type of graphic representation of assemblages is entirely dictated by the material one is working with and by the problem one wants to solve. The following steps may be a guide to the production of adequate phase composition diagrams.

1. Group the collected rocks in populations with similar bulk compositions ("normal" metapelites, metabasalts and so on).
2. Identify minerals which are present in the majority of a given group of rocks (e.g. muscovite and quartz in metapelites). Special assemblages require a special treatment (e.g. in quartz-absent corundum-bearing metapelites one may project through corundum onto an AFM plane).
3. If the excess minerals are not simply composed of oxide system components, rewrite the composition matrix in terms of the compositions of the excess phases and the desired compositions in the corners of the mole fraction triangle selected as projection plane.
4. Delete rows in the composition matrix containing the excess phases and renormalize the column vectors. Draw the diagram and keep in mind the proper distribution of one- two- and three-phase fields. Do not forget to write the compositions of the projection phases on the diagram (without this information your figure is worthless).

In Part II of this book we will make extensive use of various kinds of composition phase diagrams.

References

Carmichael RS (1989) Practical handbook of physical properties of rocks and minerals. CRC Press, Boca Raton

Greenwood HJ (1975) Thermodynamically valid projections of extensive phase relationships. Am Mineral 60:1–8

Orville PM (1969) A model for metamorphic differentiation origin of thin-layered amphibolites. Am J Sci 267:64–68

Spear FS (1988) Thermodynamic projection and extrapolation of high-variance assemblages. Contrib Mineral Petrol 98:346–351

Spear FS, Rumble D III, Ferry JM (1982) Linear algebraic manipulation of n-dimensional composition space. In: Ferry JM (ed) Characterization of metamorphism through mineral equilibria, vol 10. Mineralogical Society of America, Reviews in Mineralogy, Washington DC, pp 53–104

Spry A (1969) Metamorphic textures. Pergamon Press, Oxford

Thompson JB (1957) The graphical analysis of mineral assemblages in pelitic schists. Am
    Mineral 42:842–858
Thompson JB (1982) Composition space; an algebraic and geometric approach. In: Ferry JM
    (ed) Characterization of metamorphism through mineral equilibria, vol 10. Mineralogical
    Society of America, Reviews in Mineralogy, Washington DC, pp 1–31
Zeck HP (1974) Cataclastites, hemiclastites, holoclastites, blasto-ditto and myloblastites –
    cataclastic rocks. Am J Sci 274:1064–1073

3 Metamorphic Process

Rock metamorphism is always associated with processes and changes. Metamorphism reworks rocks in the Earth's crust and mantle. Typical effects of rock metamorphism include:

- Minerals and mineral assemblages originally not present in the rock may form, the new mineral assemblages form at the expense of old ones, consequently older minerals may disappear [e.g. a metamorphic rock may originally contain Grt + Qtz + Sil; a metamorphic event transforms this rock into one that contains Crd (cordierite) in addition to the minerals previously present in the rock].
- The relative abundance of minerals in a rock may systematically change and the new rock may have a different modal composition (metamorphism may increase the amount of Crd present in the rock and decrease the volume proportion of Grt + Qtz + Sil).
- Metamorphic minerals may systematically change their composition (e.g. the X_{Fe} of Grt and Crd may simultaneously increase during metamorphism).
- The structure of rocks in crust and mantle may be modified (e.g. randomly oriented sillimanite needles may be aligned parallel after the process).
- The composition of the bulk rock may be altered during metamorphism by adding or removing components to or from the rock from a source/sink outside the volume of the considered rock (e.g. adding K_2O dissolved in an aqueous solution to a Grt + Crd + Sill + Qtz rock may result in the formation of biotite).

The typical changes in the modal composition of rocks and in the composition of minerals that constitute the rocks are caused by chemical reactions. Metamorphism is, therefore, strongly related to the principles of chemical reactions. Mineral- and rock-forming metamorphic processes are mainly controlled by the same parameters that control chemical reactions. Metamorphic petrology studies reaction and transport processes in rocks. Metamorphic processes are caused by transient chemical, thermal and mechanical disequilibrium states in local volumes of the Earth's crust and mantle. These disequilibrium states ultimately result from large-scale geological processes and the dynamics of the Earth's planetary system as a whole. Metamorphic processes always result from disequilibrium and gradients in parameters that control reaction and transport in rocks. Metamorphic processes cease when the rocks reach an equilibrium state (or a steady state). Chemical reaction processes are always inherent in the term metamorphism. The term metamorphosis actually means transformation, modification, alteration, conversion and thus is clearly a process-related expression.

Metamorphism is a very complex incident and involves a large number of chemical and physical processes at various scales. Metamorphic processes can be viewed as a combination of (1) chemical reactions between minerals and between minerals and gasses, liquids and fluids (mainly H_2O) and (2) transport and exchange of substances and heat between domains where such reactions take place. The presence of an aqueous fluid phase in rocks undergoing metamorphism is critical to the rates of both chemical reactions and chemical transport. Consequently, an advanced understanding of metamorphism requires a great deal of insight into the quantitative description of chemical reactions and chemical transport processes, especially reversible and irreversible chemical thermodynamics.

The term metamorphism as it is related to processes, changes and reactions clearly also includes the aspect of time. Metamorphism occurs episodically and is particularly related to mountain-building episodes at convergent plate margins (collision zones) and during subsequent uplift and extension of continental crust, but also during sea-floor spreading and continental rifting.

3.1 Principles of Metamorphic Reactions

In the following we briefly explain some basic aspects of metamorphic reactions and introduce some elementary principles that are essential for a fundamental understanding of metamorphism. The treatment is not sufficient for a thorough understanding of metamorphic processes and chemical reactions in rocks. It is therefore recommended for the reader who needs to know more to study textbooks in chemical thermodynamics (e.g. Prigogine 1955; Lewis and Randal 1961; Moore 1972; Guggenheim 1986;) or textbooks that particularly deal with application of thermodynamics to mineralogy and petrology (Wood and Fraser 1976; Fraser 1977; Greenwood 1977; Powell 1978; Lasaga and Kirkpatrick 1981; Ferry 1982; Saxena and Ganguly 1987). We particularly recommend Chatterjee (1991) and the superb book by Fletcher (1993).

Now let us consider, for example, a rock that contains the minerals albite and quartz. The Ab-Qtz rock of presumed constant temperature and pressure is located at a certain depth in the crust (e.g. at point $T_h = 12\,500$ bar, 550 °C in Fig. 3.1). At that given pressure and temperature the two minerals (phases) are associated with unique values of molar Gibbs free energy. The Gibbs free energy, usually abbreviated with a G, is a thermodynamic potential with the dimension J/mol (energy/mole) and it is a function of pressure and temperature. The free energy of minerals and their mixtures are negative quantities (because they refer to the free energy of formation from the elements or oxides rather than absolute energies, e.g. G of albite at 900 K and 1 bar: -3257.489 kJ

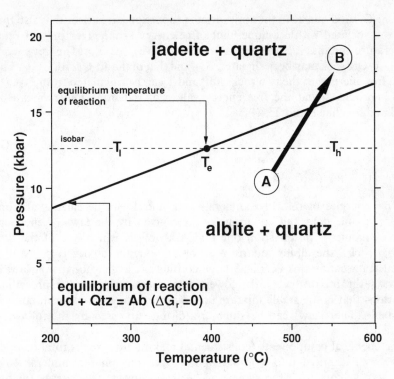

Fig. 3.1. Pressure – temperature diagram showing the quilibrium conditions of the reaction: jadeite + quartz = albite

mole^{-1}). The considered Ab + Qtz rock can be formed by mechanically mixing, for example, 1 mol albite and 1 mol quartz. All rocks represent, thermodynamically speaking, mechanical mixtures of phases. Rocks are, therefore, heterogeneous thermodynamic systems[1]. The phases, in turn, can be viewed as chemically homogeneous sub-spaces of the considered system, for example a volume of rock. Minerals, aqueous fluids, gases and melts are therefore the thermodynamic phases in rocks. The phases that make up rocks are usually chemical mixtures of a number of phase components (most minerals are solid chemical solutions and show a wide range in composition). In our example, albite and quartz shall be pure $NaAlSi_3O_8$ and SiO_2 respectively. The total free energy of the rock is the sum of the free energies of its parts that is in our case $G_{Ab} + G_{Qtz}$. The rock is characterized by a unique value of G and its total composition is $NaAlSi_4O_{10}$. However, the composition of such a mechanical mixture ($\equiv$ rock) can also be obtained by mixing (powders of) jadeite

---

[1] The thermodynamic description of heterogeneous systems has been developed and formulated mainly by W. Gibbs (Gibbs 1878, 1906). Gibbs' scientific contributions were fundamental for the development of modern quantitative petrology.

$(NaAlSi_2O_6)$ and quartz in the appropriate proportions. It is clear that also this rock is associated with a unique Gibbs free energy at the given pressure and temperature and that it corresponds to the sum of $G_{Jd} + 2\,G_{Qtz}$. The free energy of the Ab-Qtz rock may be designated G_{AQ} and that of the Jd-Qtz rock G_{JQ}. The Gibbs free energies of the two rocks at P and T can be calculated from Eqs. (1) and (2) provided that the free energy values of the three minerals can be calculated for that P and T:

$$G_{AQ} = G_{Ab} + G_{Qtz}, \tag{1}$$

$$G_{JQ} = G_{Jd} + 2\,G_{Qtz}, \tag{2}$$

The free energies of the three minerals at P and T can be calculated from thermodynamic data and equations of state given by the laws of chemical thermodynamics. The question one may ask now is which one of the two possible rocks, the albite + quartz rock or the jadeite + quartz rock, will be present at the conditions T_h (Fig. 3.1). According to the thermodynamic laws it is always the mixture with the lowest total free energy at the prevailing conditions that is the **stable** mixture (assemblage) while the other mixture is **metastable**. These laws can be summarized and expressed by the following statement:

The chemical components (constituents) making up a rocks total composition are distributed into a group of homogeneous phases [minerals and fluid(s)], that constitute the assemblage with the lowest free energy for the system at that given pressure and temperature. This assemblage is called the equilibrium phase assemblage.

At T_h G_{AQ} is more negative than G_{JQ} and thus the albite + quartz rock is stable, while the jadeite + quartz assemblage is metastable or less stable than Ab + Qtz. The free energy difference of the two rocks can be calculated by subtracting Eq. (2) from equation (1):

$$\Delta G = G_{AQ} - G_{JQ} = G_{Ab} - G_{Jd} - G_{Qtz}. \tag{3}$$

The ΔG is the free energy difference of the two rocks with identical composition but different mineralogy. The mineralogy of the two rocks is related by a reaction that can be expressed by the stoichiometric equation:

$$1\ NaAlSi_2O_6 + 1\ SiO_2 = 1\ NaAlSi_3O_8. \tag{4}$$
$$\text{jadeite}\qquad\text{quartz}\qquad\text{albite}$$

The stoichiometric coefficients in this particular reaction equation are all equal to 1. The equation suggests that albite in the feldspar + quartz rock that is stable at T_h may decompose to jadeite + quartz at some other conditions than T_h. Similarly, the reaction Eq. (4) describes the formation of 1 mol Na-feldspar from 1 mol jadeite + 1 mol quartz. The free energy change of the reaction ΔG_r is calculated from:

$$\Delta G_r = G_{Ab} - G_{Jd} - G_{Qtz}. \tag{5}$$

By convention, the stoichiometric coefficients on the right hand side of a reaction equation are always taken positive, while those on the left hand side are taken as negative. In the example made here the energy difference of the two rocks [Eq. (3)] is identified as the free energy change of the albite-forming reaction. ΔG_r is, like G of the individual phases, a function of P and T. Three basic cases may be distinguished:

$$\Delta G_r < 0. \tag{6}$$

For all pressure and temperature conditions that satisfy Eq. (6) the products of the reaction are stable and the reactants are metastable. At T_h, ΔG_r is negative and albite + quartz constitute the stable assemblage. A rock with jade-ite + quartz is metastable at T_h.

$$\Delta G_r > 0. \tag{7}$$

For all conditions that satisfy equation [7] the reactants of the reaction are stable and the products are metastable. At T_l, ΔG_r is positive and jade-ite + quartz constitute the stable assemblage (Fig. 3.1). A rock with al-bite + quartz is metastable at T_l.

$$\Delta G_r = 0. \tag{8}$$

For all conditions that satisfy Eq. (8) the reactants and the products of the reaction are stable simultaneously at the same conditions. These conditions are referred to as the equilibrium conditions of the reaction. In Fig. 3.1, the equilibrium condition of the reaction expressed by Eq. (8) is represented by a straight line. Along that line all three minerals are simultaneously stable. At a pressure of 12 500 bar, the equilibrium temperature of the reaction is 400°C and corresponds to the point T_e along the isobar of Fig. 3.1. At this unique temperature the rock may contain all three minerals in stable equilibrium. At any temperature other than T_e, the reaction is not at equilibrium and it will proceed in such a way as to produce the assemblage with the most negative free energy for a given temperature. In Fig. 3.1, the equilibrium line of the reaction divides the pT space into two half spaces with two distinct stable assemblages. The pT diagram represents in this way a free energy map of the considered system. Like a topographic map where the surface of the Earth is projected along a vertical axis, a pT-phase diagram represents a map where the lowest free energy surface of the system is projected along the G-axis onto the pT plane.

At equilibrium, the mineralogical composition of a rock is entirely dictated by its chemical composition, the pressure and the temperature. If temperature and pressure change, new assemblages may have a lower free energy, and chemical reaction will replace the old assemblage by a new, more stable assemblage. A rock always tries to reach a state of equilibrium by minimizing its free energy content. This is accomplished by readjusting the mineralogy or the composition of minerals if necessary.

Stable assemblages cannot be distinguished from metastable ones by any petrographic technique, and metastable equilibria cannot be separated from stable equilibria. This becomes obvious if we look at a rock sample collected on a rainy day that contains the three aluminosilicate minerals, kyanite, andalusite, sillimanite in addition to quartz (a rock type that is relatively common in the central Alps). All three Al-silicates are metastable in the presence of quartz and water under surface conditions relative to the hydrous Al-silicate kaolinite. In general, at some arbitrary metamorphic pT condition only one Al-silicate + quartz can be stable, the other two possible two-phase assemblages must be metastable. Metastable persistence of metamorphic minerals and assemblages is common in metamorphic rocks. This is, of course, an extremely fortunate circumstance for metamorphic petrologists. Rocks that formed in the deep crust or mantle with characteristic high-pressure and high-temperature mineral assemblages may be collected at the Earth's surface. If metastable assemblages were not common in metamorphic rocks we would find only low-pT rocks at the surface. Metastable assemblages may survive even over geological time scales (hundreds of millions of years). It is obvious from Fig. 3.1 that a rock that contains jadeite + quartz is metastable under conditions at the surface of the Earth. However, rocks containing Jd + Qtz can be found at the surface and witness that the rocks were exposed to very high pressures earlier in their geological history. Other examples of metastable persistence of minerals that are stable at high pressure or temperature are the presence of minerals such as sillimanite, coesite, aragonite, and diamond in rocks collected at the Earth's surface.

Some criteria and methods to detect **disequilibrium** in rocks do exist, however. One may distinguish two main kinds of disequilibrium; structural disequilibrium and chemical disequilibrium. Structural disequilibrium is indicated by distinct shapes and forms of crystals and spatial distribution and arrangement of groups of minerals in heterogeneous systems. Structural disequilibrium may be found in rocks that underwent chemical reactions or successive series of reactions that all ceased before equilibrium structures developed and overall chemical equilibrium was reached. Many different kinds of compositional features of metamorphic rocks and minerals may indicate chemical disequilibrium. As an illustration we may use our example again. We may have collected a rock that contains omphacite + quartz (sodium-rich clinopyroxene where jadeite represents a phase component in that pyroxene) and independent information suggests that this rock formed at about 600°C. The same rock may also contain pure albite in domains, local patches or veins. The feldspar composition in this rock represents a clear chemical disequilibrium feature. The minerals omphacite + albite + quartz never coexisted in stable or metastable equilibrium because the pyroxene contains a calcic component

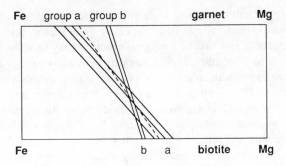

Fig. 3.2. Schematic figure showing the Fe-Mg distribution between pairs of coexisting garnet and biotite from one specific rock sample

(diopside) and the feldspar does not (no anorthite). However, at the inferred temperature of 600°C, plagioclase in a mafic omphacite-bearing rock should, in an equilibrium situation, contain some anorthite component. The presence of albite in such a rock must be related to some metamorphic process that progressed under conditions other than the equilibration of the omphacite + quartz assemblage. It is, on the other hand, important to note that the assemblage omphacite + quartz + albite represents an overall disequilibrium but the assemblages omphacite + quartz and albite + quartz may well be equilibrium assemblages that equilibrated at different times at different pT conditions. The question of equilibrium is always related to the scale of equilibrium domains. Disequilibrium may exist between large rock bodies in the crust, layers of rocks of different composition in an outcrop, between local domains of a thin section or between two minerals in mutual grain contact. Any chemically zoned mineral represents disequilibrium. At some scale there is always and at any time disequilibrium.

Chemical properties of coexisting minerals in a rock are often used to reason about the equilibrium question. As an example, a series of garnet-biotite pairs have been analyzed from a homogeneous Grt-Bt-schist and the data are schematically shown in Fig. 3.2. The data pairs can be arranged into two groups. *Group a* represents matrix biotite in contact with the rim of garnet, whilst *group b* represents analyses of biotite grains included in the core of the garnet. Overall chemical equilibrium (stable or metastable) between garnet and biotite requires that both minerals have a uniform composition in the rock. Crossing tie lines are inconsistent with overall equilibrium. However, it is evident that the Grt - Bt pairs from each group are not in conflict with the requirements of chemical equilibrium. They may constitute two different **local equilibrium** systems. The Grt-Bt pair connected with a dashed line in Fig. 3.2 shows clear crossing tie-line relationships within *group a* pairs and thus, clearly represents a disequilibrium pair. Note that the absence of crossing tie-line relationships does not necessarily prove that the rock was in a state of overall equilibrium. The lack of obvious disequilibrium phenomena can be taken as evidence for, but not proof of equilibrium.

Metastable persistence of minerals and mineral assemblages but also clear disequilibrium in rocks follows from the controlling factors and circumstances of reaction kinetics. The rate of a mineral reaction may be slower than the rate of, e.g., cooling of a volume of rock. Lacking activation energy for reaction in cooling rocks and nucleation problems of more stable minerals typically affects reaction kinetics. The kinetics of reactions in rocks is extremely sensitive to the presence or absence of H_2O. Aqueous fluids serve as solvent and reaction medium for mineral reactions. For example, if kyanite-bearing rocks are brought to pressure and temperature conditions where andalusite is more stable than kyanite, kyanite may be replaced by andalusite in rocks containing free aqueous fluid in the pore space or along grain boundaries, whilst andalusite may fail to form in fluid-absent dry rocks.

However, it is commonly reasonable to assume that during **prograde** metamorphism rocks pass through successive sequences of equilibrium mineral assemblages. These sequences can be viewed as a series of stages, each of them characterized by an equilibrium assemblage and the different stages are connected by mineral reactions. This assumption is founded on much convincing evidence that prograde metamorphism takes place under episodic or continuous water-present conditions. One would therefore expect to find disequilibrium and metastable assemblages particularly in rocks that were metamorphosed under fluid-absent or fluid-deficient conditions. Aqueous fluids are typically absent in cooling rocks after they have reached maximum metamorphic pressure and temperature conditions. Structural and chemical disequilibrium is also widespread in very high-grade rocks of the granulite facies that lost the hydrous minerals and aqueous fluids during earlier stages of prograde metamorphism. Microstructures such as reaction rims, symplectites, partial replacement, corrosion and dissolution of earlier minerals are characteristic for granulite facies rocks. They indicate that, despite relatively high temperatures (700 to 900°C), equilibrium domains were small and chemical communication and transport were hampered as a result of dry or H_2O-poor conditions.

To further illustrate some aspects of disequilibrium, we may consider large bodies of incompatible rock types in the crust. In many orogenic fold belts (Alps, Caledonides) mantle-derived ultramafic rock fragments were emplaced in the continental crust during the collision phase and stacking of nappes. Mantle fragments (harzburgites, lherzolites, serpentinites) of various dimensions can be found as lenses in granitic crust (Fig. 3.3). Forsterite (in the mantle fragment) + quartz (in the crustal rocks) is metastable relative to enstatite at any geologically accessible pressure and temperature conditions. The presence of Fo-bearing mantle rocks in Qtz-rich crustal rocks represents a large-scale disequilibrium feature. Chemical mass transfer across the contact of the incompatible rock types results in the formation of shells of reaction zones that encapsulate the mantle fragment. The minerals found in the reaction zones depend on the pT conditions at the reaction site. In our example, the reaction shells are made up of talc-rich and biotite-rich zones that are often nearly monomineralic (Fig. 3.3). Such biotite- or amphibole-rich shells that grew as a

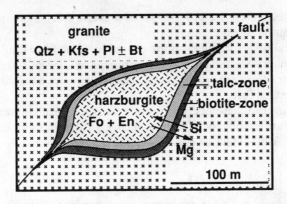

Fig. 3.3. Large-scale disequilibrium between a lens of harzburgite (ultramafic rock fragment from the upper mantle) that has been tectonically emplaced in granitic (gneissic) crust. The harzburgite lens has been enveloped by protective shells of talc and biotite. These shells were produced by chemical reaction between the incompatible lithologies and their formation required transfer of, e.g., Si from the granite to the harzburgite and Mg from the harzburgite to the granite

result of disequilibrium on a large scale and subsequent chemical transport and reaction are also known as black-wall formation. The process is one of contact metasomatism. Chemical communication, transport and reaction are most efficient if aqueous fluids are present in excess and wet the grain boundaries in the incompatible rock bodies. Exchange metasomatism proceeds until the incompatible mineralogy has been used up by the process or the rocks lose their aqueous fluid phase. In rocks with dry grain boundaries, diffusional mass transfer is extremely slow and inefficient even over geological time spans.

Let us return to the jadeite-quartz-albite example (Fig. 3.1). At the conditions of point A, a rock containing albite + quartz is more stable than a rock made of jadeite + quartz. Consider an albite + quartz rock that has reached a state of chemical and structural equilibrium at conditions of point A. The assemblage of this rock may be replaced by jadeite + quartz by reaction [4] if the rock is brought to the conditions at point B. It is obvious that a transfer from A to B would bring the Ab + Qtz rock into the field where Jd + Qtz represents the stable assemblage.

3.2 Pressure and Temperature Changes in Crust and Mantle

3.2.1 General Aspects

In the previous section it became obvious that the changes in chemical and physical properties of metamorphic rocks of constant composition are caused by changes in pressure and temperature. The next question must be: what kind of geologic mechanisms lead to changes of pressure and temperature in a given volume of crust or mantle.

In a general way, such changes are caused by some force that acts on rocks (driving force: driving the process or change). Any kind of force applied to a rock will cause some flow or transfer of a property in such a way as to reduce the size of the applied force.

Temperature differences between volumes of rocks result in transport of heat (heat flow) from the hot rock to the cold rock until both rocks reach the same temperature. A simple linear equation describes this process:

$$J_q = - L_q \nabla T. \tag{9}$$

Equation (9) states that a non-uniform temperature distribution in a system will transfer heat (Q) in the direction normal to the isotherms of the temperature field and from high temperature to low temperature in order to decrease the temperature difference. The linear equation (known as Fourier's law) describes heat flow in systems with small temperature differences (such as most geologic systems). L_Q is a material dependent constant related to the thermal conductivity of the material $\varkappa$. J_q is the rate of heat transfer per unit area (heat flow vector) parallel to the highest gradient in the temperature field ($J_q = \partial Q/\partial t$). ∇T represents the gradients in temperature in the three-dimensional space. If there are only linear temperature gradients in one direction, the ∇T in Eq. (9) reduces to $(T_{x1} - T_{x2})$, the temperature difference between two points. The general expression presented above shows that it is gradients in intensive variables in a system that ultimately cause "processes" and "changes" geological or other. As metamorphism is mainly related to chemical reactions in rocks that are largely controlled by changes in temperature, pressure and rock composition, it is useful to discuss geological aspects of heat transfer, pressure changes and chemical mass transfer in crust and mantle in some detail.

3.2.2 Heat Flow and Geotherms

Fourier's law states that heat will be transported from a place at high temperature to an area at low temperature. In the case of the Earth, and looking at it on a large scale, there is a hot interior and a cold surface. This necessarily results in transport of heat from the center to the surface. The surface of the earth is at a nearly constant temperature (about 0°C) and the core mantle boundary is probably also at a constant temperature as a result of the liquid state of the outer core that permits very rapid convection and heat transport. The general situation is depicted in Fig. 3.4. A consequence of the temperature difference between the two surfaces is a steady and continuous heat flow from the interior to the surface.

Heat flow is measured in mW m^{-2} (milliwatt per square meter); however, in the literature heat flow data are often given in HFU (heat flow units). An HFU is defined as μcal cm^{-2} s^{-1}, that is equivalent to 4.2 μJ s^{-1} cm^{-2} or 0.042 J s^{-1} m^{-2}.

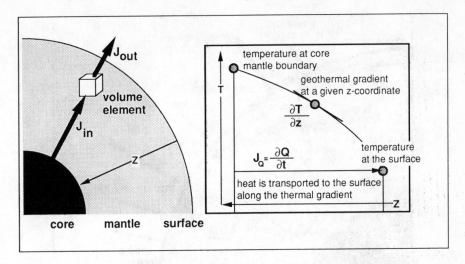

Fig. 3.4. Conductive heat transfer between the hot interior and the cold surface of the Earth

Because $J\,s^{-1}$ is equivalent to W (watts), it follows that 1 HFU is equal to 42 mW m^{-2}.

The heat flow from the interior of the Earth is on the order of 30 mW m^{-2}. However, heat flow measurements at the surface of the Earth vary between about 30 and 120 mW m^{-2}. The total heat flow at the surface is composed of a number of contributions: (1) heat flow from the interior resulting from conductive heat transport as described by Fourier's law, (2) transport of heat by convection in the mantle (Fig. 3.5), (3) transport of heat generated by decay of radioactive elements.

The continental crust consists mostly of granitoid rocks that produce about 30 mJ heat per kilogram and year. Oceanic crust, that is generally composed of basaltic rocks, produces about 5 mJ heat per kilogram and year, whereas mantle rocks produce only a small amount of radioactive heat (ca. 0.1 mJ heat per kilogram and year). The extra heat produced in the crust contributes significantly to the observed heat flow at the surface.

Heat flow through a volume element of the crust may occur under the following conditions:

1. The heat flow into the crustal volume is equal to the heat flow out from that volume. In this case the temperature in the volume element remains constant. The temperature profile along z is independent of time (**steady-state geotherm**).
2. The heat flow into the crustal volume is greater than the heat flow out from that volume. The excess heat put into the crustal volume will be used in two different ways: (a) to increase the temperature of the rock volume, (b) to drive endothermic chemical reactions in rocks.

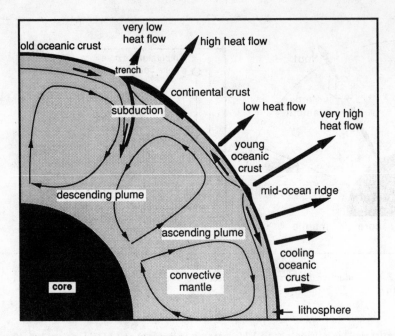

Fig. 3.5. Steady-state heat flow resulting from conductive heat transfer is modified by active tectonic processes. Heat flow at the surface varies by a factor of 4

3. The heat flow into the crustal volume is lower than the heat flow out from the volume. In this case the heat loss of the volume of rock results in a temperature decrease (the rock cools). In this situation exothermic chemical reactions may produce extra heat to some extent. These reactions have the effect to prevent the rocks from cooling.

Different heat flows at the surface also have the consequence that rocks at the same depth in the crust and upper mantle may be at different temperatures leading to lateral heat transport parallel to the earth surface (parallel to xy-surfaces). This is shown in Fig. 3.6. Along a profile from Denmark to southern Norway the observed surface heat flows have been used by Balling (1985) to model the temperature field in the crust and mantle. The temperature field is shown in Fig. 3.6. Because flow vectors are always normal to the force field from which the flow results, the heat flow trajectories will roughly look like the flow vectors shown in Fig. 3.8. It is obvious in this two dimensional section that at a given point in the crust the heat flow has a vertical and a horizontal component. Also note that the MOHO under Denmark that is underlain by mid-Paleozoic continental crust of central Europe is at about 700°C, whereas the MOHO under the Precambrian crust of the Baltic shield in the north is only at about 350°C (!). It follows that the base of continental crust may be at largely different temperatures depending on the state and evolutionary history of a lithosphere segment and on the thermal regime deeper in the mantle. Observed surface heat

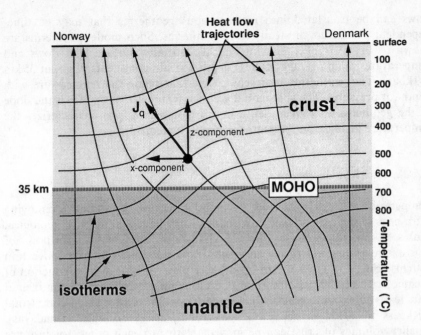

Fig. 3.6. Modelled temperature field along a cross section from Denmark to Norway. (After Balling 1985)

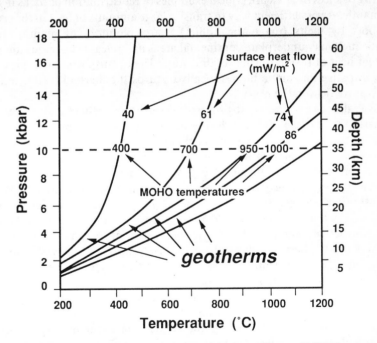

Fig. 3.7. Model geotherms with associated surface heat flow values and MOHO temperatures. (Modified after Stern et al. 1987)

flows can be translated into instantaneous geotherms that may be time-dependent geotherms or steady-state geotherms. Such model geotherms are shown in Fig. 3.7 together with typical associated surface heat flows and temperatures at an average MOHO depth beneath continents of about 35 km (10 kbar). The geotherms represent curves that relate the temperature with depth (or pressure). The geothermal gradient (expressed as dp/dT) is the slope of the geotherm at a given depth in crust or mantle and characterizes the temperature increase per pressure (or depth) increment.

3.2.2.1. Transient Geotherms

Temperature changes in the crust and mantle are caused by adding or removing extra heat to the rocks. Heat flow changes may have a number of geological causes. Deep-seated causes include changes in the relative positions of lithosphere plates and their continental rafts relative to mantle convection systems (Fig. 3.5). Collision of lithosphere plates may lead to subduction of oceanic crust and cause abnormal thermal regimes. Collision of continental rafts leads to extreme crustal thickening (from 35 km of normal continental thickness to 70 km in active collision zones, active example: Himalayas). Smaller volumes of crust can be moved relative to each other, for instance during a continent-continent collision. The formation of nappes and the stacking of slices of crust are typical examples of redistribution of rocks in crust and mantle. A very efficient way to transfer large amounts of heat in the crust is transport by melts (magmatic bodies). Large volumes of mantle-derived basaltic melt may underplate continental areas and release enormous amounts of latent heat of fusion during solidification. Heat transported to the crust by rising intrusive magmatic bodies locally disturbs the thermal regime and the temperature field in the crust.

Also, all tectonic transport processes disturb the pre-tectonic steady-state geotherm. At any one instant during active tectonic transport, the crust is characterized by a specific geotherm that relates temperature to depth. The geotherm is now changing with time. These **transient geotherms** describe the temperature-depth relationship at a given time and for a given geographic location (xy-coordinates). Some aspects of tectonic transport and its effect on the instantaneous geotherm are depicted in Fig. 3.8. The figure shows schematically the temperature field along a cross section through an Andean-type destructive margin. The down-going slab of oceanic lithosphere is relatively cold. Thermal relaxation is a slow process compared to tectonic movements. The continental lithosphere above the subducted oceanic plate is characterized by a complex thermally anomalous region that results from magmatic heat transfer and from smaller scale deformation in the crust (nappes). The geotherm at location **a** at a given instant (instantaneous geotherm) in the active history of the subduction event may look like the geotherm shown at the left hand side of Fig. 3.8. The tectonic transport may even be fast enough to create temperature inversion in a transient geotherm.

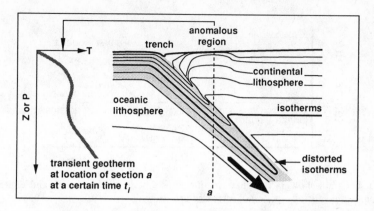

Fig. 3.8. Temperature field along a cross section through an Andean-type subduction zone and a transient geotherm at location *a*. (Modified from Brown and Mussett 1981)

Such temperature inversions will, however, be very short-lived phenomena and they will be eliminated rapidly if subduction ceases.

3.2.3 Temperature Changes and Metamorphic Reactions

If a crustal volume receives more heat than required to maintain a steady-state geotherm, the volume of rock will experience an increase in temperature and it will undergo **prograde metamorphism**. The heat capacity of the rock determines how much the temperature rises in relation to the heat added. Typical heat capacities for silicate rocks are on the order of 1 kJ per kg rock and °C. Suppose a quartzite layer in the crust receives 100 kJ per kg rock extra heat, this heat will increase the temperature of the quartzite by 100°C. Extra heat may also be received by rocks that must adjust their mineralogy by chemical reactions in order to maintain a state of minimized total free energy (as explained above). These endothermic reactions will consume a part of the heat added to the rock. Many metamorphic reactions during prograde metamorphism are dehydration reactions, which means that the reactions replace hydrous minerals (such as zeolites, micas, chlorite, amphibole) by less hydrous or anhydrous minerals and therefore release H_2O to a fluid phase. This fluid may or may not escape from the site of production. Such reactions are strongly endothermic and consume a large amount of heat [about 90 kJ per 1 mol H_2O (or CO_2)]. Consider now a layer of mica schist in the crust containing about 1 mol H_2O per kg rock. If a temperature increase of 100°C (as in the quartzite layer) results in complete destruction of the hydrous minerals in this rock, a total of 190 kJ per kg of schist is required to obtain the same temperature increase (100 kJ for the temperature increase and 90 kJ for the reaction).

The heat effect of mineral reactions in rocks also influences the temperature-time relationship in a volume of reactive rock. The case is illustrated in Fig. 3.9 using the $Qtz + Jd \Rightarrow Ab$ reaction as an example. Supply of excess heat (in

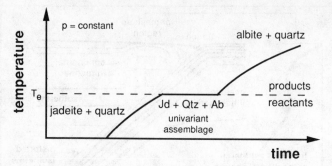

Fig. 3.9. Diagram showing the phase relationships among jadeite, quartz and albite as a function of time during a period of increased heat flow

addition to the steady state heat flow) will cause an increase in temperature until the equilibrium conditions (T_e) of the chemical reaction is reached. The heat input will then be used for the reaction[2] over a certain period of time and the temperature remains constant until one of the two reactant minerals (in this case Jd) is used up. After completion of the reaction, the temperature of the rock will continue to increase. The effect is analogous to heating water to the boiling point. The temperature of the water increases to the boiling temperature as heat is added to the water. During the boiling of the water the temperature remains constant and the heat is used to drive the reaction (phase transition) liquid water ⇒ steam. After all water is boiled off, further addition of heat will increase the temperature of the steam.

If a crustal volume receives less heat than it requires to maintain a steady-state geotherm, it will cool and be potentially affected by **retrograde metamorphism**. During cooling, it may become necessary for the rock to change its mineralogical composition in order to maintain a state of minimum free energy. In our example, for a rock consisting of albite and quartz, the reaction Ab ⇒ Qtz + Jd converts all albite into jadeite and quartz as the rock cools to T_e (Fig. 3.9) under equilibrium conditions. The heat released by the reaction buffers the temperature to a constant value (T_e) until all albite has been used up and further cooling can take place. The situation is analogous to removing heat from water. The temperature will decrease until water begins to freeze. As long as both reactants (liquid water) and products (ice) are present in the system, the temperature is confined to the freezing temperature. After conversion of all water into ice the temperature will continue to fall if further heat is removed from the ice.

---

[2] Note: the heat required for this type of H_2O-absent reaction is much less than the heat consumption for a dehydration reaction discussed above.

3.2.4 Pressure Changes in Rocks

Normally, the pressure field in crust and mantle is characterized by isobaric surfaces that are about parallel to the surface of the Earth. As outlined above, the free energy of minerals and associations of minerals (rocks) is also a function of pressure. Pressure changes in rocks are related to the change of position along the vertical space coordinate (z-axis in Fig. 3.4). The prevailing pressure at a given depth in a crustal profile on a steady-state geotherm is given by the average density of the material above the volume of interest. It can be calculated from ($dp/dz = -g\varrho$):

$$P_{(z)} = -g \sum_0^z \varrho_{(z)} \, dz + P_{(z=0)}, \tag{10}$$

where g is the acceleration due to gravity (9.81 m s^{-2}), ϱ is the density of the rock at any z (e.g. $2.7 \text{ g cm}^{-3} = 2700 \text{ kg m}^{-3}$ and taken as independent of depth) and $P_{(z=0)}$ is the pressure at $z = 0$ (e.g. $10^5 \text{ Pa} = 1 \text{ bar} = 10^5 \text{ N m}^{-2} = 10^5 \text{ kg m}^{-1} \text{ s}^{-2}$). The z-axis direction is always taken negative. The pressure, for example at 10 km (10000 m) depth, is then calculated as 264.97 MPa or 2.6497 kbar. The pressure at the base of continental crust of normal thickness (35–40 km) is about 10 kbar. The pressure at z results from the weight of the rock column above the volume of interest and is designated **lithostatic pressure**. This pressure is usually nearly isotropic. Non-isotropic pressure (stress) may occur as a result of a number of geologic feasible processes and situations as discussed below.

A pressure difference of about 100 bar at the same depth (10 km) will occur if the rock column along a nearby profile has a different average density (e.g. 2800 kg m^3). Because crustal material has very similar densities, possible pressure gradients between volumes of rocks at the same depth are typically only small fractions of the lithostatic pressure (< 10%). However, such pressure gradients may cause the flow of fluids stored in the interconnected pore space of rocks according to Darcy's law. The transport of fluids may cause chemical reactions in the rocks receiving these fluids if the fluids are not in equilibrium with the solid assemblage of the rock.

Non-isotropic pressure (stress) also results from tectonic forces and from volume changes associated with temperature changes of rocks. Stress is one of the major controlling factors of the structure of metamorphic rocks. It is also a very important force for small-scale migration and redistribution of chemical components in metamorphic rocks. The processes of pressure solution, formation of fabric (such as foliation) and formation of metamorphic banding from homogeneous protolith rocks are caused by non-isotropic pressure distribution in rocks and associated transport phenomena. The redistribution of material in stressed rocks seeks to reduce the non-isotropic pressure and return to isotropic lithostatic pressure conditions.

In general, the lithostatic pressure acting on a volume of rock changes as a result of a change of the depth of that volume of rock (change of the position along the z-direction). In almost all geological occurrences depth changes are caused by tectonic processes (that ultimately are also the result of thermal processes, e.g. mantle convection, density changes resulting from temperature changes). A lithosphere plate can be subducted in a collision zone and be transported to great depth in the mantle (or even to the core-mantle boundary) before it is resorbed by the mantle. Some of the subducted material may return to the surface before a normal steady-state geotherm is established. Metamorphic rocks of sedimentary or volcanic origin with mineral assemblages that formed at pressures in excess of 30 kbar (3×10^9 Pa = 3 GPa = 30 kbar) have been reported from various orogenic belts; 30 kbar pressure is equivalent to about 100 km subduction depth $[z = -P/(g\varrho)]$. In continental collision zones the continental crust often is thickened to the double of its normal pre-collision thickness. At the base of the thickened crust the pressure increases from 9 to 18 kbar. The changes in mineralogical composition of the rocks undergoing such dramatic pressure changes depend on the nature and composition of these rocks and will be discussed in Part II of this book. Variations in depth and associated pressure changes also can be related to subsiding sedimentary basins where a given layer **a** of sediment is successively overlain and buried by new layers **b, c**.. of sediments. The layer **a** experiences a progressive increase in pressure and temperature. If sedimentation and subsidence is slow the layer **a** may in fact follow a steady state geotherm. Metamorphism experienced by layer **a** is commonly described under the collective term **burial metamorphism**. Crustal extension or lithosphere thinning is often the cause for the required subsidence creating deep sedimentary basins. Burial metamorphism is therefore also caused by tectonic processes.

3.3 Gases and Fluids

Sedimentary rocks such as shales often contain large modal proportions of hydrous minerals. In fact, sediments deposited in marine environments can be expected to contain, under equilibrium conditions, a mineralogy that corresponds to a **maximum hydrated state**. Adding heat to the hydrous mineralogy (clays) of a sediment during a metamorphic event will drive reactions of the general form:

hydrous assemblage $\Rightarrow$ less hydrous or anhydrous assemblage $+ H_2O$.   (11)

The general reaction (11) is a description of dehydration processes taking place during prograde metamorphism. The important feature of dehydration reactions is the release of H_2O. Steam is, in contrast to solid minerals, a very compressible phase and its volume is strongly dependent on pressure and temperature. In Fig. 3.10 the molar volume of H_2O is shown as a function of pressure and temperature. At low temperatures, H_2O occurs either as a high density liquid phase or as low density steam phase depending on the prevailing pressure. The phase boundary between liquid and steam H_2O is the saturation

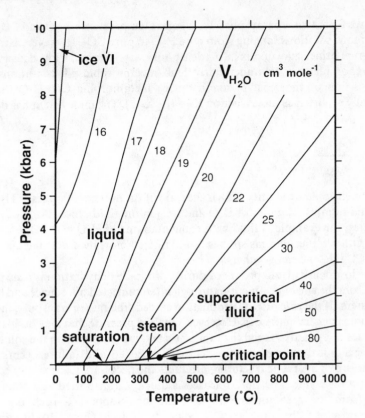

Fig. 3.10. Isochores of H_2O as a function of pressure and temperature. (Helgeson and Kirkham 1974)

or boiling curve. With increasing temperature the density contrast between steam and liquid H_2O continuously decreases along the boiling curve. At the critical point steam and liquid have the same density and the distinction between steam and liquid becomes obsolete. At temperatures above the critical point the H_2O phase is therefore a supercritical fluid phase or simply a fluid. Water in metamorphic systems at temperatures above $374\,°C$ is referred to as the (aqueous) **fluid phase or the fluid**. The critical point of H_2O is at $374\,°C$ and 217 bar (21.77 MPa).

Metamorphic fluids are usually dominated by H_2O. The H_2O of aqueous fluids in metamorphic rocks has its origin from various sources: (1) relic formation waters of sedimentary rocks, (2) dehydration water from H_2O originally stored in hydrous minerals, (3) meteoric water, and (4) magmatic water released from solidifying magmas. In the lower crust and in crustal segments with abundant carbonate rocks, CO_2 may become an additional important component in the fluid phase. Aqueous fluids also may contain significant amounts of dissolved salts (NaCl) and other solutes (e.g. aqueous silica complexes). At temperatures below $265\,°C$ (at $2\,kbar$), H_2O and CO_2

mixtures form two separate phases. At higher temperatures, H_2O and CO_2 form continuous solutions ranging from pure H_2O to pure CO_2. However, extremely NaCl-rich brines may unmix into a dense aqueous brine and a low density CO_2-rich vapor phase even under fairly high metamorphic pT conditions. The composition of the most common binary metamorphic CO_2-H_2O fluids is normally reported as mole fraction X_{H_2O} (or X_{CO_2}). The mole fraction is defined as:

$$X_{H_2O} = \frac{n_{H_2O}}{n_{H_2O} + n_{CO_2}}. \tag{12}$$

The dimensionless quantity X_{H_2O} is the ratio of the number of moles of H_2O and the total number of moles of H_2O and CO_2 in the fluid phase. An X_{H_2O} of 0.5 indicates, for example, a fluid with equal amounts of H_2O and CO_2 on a mole basis. Other molecular gas species found in metamorphic fluids include: CH_4, N_2, HCl, HF, and many others.

The low density fluid produced during prograde dehydration is transported away from the site of production through interconnected pore space and lost by the system. If the rate of H_2O production exceeds the rate of transport, then the local pore pressure increases. However, the fluid pressure that may build up this way does not greatly exceed the lithostatic pressure imposed on the solid rock. The mechanical strength of the rocks is exceeded and failure occurs. This mechanism of **hydraulic fracturing** produces the necessary transport system for dehydration water. The fluid released is quickly transported and channelled away from the area undergoing dehydration. Suppose a rock undergoes dehydration at 5 kbar and 600 °C. The produced fluid (Fig. 3.10) has a molar volume of about 22 cm^3 that corresponds to a density of about 0.82 g cm^{-3}. This is a remarkably high density for H_2O at 600 °C but it is still much lower than the typical density of rock-forming minerals of about 3 ± 0.5 g cm^{-3}. The large density contrast between fluid and solids results in strong buoyancy forces and a rapid escape of the fluid from the rocks. Most metamorphic rocks are probably free of fluid during periods without reaction with the exception of fluids trapped in isolated pore space and as inclusions in minerals (**fluid inclusions**).

3.4 Time Scale of Metamorphism

Some simple considerations allow for an estimate of the time scale of typical orogenic metamorphism (see also Walther and Orville 1982). The range of reasonable heat flow differences in the crust that can drive metamorphic processes is constrained by maximum heat flow differences observed at the surface. The highest heat flows are measured along mid-ocean ridges (120 mW m^{-2}), the lowest on old continental shields (30 mW m^{-2}). Consider now a layer of crust with a heat flow difference between bottom (75 mW m^{-2}) and top (35 mW m^{-2}) of 40 mW m^{-2}. This means that the rock column receives every second

0.04 Joules heat (per m^2). The layer shall consist of shales (heat capacity $= 1$ kJ kg^{-1} °C^{-1}) with a volatile content (H_2O and CO_2) of about 2 moles kg^{-1}. The heat received by the layer of rock will be used to increase the temperature of the rock and to drive endothermic devolatilization reactions. About 90 kJ heat are required to release 1 mole H_2O or CO_2. If complete devolatilization occurs in the temperature interval 400 to 600°C, a total of 380 kJ will be consumed by the model shale (180 kJ for the reactions and 200 kJ for the temperature increase from 400 to 600°C). It takes 9.5×10^6 seconds (0.3 years) to supply 1 kg of rock with the necessary energy. Using a density of 2.63 g cm^{-3}, 1 kg of rock occupies a volume of 380 cm^3 and it represents a column of 0.38 mm height and 1 m^2 ground surface. From this it follows that metamorphism requires about 8 years to advance by 1 cm. 8 million years (Ma) are required to metamorphose a shale layer of 10 km thickness. Similarly, if the heat flow difference is only 20 mW m^{-2}, the layer thickness 20 km and the initial temperature 200°C, it will then take about 48 Ma to metamorphose the rock from a hydrous 200°C shale to an anhydrous 600°C metashale.

These crude calculations show that typical time spans for regional scale metamorphic processes are on the order of 10 to 50 million years. Similar time scales have been derived from radiometric age determinations.

3.5 Pressure-Temperature-Time Paths and Reaction History

During tectonic transport, any given volume of rock follows its individual and unique path in space and time. Each volume of rock may experience loss or gain of heat, and changes of its position along the z-coordinate result in changes of the lithostatic pressure loaded on the rock. Figure 3.11A shows again a simple model of a destructive plate margin. The situation here depicts a continent-continent collision with the formation of continental crust twice its normal thickness. At the depth level c there is a volume element of rock (indicated by a filled square) at the time t_1 (0 Ma). In Fig. 3.11 B–E, a series of schematic pT diagrams show the position of the element of rock in pT space during the tectonic transport. The time slices are arbitrary and have been chosen in accordance with time scales of the formation of Alpine-type mountain chains. The volume element at t_1 is on a stable steady-state geotherm. At time t_2 (10 Ma) active tectonic transport moves the crust together with the volume element underneath, another continental crust of normal 35 km thickness. The increasing depth position of the volume element is accompanied by an increasing pressure on the volume element [Eq. (10)]. At the same time, the volume element begins to receive more heat than at its former position at t_1. However, because heat transport is a slow process compared with tectonic transport, dp/dT tends to be much steeper than the corresponding dp/dT slope of the initial steady-state geotherm. The consequence is shown in Fig. 3.11C. The volume element has traveled in pT space along a path that is at the high-pressure side of the initial steady-state geotherm. The rocks of the volume element are now on a

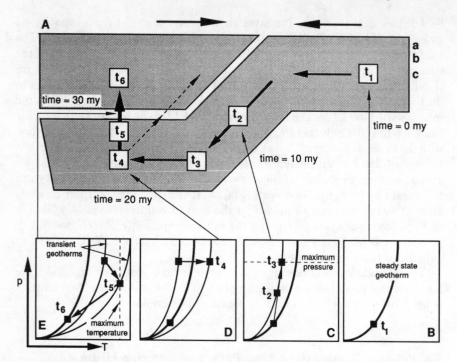

Fig. 3.11A–E. Schematic diagram showing the position of a volume element in the crust as a function of time during a continent-continent collision and the corresponding paths followed by the element in pT space

geotherm that changes its shape as time progresses, a transient geotherm. At t_3 the crust has reached twice its normal thickness (about 70 km), that is about the maximum thickness in continent-continent collision zones. The rocks have reached their maximum depth and consequently their maximum pressure point. The base of the double crust (70 km) is at a maximum pressure of about 2 GPa (20 kbar). Continued plate motion does not increase the thickness of the crust and pressure remains constant for all elements as long as thrusting is going on. On the other hand, heat transport to the rocks of the volume element in question increases the temperature as shown in Fig. 3.11D. From this point, a number of feasible mechanisms may control the path of the volume element. Continued tectonic transport may return slices and fragments of rock to shallower levels in a material countercurrent (indicated with a dashed arrow in Fig. 3.11A), or simply, after some period of time the plate convergence stops for a number of possible reasons (e.g. because frictional forces balance the force moving the plate). The thick crust starts uplift, and erosion removes the double crust and restores the original thickness. By this mechanism the volume element may return to its original depth position and, given enough time, the stable steady-state geotherm will be re-established. The path between t_4 and t_5 is characterized by decompression (transport along the z-axis). If initial uplift rates are slow

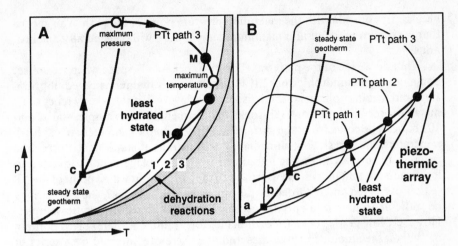

Fig. 3.12. A pTt path of a volume element at depth *c* of Fig. 3.11. The mineral assemblage stable at the tangent point of the path with dehydration reaction *2* will, in general, be preserved in metamorphic rocks. The tangent point corresponds to the least hydrated state of the rock. **B** Clockwise pTt loops for volumes of crust from different depth levels (Fig. 3–11). The mineral assemblage of the individual volumes corresponds to the least hydrated state for that volume. The pT points of the least hydrated state from all samples of a metamorphic terrain define a curve that is known as piezo-thermic array

compared with heat transport rates, the rocks will experience a continued temperature increase during uplift (as shown in Fig. 3.11E). However, at some stage along the path the rocks must start to lose more heat to the surface than they receive from below, and consequently cooling begins. The point t_5 in Fig. 3.11E represents the maximum temperature position of the path traveled by the volume element. At t_6 the volume element has returned to its former position on the steady-state geotherm.

The consecutive sequence of steps illustrated in Fig. 3.11 for a typical large-scale tectonic process and its consequences for pressure and temperature of a volume of rock is summarized in Fig. 3.12A. The volume of rock at level **c** (pTt-path 3) follow a clock wise pressure-temperature loop. Such clockwise pTt paths are characteristic features for the metamorphic evolution of rocks from most orogenic belts. Such pTt paths have been documented from such diverse mountain belts as: Scandinavian Caledonides, Alps, Appalachian, and the Himalayas. Clockwise loops may in detail show a number of additional complications and local features. Counter clockwise pTt-paths have been reported from granulite facies terrains where an event of heating from magmatic intrusions precedes crustal thickening. They may also occur in terrains that experience an initial phase of crustal extension. Very often "normal" orogenesis, is characterized by the following sequence of pTt path sections: isothermal compression, isobaric heating, isothermal decompression and isobaric cooling. Returning to Fig. 3.12A, it is evident that the maximum temperature point along the pTt path followed by a metamorphic rock does not necessarily coincide with the maximum pressure point of the path. This means that maximum pressure and maximum temperature will be generally diachronous.

What do we see of the path travelled by a rock in pTt space when we collect and study a metamorphic rock? If the rock always maintains an equilibrium state and metamorphism is strictly isochemical, we will not find any relics of its metamorphic history. However, rocks that formed at, for example, 800°C and 10 kbar can be collected at the surface and they do show characteristic high-pT mineral assemblages! This immediately tells us that at some stage during the reaction history a certain high-grade mineral assemblage is not converted to low-grade assemblages on the way back to the surface, but which point of the pT path is preserved and recorded by the mineralogy of the rock? The answer may be difficult to give in a specific geologic situation and for specific samples. However, low-grade rocks (or sediments, for example a shale) often contain modally large amounts of hydrates and are, therefore, affected by a series of discontinuous and continuous dehydration reactions during prograde metamorphism. In the pressure-temperature range typical for crustal metamorphism, dehydration reactions have equilibrium conditions as shown in Fig. 3.12A (reactions 1, 2 and 3). A rock traveling along path 3 will be affected by dehydration reactions. When crossing dehydration reaction 1 at point M it will partially lose water and continue as a rock containing modally fewer hydrates (mica, chlorite, amphibole). Dehydration of the rock will continue until the tangent point of the pT path and last dehydration reaction is reached. This point corresponds to the **least hydrated state** in the metamorphic history of the rock. It does not necessarily correspond to the maximum temperature point reached by the rock. Particularly if the pTt path shows a pronounced period of rapid uplift and decompression, the pT coordinates of the least hydrated state may be dramatically different from maximum p and T. At this point the rock contains a prograde assemblage in equilibrium with an aqueous fluid phase saturating the mineral grain boundaries and filling the pore space. The total amount of free fluid is extremely small in most metamorphic rocks, however. Typical flow porosities of metamorphic rocks at p and T are on the order of 0.2 vol% (e.g. 2 cm^3 H_2O/1000 cm^3 rock). After having passed the point of the least hydrated state, the rock crosses the dehydration reactions in the reversed direction. The reactions consume water and form hydrates from anhydrous (or less hydrous) minerals. The minute amount of free fluid will be used up readily by the first rehydration reaction. For example, 2 ‰ free H_2O in a rock of appropriate composition and mineralogy at 9 kbar and 700°C (corresponding to 0.1 mol H_2O per 1000 cm^3 of rock) can be used to form 15 cm^3 biotite per 1000 cm3 rock (1.5 vol% biotite). The rock will then be devoid of a free fluid phase which effectively prevents the mineralogy of the rock to adjust to changing pT conditions. Water is not only essential in dehydration reactions but plays a central role as a transport medium and catalyst for chemical reactions in rocks. Its absence has the effect of closing the reaction history of a rock. Even reconstructive phase transitions (e.g. kyanite = sillimanite) require the presence of water in order to proceed at finite rates even in geological times.

Consequently, by studying the phase relationships of metamorphic rocks, one will in general determine the pressure-temperature conditions corresponding to the least hydrated state.

On its way to the surface (or point c) the rock crosses reaction 1 at point N. However, the reaction will affect the rock only if water is available for the back reaction. Rehydration (retrograde metamorphism) is widespread and very common in metamorphic rocks. However, the rehydration reactions very often do not run to completion, and assemblages from the least hydrated state may survive as relics. The necessary water for retrograde metamorphism is usually introduced to the rocks by late deformation events. Deformation may occur along discrete shear zones that also act as fluid conduits. The shear zone rocks may contain low-grade mineral assemblages whereas the rocks unaffected by late deformation still show the high-grade assemblage. With some luck, it is sometimes possible to identify in a single rock sample a series of mineral assemblages that formed at consecutive stages of the reaction history. The reaction history of rocks can often be well documented for all stages that modified the assemblage of the least hydrated state.

Relics from earlier portions of the pT path (before it reached its least hydrated state) are seldom preserved in metasedimentary rocks. Occasionally, early minerals survive as isolated inclusions in refractory minerals such as garnet. Garnet in such cases effectively shields and protects the early mineral from reacting with minerals in the matrix of the rock with which it is not stable at some later stage of the rock's history. Rocks with water-deficient protoliths such as basalts, gabbros and other igneous rocks may better preserve early stages of the reaction history. In some reported case studies, meta-igneous rocks recorded the entire pTt path from the magmatic stage to subduction metamorphism and subsequent modifications at shallower crustal levels.

Rocks collected in a metamorphic terrain may originate from different depth level in the crust (Fig. 3.11, levels a, b and c). All rocks follow their individual pT path as shown in Fig. 3.12b. Analysis of phase relationships and geologic thermobarometry may therefore yield a series of different pT coordinates corresponding to the least hydrated states of the individual samples. The collection of all points from a given metamorphic terrain define a so-called **piezo-thermic array**. Its slope and shape characterize the specific terrain. The details of a piezo-thermic array of a given terrain depend on the geologic history and dynamic evolution of the area.

3.6 Chemical Reactions in Metamorphic Rocks

Chemical reactions in rocks may be classified according to a number of different criteria. Below follows a brief presentation of various kinds of reactions that modify the mineralogy or mineral chemistry of metamorphic rocks:

3.6.1 Reactions Among Solid Phase Components

These are often termed solid-solid reactions because only phase components of solid phases occur in the reaction equation. Typical solid-solid reactions are, for example:

3.6.1.1 Phase Transitions, Polymorphic Reactions

Al_2SiO_5	Ky = And ; Ky = Sil ; Sil = And
$CaCO_3$	calcite = aragonite
C	graphite = diamond
SiO_2	α-Qtz = β-Qtz ; α-Qtz = coesite,...
$KAlSi_3O_8$	microcline = sanidine

3.6.1.2 Net-Transfer Reactions

Such reactions transfer the components of reactant minerals to minerals of the product assemblage. Reactions involving anhydrous phase components only

Jd + Qtz = Ab
Gros + Qtz = An + 2 Wo
3 Fe-Crd = 2 Alm + 4 Sil + 5 Qtz

Volatile-conserving solid-solid reactions

Tlc + 4 En = Ath
Lws + 2 Qtz = Wa
2 Phl + 8 Sil + 7 Qtz = 2 Ms + Crd.

3.6.1.3 Exchange Reactions

These reactions exchange components between a set of minerals. Reactions involving anhydrous phase components only

Fe-Mg exchange between olivine and orthopyroxene: Fo + Fs = Fa + En
Fe-Mg exchange between clinopyroxene and garnet: Di + Alm = Hed + Prp

Volatile-conserving solid-solid reactions

Fe-Mg exchange between garnet and biotite: Prp + Ann = Alm + Phl
Cl-OH exchange between amphibole and biotite:
Cl-Fpa + OH-Ann = OH-Fpa + Cl-Ann

3.6.1.4 Exsolution Reactions/Solvus Reactions

High-T alkali feldspar = K-feldspar + Na-feldspar
Ternary high-T feldspar = meso-perthite + plagioclase
Mg-rich calcite = calcite + dolomite
High-T CPX = diopside + enstatite
Aluminous OPX = enstatite + garnet.

The common feature of all solid-solid reactions is that the equilibrium conditions of the reaction are independent of the composition of the fluid phase, or more generally speaking, of the chemical potentials of volatile phase components during metamorphism. It is for this reason that all solid-solid reactions are potentially useful geologic thermometers and barometers. The absence of volatile components in the reaction equation of solid-solid reactions should not be confused with general fluid-absent conditions during reaction progress. Metamorphic reactions generally require the presence of an aqueous fluid phase in order to achieve significant reaction progress even in geological time spans, attainment of chemical equilibrium in larger scale domains and chemical communication over several grain-size dimensions. Although pT coordinates of the simple reconstructive phase transition kyanite = sillimanite are independent of μ_{H_2O}, the detailed reaction mechanism may involve dissolution of kyanite in a saline aqueous fluid and precipitation of sillimanite from the fluid at nucleation sites that can be structurally unrelated to the former kyanite (e.g. Carmichael 1968). It is clear that the chlorine-hydroxyl exchange between amphibole and mica represents a process that requires the presence of

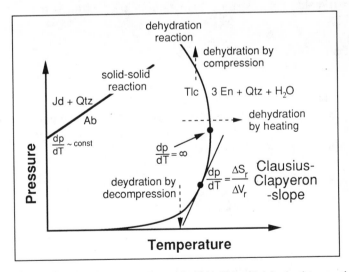

Fig. 3.13. Schematic equilibrium conditions of solid-solid and dehydration reactions in pT space and the Clausius-Clapeyron equation. ΔS_r, Entropy change; ΔV_r, volume change of the reaction

a saline aqueous fluid. However, the equilibrium of the exchange reaction is independent on the composition of that fluid.

Phase transitions, "solid-solid" net transfer and exchange reactions often show near linear equilibrium relationships in pT space. The slope of the equilibrium curves of such reactions can be readily calculated (estimated) from the Clausius-Clapeyron equation (Fig. 3.13).

3.6.2 Reactions Involving Volatiles as Reacting Species

3.6.2.1 Dehydration Reactions

Reactions involving H_2O are the most common metamorphic reactions. Low-grade metasediments contain modally abundant hydrates. Pelitic and mafic rocks from the anchizone consist mainly of clay minerals. Serpentine minerals make up lower-grade ultramafic rocks. These sheet silicates contain up to about 12 wt% H_2O. The hydrates are successively removed from the rocks by continuous and discontinuous dehydration reactions as heat is added to the rocks. The release of H_2O during prograde metamorphism of hydrate-rich protoliths ensures that a free H_2O fluid is present in the rocks either permanently or periodically during the progress of dehydration reactions. This in turn often permits discussion of metamorphism for conditions where the lithostatic pressure acting on the solids also applies to a free fluid phase. If the fluid is pure H_2O (and choosing an appropriate standard state), the condition is equivalent to $a_{H_2O} = 1$. The general shape of dehydration equilibria is shown schematically in Fig. 3.13. Using the talc breakdown reaction as an example, the general curve shape of dehydration reactions on pT diagrams can be deduced as follows:

$$Tlc = 3 \ En + 1 \ Qtz + 1 \ H_2O \ (V). \tag{13}$$

The dp/dT-slope of a tangent to the equilibrium curve is given at any point along the curve by the Clapeyron equation. All dehydration reactions have a positive ΔS_r and the curvature is largely controlled by the volume change of the reaction. The volume term can be separated into a contribution from the solids and the volume of H_2O respectively:

$$\Delta V_r = \Delta V_{solids} + \Delta V_{H_2O}. \tag{14}$$

The volume change of the solids can be calculated from:

$$\Delta V_{solids} = 3 \ V_{En} + 1 \ V_{Qtz} - 1 \ V_{Tlc}. \tag{15}$$

ΔV_{solids} is a small and negative quantity and varies very little with p and T. Consequently, the curve shape of dehydration reactions largely reflects the volume function of H_2O that is shown in Fig. 3.10. At low pressures, V_{H_2O} is large and the dp/dT slope small. In the pressure range of about 2 to 3 kbar, the molar volume of H_2O rapidly decreases and dehydration reactions tend to be strongly curved. At pressures greater than about 3 kbar and up to 10–15 kbar, the volume of H_2O is usually comparable to the volume change of the solids and ΔV_r tends to be very small. Perceptively, dp/dT slopes are very large in this pressure range. At even higher pressures, the volume of H_2O cannot compensate the negative ΔV_{solids} and ΔV_r and the dp/dT-slope also becomes negative. From the general shape of dehydration equilibria in pT space shown in Fig. 3.13, it is evident that talc, for instance, can be dehydrated by heating (temperature increase), decompression or compression.

The dehydration of zeolite minerals is often accompanied by large negative ΔV_{solids} and the dehydration equilibria have negative slopes already at moderate or low pressures. For example, the reaction analcim + quartz ⇒ albite + H_2O has a ΔV_{solids} of –20.1 cm^3 mole^{-1} and the equilibrium has a negative slope above pressures of about 2 kbar.

In contrast to dehydration reactions, dp/dT slopes of solid-solid reactions (and phase transitions) are nearly constant (Fig. 3.13).

The equilibrium constant of the talc breakdown reaction (13) can be expressed by the mass action equation:

$$K = \frac{a_{En}^3 \, a_{Qtz} \, a_{H_2O}}{a_{Tlc}} \tag{16}$$

The equilibrium constant K of a chemical reaction has a fixed value at a given p and T (a constant as the name suggests). It can be calculated from the fundamental and also for geological applications extremely useful equation:

$$- R \, T \ln K = \Delta G_r^\circ. \tag{17}$$

The free energy change of the reaction on the right hand side of Eq. (16) can be calculated for any pressure and temperature pair of interest provided that the thermodynamic data for all phase components in the reaction are known (R = universal gas constant; T = temperature in K). For our talc reaction, Eq. (17) takes the form:

$$\ln K = \{\ln a_{H_2O} + [\ln a_{En}^3 + \ln a_{Qtz} - \ln a_{Tlc}]\} = - \Delta G_r^\circ/(R \, T). \tag{18}$$

By considering pure endmember solids (a = 1), the expression in square brackets becomes zero and the equilibrium is dependent on the activity of H_2O. The equilibrium is shown in Fig. 3.13 for the case $p_{H_2O} = p_{lithostatic}$ (pure H_2O fluid present, $a_{H_2O} = 1$). Equation (18) is represented graphically in Fig. 3.14 for three values of a_{H_2O}. It can be seen from Fig. 3.14, that the equilibrium conditions for dehydration reactions are displaced to lower temperatures (at p = constant) by decreasing a_{H_2O}. Solutions of eq. (18) for different a_{H_2O} are shown in Fig. 3.15. It follows from Fig. 3.14 that the maximum temperature for the talc-breakdown

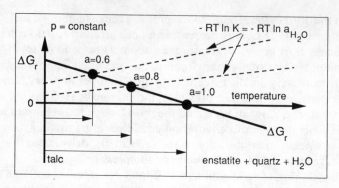

Fig. 3.14. Solutions of the equation $\Delta G_r = - RT \ln K$ for dehydration reactions and three different values of a_{H_2O}

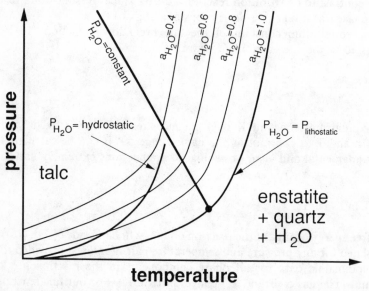

Fig. 3.15. Schematic equilibrium conditions of dehydration reactions for various a_{H_2O} conditions

reaction is given by the condition $a_{H_2O} = 1$ that is equivalent to $p_{H_2O} = p_{lithostatic}$ and the presence of a pure H_2O fluid.

It also follows from Fig. 3.14, that talc dehydrates to enstatite and quartz at lower temperatures if $a_{H_2O} < 1.0$ at any pressure. The pT space can, therefore, be contoured with a series of dehydration curves with constant a_{H_2O}. Geologically, the condition of $a_{H_2O} < 1.0$ can be realized in basically two different situations: (1) the fluid is not a pure H_2O fluid but is rather a solution between H_2O and some other components not taking part in the reaction (e.g. CH_4, N_2, CO_2),

(2) there is no free fluid phase present at all. If p_{H_2O} is held constant, dehydration occurs along a curve with a negative slope intersecting the $a_{H_2O} = 1$ curve where $p_{H_2O} = constant = p_{lithostatic}$. Towards higher pressure, a_{H_2O} decreases along the curve. This condition is of little geologic significance. The equilibrium curve for the condition $p_{H_2O} = p_{hydrostatic}$ has a shape similar as for $p_{H_2O} = p_{lithostatic}$. The two curves converge at very low pressures. Note, however, that in this situation $a_{H_2O} = 1$ in the fluid along the curve and it does not vary with increasing pressure. The pressure is different for the solids and the fluid. This consequently means that such a system is not in mechanical equilibrium. Hydrostatic pressure conditions may occur in very porous or strongly fractured rocks at shallow crustal levels. Typical geologic environments where such conditions may occur include hydrothermal vein formation, oceanic metamorphism, shallow level contact aureoles. Hydrostatic pressures usually cannot be maintained at typical metamorphic pressures of some kbar. Pressure solution and local chemical redistribution of material rapidly isolates the fluid phase and the system returns to mechanical equilibrium. Geologically then, the situation $p_{H_2O} = p_{lithostatic}$ is the most important one and the curves $a_{H_2O} < 1$ may be relevant in rocks with an impure aqueous fluid or in "dry" environments.

3.6.2.2 Decarbonatization Reactions

This type of reaction describes the decomposition of carbonate minerals such as calcite and dolomite. For example: adding heat to a rock containing calcite and quartz will eventually cause the following reaction: $Cal + Qtz = Wo + CO_2$. The form of the equilibrium conditions in pT space is analogous to the one of the dehydration reactions. Also, the effects of variations in a_{CO_2} (p_{CO_2}) are analogous to the ones discussed above for dehydration reactions. In metamorphic carbonate rocks the fluid phase rarely consists of pure CO_2, it is usually rather a mixture of CO_2 and H_2O (and other volatile species).

3.6.2.3 Mixed Volatile Reactions

In rocks containing both hydrates (sheet silicates and amphiboles) and carbonates, reactions are important that produce or consume both CO_2 and H_2O simultaneously. In such rock types the fluid phase contains at least the two volatile species, CO_2 and H_2O. Under certain conditions, for instance if graphite is present in the rocks, other species such as H_2 and CH_4 may be present in significant amounts. However, the fluid composition in carbonate-bearing rocks can be assumed to be a binary mixture of CO_2 and H_2O in most cases. One compositional variable is therefore sufficient to describe the fluid composition. Usually this variable is defined as the mole fraction of CO_2 in the fluid (X_{CO_2}). X_{CO_2} is zero in pure H_2O fluids and equal to one in pure CO_2 fluids. Consider the following characteristic mixed volatile reaction:

$$margarite + 2\ quartz + calcite = 2\ anorthite + 1\ CO_2 + 1\ H_2O. \tag{19}$$

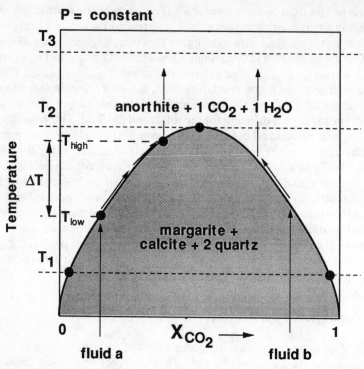

Fig. 3.16. Schematic equilibrium conditions of a mixed volatile reaction (isobaric TX diagram)

For pure solid phases the equilibrium constant is: $K_{p,T} = a_{CO_2} \, a_{H_2O}$. The equilibrium conditions are dependent on the activities of CO_2 and H_2O in addition to pressure and temperature. However, if we assume that the fluid is a binary CO_2-H_2O mixture (and using an ideal solution model for the fluid) the equilibrium constant reduces to: $K = X_{CO_2} \, (1 - X_{CO_2})$ or $K = X_{CO_2} - X_{CO_2}^2$. Depending on the actual value of the equilibrium constant at p and T, the quadratic equation may have no, one or two solutions. The equilibrium conditions of the margarite breakdown reaction is therefore a complex surface in the pTX_{CO_2} space. Because it is inconvenient to work graphically in a three-dimensional space, mixed volatile reactions are most often represented on isobaric TX diagrams or isothermal pX diagrams depending on the geological problem one wants to solve. It is also possible to select a certain metamorphic pT geotherm that best describes the regional metamorphic area under consideration and construct (pT)-X_{CO_2} sections (examples are presented in Chap. 6). For instance, let us keep the pressure constant. We can then solve the quadratic equation above for a series of temperatures and display the solution on a isobaric TX section (Fig. 3.16). At T_1 the four-mineral assemblage margarite + quartz + calcite + anorthite may coexist with either very H_2O-rich or very CO_2-rich fluids. At one unique temperature, T_2, the equation has only one solution. The temperature T_2 represents the maximum temperature for the reactant assemblage; above it, no margarite, calcite and quartz may coexist in rocks. At T_3 no (real) solution for the equation exists and the product assemblage may coexist with binary CO_2-H_2O fluids of any composition. The heavy solid curve in Fig. 3.16 connects all solutions of the equilibrium constant equation and represents the reaction equilibrium of the margarite breakdown reaction in quartz- and calcite-bearing rocks. The curve separates a TX field where the reactants are stable (shaded) from an area where the products are stable. At any given temperature, $< T_2$, the products are stable in H_2O-rich or CO_2-rich fluids, respectively. The reactants may coexist with fluids of intermediate compositions.

Consider now a margarite-bearing rock containing excess quartz and calcite with a pore fluid of composition **a**. Adding heat to this rock raises its temperature until the reaction boundary is reached at T_{low}. Further increase in the temperature causes the production of anorthite and a fluid with equal proportions of CO_2 and H_2O. Because the produced fluid is more CO_2-rich than the original pore fluid of the rock (fluid **a**), the total fluid composition is driven to more CO_2-rich compositions. The original fluid becomes more and more diluted with the fluid produced by the reaction; ultimately the fluid is entirely dominated by the reaction fluid and has the composition $X_{CO_2} = 0.5$ (1 mol CO_2 and 1 mol H_2O). This will be the case at T_2, the maximum temperature of the reactant assemblage. However, the reactant assemblage may become exhausted with respect to one or more reactant minerals before reaching T_2. For example, if the original rock contains 1 mol margarite, 3 mol calcite and 5 mol quartz per 1000 cm^3, the reaction will consume all margarite during reaction progress at some temperature along the isobaric univariant curve (T_{high}). The rock consists now of anorthite, calcite and quartz, and a fluid of the composition at T_{high}. Further addition of heat will increase the temperature of the rock but the fluid composition will remain unchanged. The univariant assemblage margarite + anorthite + calcite + quartz, therefore, occurs in a temperature interval ΔT and coexists with fluids of increasing CO_2 content. The temperature interval ΔT is also associated with gradual changes in modal composition in the rock. At T_{low} the first infinite small amount of anorthite appears and it is continuously increasing in modal abundance as the reaction progresses towards T_{high} where the last small amount of margarite disappears. The fluid composition at the maximum temperature, T_2, depends on the reaction stoichiometry. Consider the reaction:

$$11 \text{ dolomite} + \text{tremolite} = 13 \text{ calcite} + 8 \text{ forsterite} + 9 \text{ } CO_2 + 1 \text{ } H_2O. \qquad (20)$$

Here, the produced fluid has the composition $X_{CO_2} = 0.9$ (9 moles CO_2 and 1 mole H_2O) and the temperature maximum of the equilibrium is at the same fluid composition. The general shape of reaction equilibrium for reactions of the type: $a = b + m \text{ } H_2O + n \text{ } CO_2$ (both fluid species are products of the reaction) is shown in Fig. 3.17 together with mixed volatile reactions with other reaction stoichiometries.

P = constant

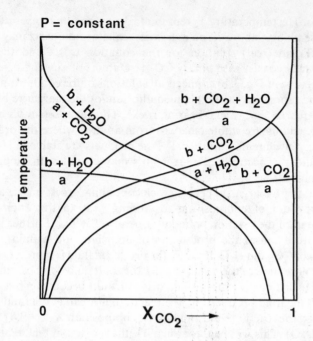

Fig. 3.17. Schematic equilibrium conditions of mixed volatile reactions with different reaction stoichiometries (isobaric TX diagram)

The equilibrium constant of the reaction:

$$2 \text{ zoisite} + 1 \text{ } CO_2 = 3 \text{ anorthite} + \text{calcite} + 1 \text{ } H_2O \tag{21}$$

can be written as: $K = (1/X_{CO_2}) - 1$. It varies between zero and $+\infty$ (ln K $-\infty$, $+\infty$). Consequently, the univariant assemblage coexists with CO_2-rich fluids at low temperatures and with H_2O-rich fluids at high temperatures. The general shape of reactions $a + CO_2 = b + H_2O$ is shown in Fig. 3.17. The reaction:

$$8 \text{ quartz} + 5 \text{ dolomite} + 1 \text{ } H_2O = \text{tremolite} + 3 \text{ calcite} + 7 \text{ } CO_2 \tag{22}$$

represents a reaction of the type: $a + H_2O = b + CO_2$ and has, compared to reaction [21], an inverse shape in TX diagrams (Fig. 3.17). It follows from the earlier discussion on pure dehydration and decarbonatization reactions (see Figs. 3.14 and 3.15) that they must have general curve shapes as shown in Fig. 3.17. The figures presented so far also permit a deduction of the general shape of mixed volatile reaction equilibria in isothermal pX sections (you may try to draw a pX figure at constant temperature similar to Fig. 3.17).

Fig. 3.18A shows the phase relationships in siliceous dolomites at constant pressure (2 kbar). Calcite and dolomite are present in excess (dolomite is, however, not stable in the brucite and periclase field respectively). The curves shown in Fig. 3.18A represent reaction equilibria of a number of mixed volatile

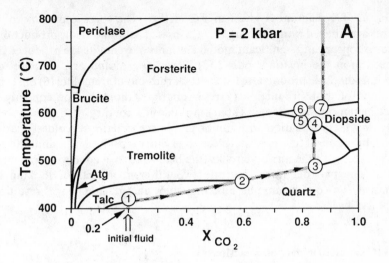

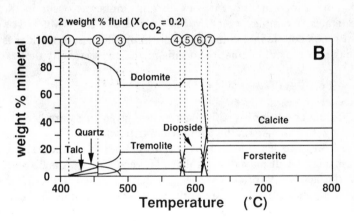

Fig. 3.18A, B. A quantitative model for isobaric metamorphism of siliceous dolomites in terms of a quantitative isobaric TX diagram (**A**) and a modal composition versus temperature diagram (**B**)

reactions. The shape of the curves corresponds to those shown in Fig. 3.17. However, the equilibrium curves on diagrams showing phase relationships among many phases usually are divided up into sections of different stability level. The stability level sections of individual equilibrium curves are bounded by intersections with other equilibria (invariant points, see also Sect. 3.8) or with the boundaries of the selected parameter window. In the **phase diagram** of Fig. 3.18A only stable sections of reaction equilibria are shown; metastable equilibria have been omitted. Fig. 3.18B shows the modal changes during progressive isobaric metamorphism of a dolomite containing initially 10 wt% quartz and 2 wt% fluid (with the composition 20 mol% CO_2 and 80 mol% H_2O). The reaction path taken by that rock during metamorphism in the equilibrium TX diagram is given in Fig. 3.18A. The specific metamorphic incidents are labelled from 1 through 7. (1) Onset of continuous talc production; (2) talc disappears and significant modal amounts of tremolite are produced by the reaction at the invariant point; (3) last quartz disappears; (4) onset of continuous diopside production; (5) the rock runs out of tremolite; (6) onset of continuous forsterite production; (7) rock runs out of diopside. Consequently, a contact metamorphic aureole (2 kbar) in dolomitic country rock will show a regular systematic zonation with minor talc in the outermost (coldest) parts, followed by a zone of tremolite marble, a zone with diopside marble and, closest to the contact with the intrusive rocks (the heat source), a zone with forsterite marbles. The two diagrams illustrate two different aspects of looking at metamorphism: equilibrium phase relationships and modal changes resulting from reactions in rocks.

3.6.2.4 Oxidation/Reduction Reactions

A number of cations making up rock forming minerals occur in different oxidation states. Examples are: Fe^{2+}/Fe^{3+}, Cu^+/Cu^{2+}, Mn^{2+}/Mn^{3+}. The most important redox couple in common rock-forming silicates and oxides is Fe^{2+}/Fe^{3+}. The two iron oxides hematite and magnetite are related by the reaction:

$$6 \, Fe_2O_3 = 4 \, Fe_3O_4 + O_2. \tag{23}$$

The equilibrium conditions of this reaction are given (for pure hematite and magnetite) by: $K = a_{O_2}$. Again the equilibrium depends on three variables: p, T and a_{O_2}. The pressure dependence is usually very small for redox reactions. The equilibrium may be displayed on isobaric T versus $\ln a_{O_2}$ diagrams. In the literature the activity is often replaced by $\ln f_{O_2}$ (fugacity of O_2), which is numerically identical to $\ln a_{O_2}$ if an appropriate standard state is chosen. The fugacity is close to the partial pressure of O_2 under low pressure conditions. However, fugacity can be related to the partial pressure of O_2 at any pressure. The partial pressure of O_2 in crustal rocks is extremely small (on the order of 10^{-20}–10^{-40} bar). However, coexistence of hematite and magnetite may also be formulated in the presence of water as:

$$6 \, Fe_2O_3 + 2 \, H_2 = 4 \, Fe_3O_4 + 2 \, H_2O \tag{24}$$

and the equilibrium can be shown on T versus $\ln a_{H_2}$ diagrams. The assemblage biotite + K-feldspar + magnetite is common in high-grade metamorphic and in igneous rocks. At equilibrium of the reaction:

$$2 \, KFe_3AlSi_3O_{10}(OH)_2 + O_2 = 2 \, KAlSi_3O_8 + 2 \, Fe_3O_4 + 2 \, H_2O \tag{25}$$

$$\text{biotite} \qquad\qquad \text{K-feldspar} \quad \text{magnetite}$$

the assemblage is dependent on O_2 (and H_2O), or vice versa the assemblage defines the activity of O_2 in the presence of water. Redox reactions do not necessarily involve volatiles. Redox reactions may also be formulated in ionic form (example Mag - Hem reaction):

$$2 \ Fe_3O_4 + H_2O = 3 \ Fe_2O_3 + 2 \ H^+ + 2 \ e^-. \tag{26}$$

Oxidation of magnetite to hematite in the presence of water produces two H^+ and two electrons. The magnetite - hematite reaction (26) depends on the redox potential and on the activity of the hydrogen ion (p_H). Graphical representation of equilibria of the type (26) is often done by means of Eh (oxidation potential) versus p_H diagrams (see Garrels and Christ 1965).

3.6.2.5 Sulphidation Reactions

Sulfides are widespread accessory minerals in metamorphic rocks. Most common is pyrrhotite (FeS) and pyrite (FeS_2). If both common Fe-sulfides occur in a metamorphic rock, the stable coexistence of pyrite and pyrrhotite requires equilibrium of the reaction:

$$2 \ FeS_2 = 2 \ FeS + S_2. \tag{27}$$

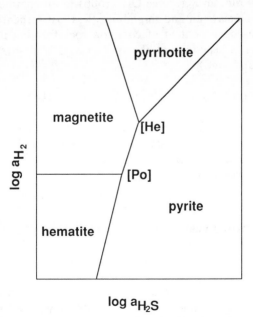

Fig. 3.19. Qualitative phase relationships among hematite, magnetite, pyrrothite and pyrite in terms of an H_2 versus H_2S activity diagram

The dominant sulfur species in metamorphic fluids are either H_2S or SO_2. If one wishes to discuss sulfide-involving reactions in terms of the most abundant species one may rewrite Eq. (27) in the presence of water ($2\ H_2O = 2\ H_2 + O_2$; $2\ H_2S = 2\ H_2 + S_2$) by:

$$FeS_2 + H_2 = FeS + H_2S. \tag{28}$$

Graphic representation of sulfide-involving reactions depends on the actual problem but partial pressure (activity, fugacity) of a sulfur species in the fluid versus temperature diagrams are popular. If sulfide reactions are combined with redox reactions, oxygen (hydrogen) versus sulfur diagrams are useful. As an example, Fig. 3.19 shows the phase relationships among some iron oxides and sulfides in terms of a_{H_2} and a_{H_2S}. The figure shows that the presence of one Fe-sulfide and one Fe-oxide in a metamorphic rock (e.g. magnetite and pyrrhotite) buffers the activities of H_2 and H_2S to distinct values along a univariant line at a given pressure and temperature in the presence of water. The rocks may be zoned with respect to oxides and sulfides depending on gradients in a_{H_2S} (e.g. under reducing conditions, rocks may contain magnetite, pyrrhotite or pyrite with increasing a_{H_2S}).

3.6.2.6 Reactions Involving Halogens

Fluorine and chlorine may replace OH groups in all common rock-forming hydrates, notably micas, talc and amphiboles (less so in serpentine minerals and chlorites). The halogen content of common rock-forming minerals may be related to exchange reactions of the type (e.g. fluorine-hydroxyl exchange between biotite and fluid):

$$KMg_3AlSi_3O_{10}(OH)_2 + 2\ HF = KMg_3AlSi_3O_{10}(F)_2 + 2\ H_2O. \tag{29}$$

Also, in this case, equilibria of the type (29) are commonly displayed and discussed in terms of partial pressure (fugacity, activity) of a gaseous halogen species (HCl, HF) versus temperature diagrams or in terms of μ-μ diagrams (activity-activity diagrams).

3.6.2.7 Complex Mixed Volatile Reactions and Fluids

It is a common situation in metamorphic rocks that they contain simultaneously halogen-bearing hydrates, carbonates, sulfides, and oxides. Dealing with such rocks consequently means that all types of metamorphic reactions must be considered simultaneously. For example, consider a calcareous mica schist containing the following minerals: calcite, quartz, K-feldspar, plagioclase, muscovite, biotite, graphite and pyrite. Among the minerals and the nine potentially important fluid species the following linearly independent equilibria can be written:

a) Equilibria involving silicates, sulfides and carbonates

$$KAl_2AlSi_3O_{10}(OH)_2 + CaCO_3 + 2 SiO_2 = KAlSi_3O_8 + CaAl_2Si_2O_8$$
$$+ H_2O + CO_2 \tag{a1}$$
$$KFe_3AlSi_3O_{10}(OH)_2 + 3 S_2 = 2 KAlSi_3O_8 + 3 FeS_2 + H_2O \tag{a2}$$

b) Equilibria involving graphite

$$C + O_2 = CO_2 \tag{b3}$$
$$C + 2 H_2 = CH_4 \tag{b4}$$
$$2 C + O_2 = 2 CO \tag{b5}$$

c) Equilibria involving fluid species only

$$2 H_2O = 2 H_2 + O_2 \tag{c6}$$
$$S_2 + 2 H_2 = 2 H_2S \tag{c7}$$
$$S_2 + 2 O_2 = 2 SO_2 \tag{c8}$$

For all eight equilibria a mass action equation can be formulated [see, for example Eqs. (16) and (17)]. In addition, in the presence of a free fluid phase in the rock the following mass balance equation (total pressure equation) can be written:

$$P_{total} = P_{fluid} = P_{H_2O} + P_{CO_2} + P_{O_2} + P_{H_2} + P_{CH_4} + P_{CO} + P_{S_2} + P_{H_2S} + P_{SO_2}. \tag{d9}$$

Equation (d9) states that the sum of the partial pressures of the fluid species is equal to the total lithostatic rock pressure. The set of nine equations can be solved for the nine unknown partial pressures and hence the fluid composition. The mineral assemblage of the calcareous mica schist defines and completely controls the composition of the coexisting fluid phase at any given pressure and temperature. In most metamorphic fluids the partial pressures of O_2 and CO are extremely low compared with the total pressure. H_2 is usually low also in most metamorphic fluids. The dominant sulfur species in the fluid is H_2S at low temperatures and SO_2 at high temperatures. CH_4 may be high in some fluids at lower temperatures. CO_2 is dominant at higher temperatures.

3.6.2.8 Reactions Involving Minerals and Dissolved Components in Aqueous Solutions

The fluids that are associated with rock metamorphism in crust and mantle inevitably migrate from the source area. They come in contact with other rocks with which they may not be in equilibrium. The chemical reactions that try to establish equilibrium between the solid rock assemblage and the fluid are known under the term fluid-rock interaction. Fluid-rock interaction is important in the formation of hydrothermal ore deposits, contact metasomatism, large-scale infiltration and reaction in orogenesis, in shear zone metasomatism and in geothermal fields (to name a few examples). As an example, imagine a crustal volume containing rocks with the assemblage margarite + calcite + quartz (Fig. 3.16). A foreign externally derived fluid with the composition $X_{CO_2} = 0.01$ arrives in this rock at T_{high}. The fluid is in equilibrium with the product assemblage of reaction (19) rather than with the reactant assemblage that is actually present. The consequence of fluid infiltration is that the fluid will drive the reaction until it reaches equilibrium with the solid phase assemblage.

Metamorphic fluids are often saline brines containing high concentrations of charged and neutral metal ions and complexes. Therefore, infiltrating fluids may also have a cation composition that is not in equilibrium with the solid phase assemblage receiving this fluid. An essential aspect of fluid-rock interaction is the reaction of phase components of the solid mineral assemblage of the rock and dissolved metal species in the foreign fluid. Some examples:

Formation of talc schists by interaction of serpentinite with quartz-saturated fluids:

$$Mg_3Si_2O_5(OH)_4 + 2\ SiO_{2_{aq}} = Mg_3Si_4O_{10}(OH)_2 + H_2O, \tag{30}$$

where $SiO_{2_{aq}}$ stands for an uncharged hydrous silica complex in the aqueous fluid. SiO_2 saturation of the aqueous fluid infiltrating the serpentinite may result from interaction of the fluid with granite or granite-gneiss enclosing a serpentinite body.

Formation of soapstone (Tlc + Mgs rock) by interaction of serpentinite with CO_2-bearing fluids:

$$2\ Mg_3Si_2O_5(OH)_4 + 3\ CO_2 = Mg_3Si_4O_{10}(OH)_2 + 3\ MgCO_3 + 3\ H_2O. \tag{31}$$

Sericitization of K-feldspar is a widespread process in retrograde metamorphism:

$$3\ KAlSi_3O_8 + 2\ H^+ = KAl_3Si_3O_{10}(OH)_2 + 2\ K^+ + 6\ SiO_{2_{aq}}. \tag{32}$$

K-feldspar   muscovite (sericite)

This last type of reaction is of great importance in geology. It is referred to as a **hydrolysis reaction.** Any other kind of metamorphic mineral reaction can be viewed as a linear combination of a series of hydrolysis reactions. For example, the phase transition kyanite = sillimanite can be viewed as a result of the two hydrolysis reactions:

a) kyanite hydrolysis (dissolution): $Al_2SiO_{5_{kyanite}} + H^+ = 2\ Al^{3+} + SiO_{2_{aq}} + H_2O$
b) sillimanite precipitation: $2\ Al^{3+} + SiO_{2_{aq}} + H_2O = Al_2SiO_{5_{sillimanite}} + H^+$
c) net reaction (a+b): $Al_2SiO_{5_{kyanite}} = Al_2SiO_{5_{sillimanite}}$

Al^{3+} is the dominant Al species in extremely acid aqueous fluid, but similar reactions can be written with other Al species such as $Al(OH)^{2+}$, $Al(OH)_2^+$, $Al(OH)_3^\circ$, $Al(OH)_4^-$ (depending on the pH of the fluid).

Albitization of K-feldspar (ion exchange reaction) is a prevailing process in low-grade alteration of granites and granite-gneisses:

$$KAlSi_3O_{8K\text{-feldspar}} + Na^+ = NaAlSi_3O_{8Na\text{-feldspar}} + K^+. \tag{33}$$

The importance of ionic reactions in the formation of hydrothermal ore deposits can be illustrated by the dependence of the solubility of covellite (CuS) on p_H.

$$CuS + 2 H^+ = Cu^{2+} + H_2S. \tag{34}$$

Suppose there is an acid metamorphic fluid with some Cu^{2+}. Covellite will precipitate from that fluid as a result of increasing p_H caused by feldspar alteration (reaction [32]). Alteration processes such as reaction (32) (micas from feldspars, chlorite from amphiboles) increase the p_H and lead to precipitation of metal oxides and sulfides.

Graphically, ionic (complex) reactions can be represented by means of p_H versus gas pressure (activity, fugacity) diagrams, activity-activity diagrams (or μ-μ diagrams), activity-temperature diagrams etc., depending on the actual geological problem.

3.7 Reaction Progress

Chemical reactions in rocks proceed in response to gradients in intensive variables (e.g. increase in temperature). The progress of a chemical reaction can be illustrated by using the following wollastonite-producing reaction as an example:

$$CaCO_3 + SiO_2 = CaSiO_3 + CO_2. \tag{35}$$

Calcite and quartz may react and produce wollastonite and release CO_2 gas. The equilibrium conditions of the reaction are shown in Fig. 3.20 as a function of temperature and the composition of a binary CO_2-H_2O fluid at a total pressure of 2 kbar. A calcite- and quartz-bearing rock with an initial pore space of 1% may contain a fluid of the composition $X_{CO_2} = 0.01$. Adding heat to this rock brings it onto the reaction boundary of the wollastonite-producing reaction at a temperature of about 450°C. Further addition of heat will produce wollastonite and the fluid becomes enriched in CO_2 as a result of CO_2 released by the reaction. Suppose the rock finally reaches a temperature of 600°C and contains the isobaric univariant assemblage calcite + quartz + wollastonite. It is a simple matter to calculate the modal amount of wollastonite produced by the reaction:

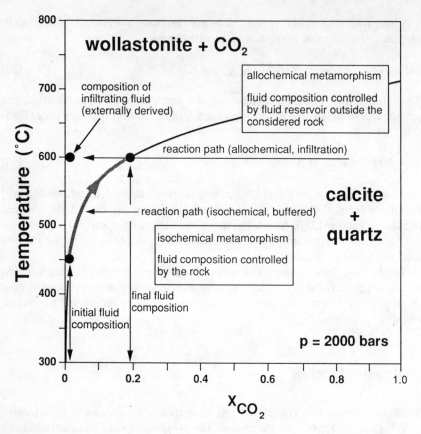

Fig. 3.20. Quantitative temperature versus fluid composition phase diagram showing the equilibrium conditions of the wollastonite reaction at 2 kbar

Reference volume: 1000 cm^3 rock

1% pore space = 10 cm^3 fluid per 1000 cm^3

Cal + Qtz reaches equilibrium with Wo at 450°C (at 2 kbar) and $X_{CO_2} = 0.01$ (see Fig. 3.20)

Cal + Qtz + Wo assemblage at 600°C (at 2 kbar) and $X_{CO_2} = 0.19$ (see Fig. 3.20)

Molar volume of H$_2$O at 450°C, 2 kbar = 24 cm^3 mol^{-1} (Fig. 3.10)

Initial fluid 99 mol% H$_2$O + 1 mol% CO$_2$ (assume for simplicity pure H$_2$O)

Number of moles of H$_2$O in the initial fluid: $n^\circ_{H_2O} = {}^{10}/_{24} = 0.42$ mol H$_2$O

Initial $X^\circ_{CO_2} = n^\circ_{CO_2}/(n^\circ_{CO_2} + 0.42)$

Final $X^f_{CO_2} = n^f_{CO_2}/(n^f_{CO_2} + 0.42)$

Solving for:

Number of moles of CO$_2$ in the initial fluid: $n^\circ_{CO_2} = 0.004209$ mol

Number of moles of CO_2 in the final fluid: $n^f_{CO_2} = 0.09655$ mol
CO_2 produced by the reaction: $\Delta n_{CO_2} = n^f_{CO_2} - n^o_{CO_2}$
$= 0.09655 - 0.004209 = 0.09234$ mol
With reference to equation [35]: 0.09234 mol CO_2 is equivalent to 0.09234
mol wollastonite produced by the reaction per 1000 cm^3 rock.
The molar volume of wollastonite is: $V^o = 40$ cm^3 mol^{-1}
Wollastonite produced by buffered reaction: 3.7 cm^3 (0.367 vol%)

The amount of wollastonite is very small and will hardly be detected under the petrographic microscope.

Let us now consider a calcite- and quartz-bearing rock with 20 vol% wollastonite that equilibrated at 600°C and 2 kbar. From the calculations above it is evident that isochemical metamorphism can produce only a very small modal amount of wollastonite (at 600°C and 2 kbar). Consequently, a very wollastonite-rich rock that formed at these conditions must be the result of interaction with a foreign, externally derived H_2O-rich fluid that pushed the rock across the reaction equilibrium into the wollastonite field shown in Fig. 3.20. The question is how much H_2O must be added to the rock in order to produce the observed modal proportion of wollastonite? This can be readily calculated:

20 vol% wollastonite ($V^o = 40$ cm^3 mol^{-1}) $= 0.5$ mol/100 cm^3
The rock contains 5 mol wollastonite/1000 cm^3 rock. It follows from the reaction stoichiometry of reaction (35) that the reaction also produced 5 mol of CO_2.
However, the final $X^f_{CO_2} = n^f_{CO_2}/(n^f_{CO_2} + n_{H_2O})$ is equal to 0.19 (Fig. 3.20).
Solving for: $n_{H_2O} = n^f_{CO_2} (1 - X^f_{CO_2}) (1/X^f_{CO_2})$
gives: $n_{H_2O} = 5 \times 0.81 \times (1/_{0.19}) = 21.315$ mol H_2O
The molar volume of H_2O is: V^o (600°C, 2 kbar) $= 31$ cm^3 (Fig. 3.10)
The total volume of H_2O that reacted with the rock is:
$V_{H_2O} = 660$ cm^3/1000 cm^3 rock

In conclusion, a rock with 20 vol.% wollastonite that formed from calcite + quartz at 600°C and 2 kbar requires interaction with 660 cm^3 H_2O per 1000 cm^3 rock. This represents a minimum value. If the composition of the interacting fluid is more CO_2-rich, more fluid is required to produce the observed amount wollastonite in the rock. The deduced amount of interacting external fluid may be expressed in terms of a time-integrated fluid/rock ratio:
fluid/rock ratio $= 660$ cm^3 /1000 cm^3 $= 0.66$

3.8 Phase Diagrams

3.8.1 Significance of Phase Diagrams

Phase diagrams display the equilibrium stability relationships among phases and phase assemblages in terms of intensive, extensive or mixed variables. A widely used type of phase diagram in metamorphic petrology is the pT diagram. It shows the equilibrium relationships among minerals (and fluids) as a function of the intensive variables pressure and temperature. The chosen ranges of the parameter values define a pT window that is appropriate for the problem of interest. The mineral and fluid compositions may be constant or variable on such a diagram. Composition phase diagrams show the phase relationships in terms of extensive variables at specified values of pressure and temperature. AFM and ACF diagrams are examples of composition phase diagrams. Phase diagram software is routinely used to compute and display petrologic phase diagrams from published sets of thermodynamic data of phase components. All phase diagrams in this book were computed by using two different programs: PTXA (or GEØ-CALC) is a convenient very user-friendly program that runs on DOS PC's and MAC's (Berman et al. 1987; Brown et al. 1988). It can be used for diagrams that display any combination of the variables pressure, temperature, composition of a binary CO_2-H_2O fluid, and activity of phase components. PERPLEX is an extremely powerful multipurpose phase diagram generator that permits the construction of remarkably complex phase diagrams if necessary (Connolly and Kerrick 1987; Connolly 1990). PERPLEX has been developed for UNIX-based work stations, but it also runs on MAC's and PC's. A short review of phase diagram programs is found in Bucher-Nurminen (1990). Other petrologic programs to be mentioned here include: SUPCRT92 by Johnson et al. (1991); PTPATH by Spear et al. (1991) and THERMO by Perkins et al. (1987). Compilations of thermodynamic data for substances of geologic interest are found in: Clark (1966); Burnham et al. (1969); Stull and Prophet (1971); Robie et al. (1978); Helgeson et al. (1978); Holland and Powell (1985, 1990); Powell and Holland (1985, 1988); Berman (1988); Johnson et al. (1991). The thermodynamic data sets of Berman (1988) [RB88] and Holland and Powell (1990) [HP90] are classified as so-called internally consistent data sets because of the specific data retrieval technique used in deriving the data. These two sets are particularly well suited for use in conjunction with the two mentioned programs and are widely used by metamorphic petrologists. Thermodynamic data can be derived from calorimetric measurements and, especially, from experimental phase equilibrium data. Most phase diagrams shown in this book have been generated by PERPLEX using the RB88 or the HP90 data base (or updated versions of them).

Phase diagrams are indispensable tools for the description, analysis and interpretation of metamorphic rocks and metamorphism. The ability to read and understand phase diagrams is central and essential in metamorphic petrology. The capability to calculate and construct phase diagrams is important for all those who are in need of phase diagrams for phase assemblages and phase compositions that are unique for the geologic problem in question. It may also be necessary to modify published phase diagrams and adapt them to the actual problem (e.g. change the pT frame or other parameter values). It is not the intention of this book to give a comprehensive treatment of the computation and construction of phase diagrams, however, it is probably useful to briefly introduce two important aspects of phase diagrams: the phase rule and the Schreinemakers rules.

3.8.2 The Phase Rule

The state of a heterogeneous system, such as a rock, depends on a number of state variables (such as pressure, temperature and chemical potentials). The requirements of heterogeneous equilibrium impose some restrictions on these variables (equilibrium constraints). The phase rule relates the number of variables in a system and the number of equations that can be written among them at equilibrium. In a system with a total number of c components and composed of p phases the number of variables is:

- for each phase: T, p, and $c - 1$ compositional variables $= c + 1$
- for the system of p phases: $p(c + 1)$

Equilibrium constraints:

The temperature of all phases must be the same (thermal equilibrium) $\Rightarrow$
- $p - 1$ equations of the type: T of phase α = T of phase β.
The pressure on all phases must be the same (mechanical equilibrium) $\Rightarrow$
- $p - 1$ equations of the type p on phase α = p on phase β.
The chemical potential of component i must be the same in all phases (chemical equilibrium) $\Rightarrow$
- $c(p - 1)$ equations of the type $\mu_{i\alpha} = \mu_{i\beta}$; $p - 1$ equations for each component from 1 to c.
Hence, the number of variables n_v is: $p(c + 1)$, and the number of equations n_e: $2(p - 1) + c(p - 1)$.

Now let us define a variable f as the difference between n_v and n_e. f simply expresses how many variables can be changed independently in a system at equilibrium.
$f = n_v - n_e = [p(c + 1) - 2(p - 1) - c(p - 1)]$; or simply

$$f = c + 2 - p. \tag{36}$$

This equation is known as the **phase rule**. f is often referred to as **variance** or **degrees of freedom** of the system. A consequence of the phase rule is that the number of phases cannot exceed the number of components by more than 2. In a one-component system (e.g. Al_2SiO_5) consisting of one phase (e.g. kyanite), there are two variables which can be varied independently (e.g. p and

T). If there are two phases present (sillimanite and kyanite), only one variable can be specified independently. If there are three phases present (andalusite, sillimanite and kyanite), the number of variables equals the number of equations ($f = 0$) and the system of equations has one unique solution. In other words, coexistence of all three aluminosilicates completely defines the state of the system; it is only possible at a unique value of pressure and temperature (the triple point, invariant point). Another well-known one-component system is the system H_2O. Ice has $f = 2$, melting ice has $f = 1$ and ice, water and steam ($f = 0$) may be in equilibrium at the unique p and T values of the triple point.

Phase Rule in Reactive Systems

Consider a system with N phase components (chemical species), some inert, some at reaction equilibrium. R is the number of independent reactions and p the number of phases.

number of variables: $p(N + 1)$
number of equations: $(N + 2)(p - 1) + R$
R = number of conditions of reaction equilibrium $\Sigma \nu \mu = 0$

$$f = n_v - n_e = N - R + 2 - p \tag{37}$$

An example; the mineral assemblage $Grt + Crd + Sil + Qtz$ consists of four phases and we may consider seven phase components (garnet: Grs, Prp, Alm ; cordierite: Fe-Crd, Mg-Crd; sillimanite: Sil; quartz: Qtz). The seven phase components are related by two independent reactions (net transfer: $2\, Alm + 5\, Qtz + 4\, Sil = 3\, Fe\text{-}Crd$; exchange: $FeMg_{-1}\, (Grt) = FeMg_{-1}\, (Crd)$). The variance of the assemblage calculated from Eq. (37) is 3. This means that the system of equations has a unique solution if three variables can be specified independently. If, for example, the Alm and Prp content of the garnet and the Fe-Crd content of the cordierite were measured in a rock containing the four minerals in equilibrium, then the system is uniquely defined and the equilibrium pressure and temperature can be calculated.

By defining $c = N - R$, the two forms of the phase rule become formally identical. The number of components in the sense of the phase rule is therefore to be taken as the total number of phase components (chemical species) minus the number of independent reactions between them. In the example made above, the number of system components is five (seven phase components – two independent reactions). Simple oxide components are often selected as system components. The five components $CaO\text{-}MgO\text{-}FeO\text{-}Al_2O_3\text{-}SiO_2$ (CMFAS) define the composition space of the $Grt + Crd + Sil + Qtz$ rock used as an example here. For a comprehensive treatment of the composition space and reaction space see Thompson (1982a, b). The phase rule deriviation has been adapted from Denbigh (1971).

3.8.3 Construction of Phase Diagrams for Multicomponent Systems After the Method of Schreinemakers

A binary system of four phases is, according to the phase rule, invariant. The four minerals may occur at a unique pair of temperature and pressure (or any other two intensive variables). Graphically the four-phase assemblage occurs at a point (invariant point) on a pT diagram. From this point four univariant assemblages represented by univariant lines (curves) radiate into the pT plane (surface). The univariant lines are characterized by the absence of one of the phases occurring in the invariant point and divide the pT plane in four divariant sectors. Each of these sectors is characterized by one unique two-phase assemblage that occurs in one, but only one, sector. The geometric distribution of univariant and divariant assemblages follows strict rules that may be derived from chemical thermodynamics. These rules were formulated by Schreinemakers in a series of articles around 1915 and are known today as Schreinemakers rules. Figure 3.21A shows the geometry of the general binary four-phase system and Fig. 3.21B represents a specific geologic example. For three-component five-phase systems and complications arising from compositional degeneracies the reader is referred to Zen (1966).

The general geometry of the binary system (Fig. 3.21A) shows an invariant assemblage of four phases with the compositions 1, 2, 3 and 4 in the two-component system a-b. The space around the invariant point is divided into four sectors by four univariant curves. Each of these curves is characterized by the absence of one phase of the invariant four-phase assemblage. The name of the absent phase is put into round brackets. The sectors contain either two, one or no metastable extension of the four univariant curves. The sector between the curves (2) and (3) contains two metastable extensions and is characterized by the presence of the divariant assemblage 1 + 4 which occurs only in this sector. The assemblage contains the most a-rich and the most b-rich phase, respectively. If the sector boundary (3) is crossed, 1 + 4 form phase 2, that is a new phase of intermediate composition [3 does not participate in the reaction, (3) is absent]. If the sector boundary (1) is crossed, 3 forms from 2 + 4 (1 is absent). In the sector between (1) and (4) three two-phase assemblages may be present depending on the relative proportions of the components a and b in the rock. However, only one of the three assemblages is unique for the sector, namely 2 + 3. Crossing the sector boundary (4) the phase 2 decomposes to 1 + 3 (4 does not participate in the reaction). Finally, crossing the univariant curve (2) removes 3 by the reaction 3 = 1 + 4.

Fig. 3.21B shows a specific geologic example. In rocks containing excess quartz and K-feldspar and for H_2O-present situations the four minerals garnet, cordierite, biotite and sillimanite may be represented in a binary chemography. The chemography is a projection from quartz, K-feldspar and H_2O. The geometry shown for the general binary case in Fig. 3.21A may be readily translated into Fig. 3.21B. The unique divariant assemblage for each sector is shaded in Fig. 3.21B. Note, however, that some assemblages (e.g. garnet + biotite; Fig. 3.21b) occur in more than one sector. Any assemblage occurring in two

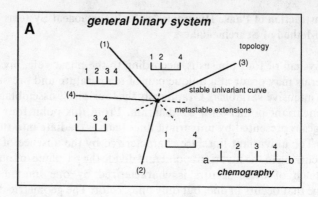

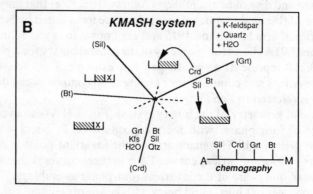

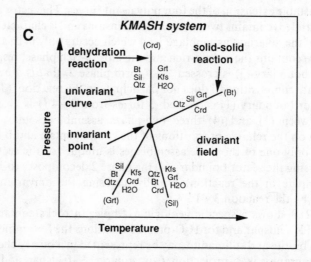

Fig. 3.21A–C. Schreinemakers diagram for general binary four phase assemblages (**A**), an example assemblage Sil, Fe-Crd, Fe-Grt, and Fe-Bt projected from K-feldspar, quartz and H_2O (**B**), and the translation of the relationships shown in figure 3–21b into pT space (**C**)

sectors (in the binary case) contain the phase which is absent at the sector boundary (biotite in our case). This is so because biotite + garnet is not affected by the biotite-absent reaction (Crd = Sil + Grt). Also note that, for instance, in the cordierite-absent reaction (Bt + Sil = Grt), the assemblage containing biotite must be in the sector that is not bounded by the biotite-absent reaction, and garnet must be in the sector that is not bounded by the garnet-absent reaction. This is a general rule and must be obeyed by all univariant assemblages along any of the curves.

Once the correct sequence and geometry of univariant assemblages is established around an invariant point the sequence may be translated into pressure-temperature space (Fig. 21C).

Three of the four reactions in our example are dehydration reactions. It is fairly safe to assume that H_2O is released from biotite as temperature increases. The reactions (Crd), (Grt) and (Sil) should therefore be arranged in such a way that H_2O appears on the high-temperature side of the univariant curve. In addition, dehydration reactions commonly have steep slopes on PT-diagrams. Two solutions are possible; (1) the (Crd) reaction extends to high pressures from the invariant point (the reactions (Grt) and (Sil) extend to the low-pressure side), or (2) (Crd) extends to low pressures. Both solutions are geometrically correct. The version (1) is shown in Fig. 3.21C because cordierite is a mineral that typically occurs in low- to medium-pressure metapelites. In solution (2) cordierite is restricted to high pressures, which is not consistent with geological experience. The slope of the solid-solid reaction (Bt) cannot be constrained in a way similar to that for the dehydration reactions. However, it must extend from the invariant point into the sector between (Sil) and the metastable extension of (Grt). This follows from Fig. 3.21b. Therefore, it may have a positive or a negative slope. The equilibrium of the reaction is in reality rather independent of temperature and it represents an example of a geologic barometer.

The geometric arrangement of univariant curves around invariant points on any kind of phase diagram must be strictly consistent with Schreinemakers rules. Part II, on progressive metamorphism of various rock types, provides many examples of Schreinemakers bundles in two- and three-component systems on various types of phase diagrams.

References and Further Reading

Balling NP (1976) Geothermal models for the crust and uppermost mantle of the Fennoscandian Shield in South Norway and the Danish Embayment. J Geophys Res 42:237–256

Balling NP (1985) Thermal structure of the lithosphere beneath the Norwegian-Danish basin and the southern baltic shield: a major transition zone. Terra Cognita 5:377–378

Bell TH, Cuff C (1989) Dissolution, solution transfer, diffusion versus fluid flow and volume loss during deformation/metamorphism. J Metamorph Geol 7:425–448

Bell TH, Hayward N (1991) Episodic metamorphic reactions during orogenesis: the control of deformation partitioning on reaction sites and reaction duration. J Metamorph Geol 9:619–640

Berman RG (1988) Internally-consistent thermodynamic data for minerals in the system: Na_2O- K_2O- $CaO-$ $MgO-$ $FeO-$ Fe_2O_3- Al_2O_3- SiO_2- TiO_2- H_2O- CO_2: J Petro, 29, 2:445–522

Berman RG, Brown TH, Perkins EH (1987) GEØ-CALC: software for calculation and display of pressure – temperature – composition phase diagrams. Am Mineral 72, 7–8:861

Brady JB (1988) The role of volatiles in the thermal history of metamorphic terranes. J Petrol 29:1187–1213

Brodie KH, Rutter EH (1985) On the relationship between deformation and metamorphism with special reference to the behaviour of basic rocks. In: Thompson AB, Rubie DC (eds) Advances in physical geochemistry. Springer, Berlin Heidelberg New York, pp 138–179

Brown GC, Mussett AE (1981) The inaccessible earth. George Allen & Unwin, London, 235 pp

Brown TH, Berman RG, Perkins EH (1988) GEØ-CALC: software package for calculation and display of pressure-temperature-composition phase diagrams using an IBM or compatible computer. Comput Geosci 14:279–289

Bucher-Nurminen K (1990) Geological phase diagram software. Terra Nova 2:401–410

Burnham CW, Holloway JR, Davis NF (1969) Thermodynamic properties of water to 1000°C and 10000 bars. Geol Soc Am Spec Pap 132:96

Carmichael DM (1968) On the mechanism of prograde metamorphic reactions in quartz-bearing pelitic rocks. Contributions to Mineralogy and Petrology 20:244–267

Cermak V, Rybach L (1987) Terrestial heat flow and the lithosphere structure. Terra Cognita 7, 4:685–687

Chatterjee ND (1991) Applied mineralogical thermodynamics. Springer, Berlin Heidelberg New York, 321 pp

Clark SP (1966) Handbook of physical constants. Geol Soc Am Mem, Boulder, 587 pp

Connolly JAD (1990) Multivariable phase diagrams. An algorithm based on generalized thermodynamics. Am J Sci 290:666–718

Connolly JAD, Kerrick DM (1987) An algorithm and computer program for calculating computer phase diagrams. CALPHAD 11:1–55

Davy P, Gillet P (1986) The stacking of thrust slices in collision zones and its thermal consequences. Tectonics 5:913–929

Day HW (1972) Geometrical analysis of phase equilibria in ternary system of six phases. Am J Sci 272:711–734

Day HW, Chamberlain CP (1989) Implications of thermal and baric structure for controls on metamorphism, northern New England, USA. In: Daly, S, Cliff RA, Yardley, BWD, (eds) Evolution of metamorphic belts. Geological Society Special Publication, Blackwell, Oxford, pp 215–222

Denbigh K (1971) The principles of chemical equilibrium. Cambridge University Press, London, 494 pp

Durney DW (1972) Solution transfer, an important geological deformation mechanism. Nature 235:315–317

Dyar MD, Guidotti CV, Holdaway MJ et al. (1993) Nonstoichiometric hydrogen contents in common rock-forming hydroxyl silicates. Geochim Cosmochim Acta 57:2913–2918

England PC, Richardson SW (1977) The influence of erosion upon the mineral facies of rocks from different metamorphic environments. J Geol Soc Lond 134:201–213

England PC, Thompson AB (1984) Pressure-temperature-time paths of regional metamorphism I. Heat transfer during the evolution of regions of thickened continental crust. J Petrol 25, 4:894–928

Erambert M, Austrheim H (1993) The effect of fluid and deformation on zoning and inclusion patterns in poly-metamorphic garnets. Contrib Mineral Petrol 115:204–214

Eugster HP (1977) Compositions and thermodynamics of metamorphic solutions. In: Fraser DG (ed) Thermodynamics in geology. Dordrecht, pp 183–202

Eugster (1986) Minerals in hot water. Am Mineral 71, 5–6:655–673

Eugster HP, Gunter WD (1981) The compositions of supercritical metamorphic solutions. Bull Minéral 104:817–826

Ferry JM (1983) Application of the reaction progress variable in metamorphic petrology. J Petrol 24:343–376

Ferry JM (1980) A case study of the amount and distribution of heat and fluid during metamorphism. Contrib Mineral Petrol 71:373–385

Ferry JM (ed) (1982) Characterization of metamorphism through mineral equilibria. Reviews in Mineralogy, vol 10. Mineralogical Society of America, Washington DC, 397 pp

Fletcher P (1993) Chemical thermodynamics for earth scientists. Longman Scientific and Technical, Essex, 464 pp

Fraser DG (ed) (1977) Thermodynamics in geology. NATO Advanced Study Institutes Series 30, Reidel, Dordrecht, 410 pp

Froese E (1977) Oxidation and sulphidation reactions. In: Greenwood HJ (ed) Application of thermodynamics to petrology and ore deposits. Short course. Mineralogical Society of Canada, Vancouver, pp 84–98

Frost BR (1988) A review of graphite- sulfide- oxide- silicate equilibria in metamorphic rocks. Rend Soc Ita Mineral Petrol 43:25–40

Frost BR, Tracy RJ (1991) P-T paths from zoned garnets: some minimum criteria. Am J Sci 291:917–939

Fyfe WS, Turner FJ and Verhoogen J (1958) Effect of temperature on equilibrium entropy of solids. In: Metamorpic reactions and metamorphic facies. Geol Soc Am Bull 73:25–34.

Garrels RM, Christ CL (1965) Solutions, minerals and equilibria. Freeman, Cooper, San Francisco, 450 pp

Gibbs JW (1878) On the equilibrium of heterogeneous substances. Am J Sci (Trans Conn Acad) 16, 3:343–524

Gibbs JW (1906) The scientific papers of J. Willard Gibbs. Thermodynamics. Longmans, Greed, London

Greenwood HJ (1968) Matrix methods and the phase rule in petrology. 23rd Int Geol Congr, pp 267–279

Greenwood HJ (1975) Buffering of pore fluids by metamorphic reactions. Am J Sci 275:573–594

Greenwood HJ (ed) (1977) Application of thermodynamics to petrology and ore deposits. Short course. Mineralogical Society of Canada, Vancouver, 230 pp

Guggenheim EA (1986) Thermodynamics. North-Holland Physics Pub, Amsterdam, 390 pp

Harker A (1932) Metamorphism. A study of the transformation of rock masses. Methuen, London

Helgeson HC, Kirkham DH (1974) Theoretical prediction of the thermodynamic behaviour of aqueous electrolytes at high pressures and temperatures. I. Summary of the thermodynamic/electrostatic properties of the solvent. Am J Sci 274:1089–1098

Helgeson HC, Delany JM, Nesbitt HW et al. (1978) Summary and critique of the thermodynamic properties of rock-forming minerals. Am J Sci 278-A:229

Holland TJB (1985) An internally consistent dataset with uncertainties and correlations: 2. Data and results. J Metamorph Geol 3:343–370

Holland TJB, Powell R (1990) An enlarged and updated internally consistent thermodynamic dataset with uncertainties and correlations: the system K_2O- Na_2O- CaO- MgO- MnO- FeO- Fe_2O_3- Al_2O_3- TiO_2- SiO_2- C- H_2- O_2. J Metamorph Geol 8:89–124

Johnson JW, Oelkers EH, Helgeson HC (1992) SUPCRT92: a software package for calculating the standard molal thermodynamic properties of minerals, gases, aqueous species, and reactions from 1 to 5000 bars and 0 to 1000°C. Computers and Geosciences 18:899–947

Johnson SE (1993) Testing models for the development of spiral-shaped inclusion trails in garnet porphyroblasts: to rotate or not to rotate, that is the question. J Metamorph Geol 11:635–659

Johnson SE (1993) Unravelling the spirals: a serial thin-section study and three-dimensional computer-aided reconstruction of spiral-shaped inclusion trails in garnet porphyroblasts. J Metamorph Geol 11:621–634

Jones KA, Brown M (1990) High-temperature 'clockwise' P-T paths and melting in the development of regional migmatites: an example from southern Brittany, France. J Metamorph Geol 8:551-578

Kukkonen IT, Cermk V, Hurtig E (1993) Vertical variation of heat flow density in the continental crust. Terra 5:389-398

Lasaga AC, Kirkpatrick RJ (eds) (1981) Kinetics of geochemical processes. Reviews in Mineralogy. Mineralogical Society of America, Washington DC, 398 pp

Lasaga AC, Rye DM (1993) Fluid flow and chemical reaction kinetics in metamorphic systems. Am J Sci 293:361-404

Lewis GN, Randall M (1961) Thermodynamics. (Revised by Pitzer KS, Brewer L. McGraw-Hill, New York, 723 pp

Miyashiro A (1961) Evolution of metamorphic belts. J Petrol 2:277-311

Miyashiro A (1964) Oxidation and reduction in the earths crust, with special reference to the role of graphite. Geochim Cosmochim Acta 28:717-729

Miyashiro A, Aki K, Celas Sengor AM (1979) Orogeny. John Wiley, New York

Moore WL (1972) Physical chemistry. Longman, London, 977 pp

Norris RJ, Henley RW (1976) Dewatering of a metamorphic pile. Geology 4:333-336

Norton D, Knight J (1977) Transport phenomena in hydrothermal systems: cooling plutons. Am J Sci 277:937-981

Oxburgh ER (1974) The plain man's guide to plate tectonics. Proc Geol Assoc 85:299-358

Oxburgh ER, England PC (1980) Heat flow and the metamorphic evolution of the eastern Alps. Eclogae Geol Helv 73, 2:379-398

Oxburgh ER, Turcotte DL (1971) Origin of paired metamorphic belts and crustal dilation in island arc regions. J Geophys Res 76:1315-1327

Oxburgh ER (1974) Thermal gradients and regional metamorphism in overshrust terrains with special reference to the eastern Alps. Schweiz Mineral Petrogr Mitt 54:642-662

Perkins D, Essene EJ, Wall VJ (1987) THERMO: a computer program for calculation of mixed-volatile equilibria. Am Mineral 72, 2-3:446-447

Platt JP (1986) Dynamics of orogenic wedges and the uplift of high-pressure metamorphic rocks. Geol Society Am Bull 97:1037-1053

Powell R (1978) Equilibrium thermodynamics in petrology, an introduction. Harper and Row, Hagerstown, 284 pp

Powell R, Holland TJB (1985) An internally consistent dataset with uncertainties and correlations. 1. Methods and a worked example. J Metamorph Geol 3:327-342

Powell R (1988) An internally consistent dataset with uncertainties and correlations. 3. Applications to geobarometry, worked examples and a computer program. J Metamorph Geol 6, 2:173-204

Prigogine I (1955) Thermodynamics of irreversible prosesses. John Wiley, New York, 147 pp

Robie RA, Hemingway BS, Fisher JR (1978) Thermodynamic properties of minerals and related substances at 298.15 K and 1 bar (10^5 Pascals) pressure and at higher temperatures. US Geol Surv Bull 1452:456 pp

Robinson D (1987) Transition from diagenesis to metamorphism in extensional and collision settings. Geology 15:866-869

Saxena S, Ganguly J (1987) Mixtures and mineral reactions. Minerals and rocks. Springer, Berlin Heidelberg New York, 260 pp

Sclater JG, Jaupart C, Galson D (1980) The heat flow through oceanic and continental crust and the heat loss of the Earth. Rev Geophys Space Phys 18:269-311

Seyfried WE (1987) Experimental and theoretical constraints on hydrothermal alteration processes at mid-ocean ridges. Annu Rev Earth Planet Sci Lett 15:317-335

Skippen GB, Carmichael DM (1977) Mixed-volatile equilibria. In: Greenwood HJ (ed) Application of thermodynamics to petrology and ore deposits. Short course. Mineralogical Society of Canada, Vancouver, pp 109-125

Sleep NH (1979) A thermal constraint on the duration of folding with reference to Acadian geology New England, USA. J Geol 87:583-589

Spear FS, Rumble D, III, Ferry JM (1982) Linear algebraic manipulation of n-dimensional composition space. In: Ferry JM (ed) Characterization of metamorphism through mineral equilibria. Reviews in Mineralogy. Mineralogical Society of America, Washington DC, pp 53–104

Spear FS, Selverstone J, Hickmont D et al. (1984) P-T paths from garnet zoning: a new technique for deciphering tectonic processes in crystalline terranes. Geology 12:87–90

Spear FS, Peacock SM, Kohn MJ et al. (1991) Computer programs for petrologic P-T-t path calculations. A Mineral 76, 11–12:2009–2012

Spear FS, Peacock SM (1989) Metamorphic pressure-temperature-time paths. Short course in geology 7. American Geophysical Union, Washington DC, pp 102

Spooner ETC, Fyfe WS (1973) Sub-sea-floor metamorphism, heat and mass transfer. Contrib Mineral Petrol 42:287–304

Stern T, Smith EGC, Davey FJ, Muirhead KJ (1987) Crustal and upper mantle structure of the northwestern North Island, New Zealand, from seismic refraction data. Geophysical Journal of the Royal Astronomical Society 91:913–936

Stull DR, Prophet H (eds) (1971) JANAF thermochemical tables. Matl Standards Ref Data Ser. National Bureau of Standards, Washington, 1141 pp

Thompson AB (1981) The pressure-temperature (P,T) plane viewed by geophysicists and petrologists. Terra Cognita 1:11–20

Thompson AB, Tracy RJ, Lyttle PT et al. (1977) Prograde reaction histories deduced from compositional zonation and mineral inclusions in garnet from the Gassetts schist, Vermont. Am J Sci 277:1152–1167

Thompson JB (1982a) Composition space; an algebraic and geometric approach. In: Ferry JM (ed) Characterization of metamorphism through mineral equilibria. Reviews in Mineralogy, Mineralogical Society of America, Washington DC, pp 1–31

Thompson JB (1982b) Reaction space; an algebraic and geometric approach. In: Ferry JM (ed) Characterization of metamorphism through mineral equilibria. Reviews in Mineralogy, Mineralogical Society of America, Washington DC, pp 33–52

Walther JV, Orville PM (1982) Rates of metamorphism and volatile production and transport in regional metamorphism. Contrib Mineral Petrol 79:252–257

Wood BJ, Fraser DG (1976) Elementary thermodynamics for geologists. Oxford University Press, Oxford, 303 pp

Wood BJ, Walther JV (1984) Rates of hydrothermal reactions. Science 222:413–415

Zen E-An (1966) Construction of pressure-temperature diagrams for multi-component systems after the method of Schreinemakers – a geometrical approach. US Geol Surv Bull 1225:56 pp

Zwart HJ (1962) On the determination of polymetamorphic mineral associations, and its application to the Bosot area (central Pyrénées). Geol Rundsch 52:38–65

4 Metamorphic Grade

4.1 General Considerations

According to Turner (1981, p. 85), the term **metamorphic grade** or **grade of metamorphism** was introduced by Tilley (1924) "to signify the degree or state of metamorphism" (p. 168) and, more specifically, "the particular pressure-temperature conditions under which the rocks have arisen" (p. 167). Since reliable P-T values were not known for metamorphic rocks at that time, and since temperature was generally accepted as the most important factor of metamorphism (cf. Chap. 3), it became current usage to equate grade rather loosely with temperature. As a recent example, Winkler (1979, p. 7) suggested a broad fourfold division of the P-T field of metamorphism primarily based on temperature, which he named very low-, low-, medium-, and high-grade metamorphism. Even though Winkler noted that information on pressure should be stated as well, a subdivision of metamorphic grade with respect to pressure seemed to be less important. This is well understandable because most of his P-T diagrams were limited to P < 12 kbar, whilst the present-day P-T space of metamorphism has to be extended to much higher pressures (Fig. 1.1).

In this book, the term metamorphic grade will be used as a qualitative indicator of the physical conditions that have been operating, with elevated P-T conditions being characteristic of higher grade. It is a useful term for comparison of rocks within a single progressive metamorphic area. When applied to rocks in different regions, its meaning is not always clear, however, and its exact nature should be clearly specified.

In the following, various concepts to determine metamorphic grade will be discussed.

4.2 Index Minerals and Mineral Zones

Mineral zones were mapped for the first time by Barrow (1893, 1912) in pelitic rocks of the Scottish Highlands. He recognized the systematic entrance of new minerals proceeding upgrade; these minerals were designated **index minerals**. The following succession of index minerals with increasing metamorphic grade

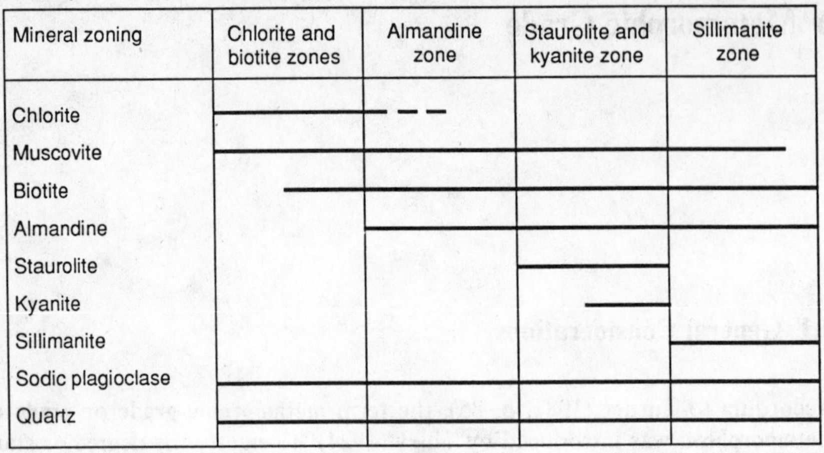

Mineral zoning	Chlorite and biotite zones	Almandine zone	Staurolite and kyanite zone	Sillimanite zone
Chlorite				
Muscovite				
Biotite				
Almandine				
Staurolite				
Kyanite				
Sillimanite				
Sodic plagioclase				
Quartz				

Fig. 4.1. Distribution of some metamorphic minerals of pelitic rocks from the Barrovian zones of the Scottish Highlands

has been distinguished:

chlorite ⇒ biotite ⇒ almandine-garnet ⇒ staurolite ⇒ kyanite ⇒ sillimanite.

The individual minerals are systematically distributed in distinct regional zones in the field, and corresponding **mineral zones** are defined as follows. The low grade limit is determined by a line on a map joining points of the first appearance of a certain index mineral, after which the zone is named. The high grade limit is given by a similar line for the following index mineral. A line separating two adjacent mineral zones will be termed a **mineral zone boundary** (and not an isograd as discussed below). The biotite zone, for instance, is that occurring between the biotite and almandine-garnet mineral zone boundaries. Note that an index mineral generally persists to higher grades than the zone which it characterizes, but is sometimes restricted to a single mineral zone (e.g. kyanite in Fig. 4.1).

The zonal sequence elaborated by Barrow are called **Barrovian zones**. These mineral zones have since been found in many other areas and are characteristic for medium-pressure metapelites. Sequences of mineral zones other than those proposed by Barrow have been identified in other areas. In the Buchan region of NE Scotland, the **Buchan zones** are defined by the index minerals staurolite, cordierite, andalusite and sillimanite. The Buchan zones represent a different metamorphic gradient involving relatively lower pressures than those represented by the Barrovian zones.

Two zonal schemes have been mentioned above in pelites from orogenic metamorphic terrains. It should be made clear, however, that mineral zones can be mapped, in principle, in many types of rock belonging to any type of metamorphism.

The mapping of mineral zones in determining metamorphic grade has the great advantage that it is a simple and rapid method, because the distribution of index minerals becomes obvious from inspection of hand specimens and thin sections. When working in an area for the first time, mineral zoning provides a first insight into the pattern of metamorphism. On the other hand, several shortcomings should be mentioned as well. First, by mapping mineral zones, one single mineral in each rock is considered instead of a mineral assemblage; the latter, however, contains more petrogenetic information. Second, variations in rock composition are not adequately taken into acount, and some index mineral will appear at a higher or lower grade in layers of different composition. In order to circumvent these drawbacks, today metamorphic zones are distinguished by associations of two, three or even more minerals rather than by index minerals.

4.3 Metamorphic Facies

The concept of metamorphic facies was introduced by Eskola (1915). He emphasized that mineral assemblages rather than individual minerals are the genetically important characteristics of metamorphic rocks, and a regular relationship between mineral assemblages and rock composition at a given metamorphic grade was proposed. This principle said that in a group of metamorphic rocks that have reached chemical equilibrium under the same definite P-T conditions, the mineral assemblages of the rocks depend only on their bulk chemical composition.

Eskola (1915, p. 115) defined a **metamorphic facies** as follows: "A metamorphic facies includes rocks which.. may be supposed to have been metamorphosed under identical conditions. As belonging to a certain facies we regard rocks which, if having an identical chemical composition, are composed of the same minerals." This definition has been a problem ever since Eskola first introduced the idea as discussed below:

1. A metamorphic facies includes mineral assemblages of a set of associated rocks covering a wide range of composition, all formed under the same broad metamorphic conditions (P and T after Eskola). Therefore, a metamorphic facies does not refer to a single rock type, even though many facies are named after some characteristic metabasic rocks (e.g. greenschist, amphibolite, eclogite). Furthermore, some mineral assemblages have a large stability range and may occur in several metamorphic facies, whilst other assemblages have a more restricted stability range and may be diagnostic for one facies only. In addition, some rock types do not show diagnostic assemblages at some particular metamorphic grade, e.g. many metapelites under subgreenschist facies conditions or metacarbonates under eclogite facies conditions. From these considerations it is obvious that we should search for **diagnostic mineral assemblages**, and it may be sufficient to recognize a metamorphic facies with the aid of only one such assemblages. In areas devoid of rock composition suitable

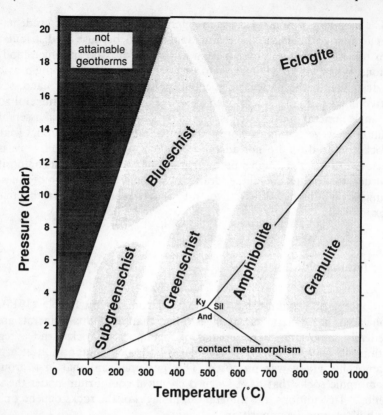

Fig. 4.2. Approximate positions of metamorphic facies on a pressure-temperature diagram

for forming these diagnostic assemblages, on the other hand, assignment to a facies cannot be made.

2. Remember that Eskola proposed the idea of metamorphic facies before there were any significant experimental or thermodynamic data on the stability of metamorphic minerals. Under these circumstances, it is understandable that the only factors which this early scheme considered to be varying during metamorphism were temperature and lithostatic pressure. Additional variables, e.g. the composition of fluids present, if any, were not yet recognized, and it was assumed that $P_{H_2O} = P_{total}$. Recent experimental and theoretical studies have shown that mineral assemblages including, e.g. lawsonite, prehnite or zeolites, are stable only in the presence of a very water-rich fluid composition, i.e. are absent at appreciable concentrations of CO_2. Furthermore, changing fluid composition may affect the P-T stability of mineral assemblages (Chap. 3). In addition, the low pressure limit of the eclogite facies is critically dependent upon the activity of H_2O in the rock systems undergoing metamorphism. This means that a more complete representation of metamorphic facies should involve relationships in P-T-X space, but detailed information on fluid composition is often missing.

Table 4.1. Diagnostic minerals and assemblages from the various metamorphic facies

Facies	Diagnostic minerals and assemblages
Subgreenschist	Laumontite, prehnite + pumpellyite, prehnite + actinolite, pumpellyite + actinolite, pyrophyllite
Greenschist	Actinolite + chlorite + epidote + albite chloritoid
Amphibolite	Hornblende + plagioclase staurolite
Granulite	Orthopyroxene + clinopyroxene + plagioclase, sapphirine, osumilite, kornerupine *no* staurolite, *no* muscovite
Blueschist	Glaucophane, lawsonite, jadeitic pyroxene, aragonite Mg–Fe–carpholite *no* biotite
Eclogite	Omphacite + garnet no plagioclase

3. Eskola's definition allows for an unlimited number of metamorphic facies. During his lifetime, Eskola increased the number of facies he recognized from five to eight. As more information on mineral assemblages became available, more metamorphic facies have been added by other writers and many have been divided into "subfacies" or zones. At some point, facies classification became impracticable. Some authors proposed retaining broad divisions, but others advocated the abolition of metamorphic facies.

Notwithstanding the inherent weaknesses of the concept, the metamorphic facies scheme continues to be used. It is convenient as a broad genetic classification of metamorphic rocks, in terms of two major variables: lithostatic pressure and temperature. It is especially useful for regional or reconnaissance studies in metamorphic regions, but is too broad for detailed metamorphic studies. As a good example, the large-scale metamorphic maps of Europe (Zwart 1973) may be mentioned.

The metamorphic facies scheme proposed in this book is shown in Fig. 4.2, and some diagnostic mineral assemblages are listed in Table 4.1. Contrary to other schemes, no facies for contact metamorphism are distinguished here, because such facies are only of limited extent and metamorphic zones are sufficient to designate variations in metamorphic grade. The simple facies scheme of Figure 4.2 closely follows Eskola's classic treatment, and each metamorphic facies is briefly characterized below.

4.3.1 Subgreenschist Facies

Eskola remained unconvinced that certain feebly recrystallized zeolite-bearing rocks represented equilibrium-phase assemblages; hence he chose not to establish a separate metamorphic facies. However, later work by Coombs in New Zealand has revealed the ubiquitous, systematic occurrence of such lithologies, and accordingly has proposed two very low-grade metamorphic facies, the zeolite and prehnite-pumpellyite facies. Recently, more complex subdivisions have been suggested, as discussed in Chapter 6. These various facies of incipient metamorphism can be recognized in only a small range of rock types; for this reason, and in order to maintain a simple scheme, they are lumped together here under the name subgreenschist facies.

4.3.2 Greenschist Facies

The name is derived from greenschist, a metabasite which commonly shows the mineral assemblage actinolite + chlorite + epidote + albite ± quartz ± hornblende ± garnet. The chlorite, biotite and garnet Barrovian zones belong to this facies, and more Al-rich metapelites may contain chloritoid. This facies covers a relatively small range of temperature at low to intermediate pressures. The epidote-amphibolite facies distinguished by some authors is included here in the greenschist facies.

4.3.3 Amphibolite Facies

The name stems from amphibolite, a metabasite containing plagioclase + hornblende ± quartz ± garnet ± clinopyroxene. The staurolite, kyanite and sillimanite Barrovian zones belong to this facies. Tremolite-bearing marbles are also characteristic.

4.3.4 Granulite Facies

This embraces highly dehydrated mineral assemblages at maximum temperatures of orogenic metamorphism. In metabasites, ortho- and clinopyroxenes take the place of hornblende, and this facies is sometimes called the two-pyroxene facies. Since granulite facies rocks are strongly dehydrated, their formation should be promoted by lower water pressure; indeed, large amounts of CO_2 are often reported from fluid inclusions of such rocks.

4.3.5 Blueschist Facies

The name is derived from blue sodic amphiboles, mainly glaucophane and minor crossite; they are typically found in mineral assemblages together with lawsonite, jadeitic pyroxene and aragonite. Blueschist facies assemblages are formed at low temperatures and relatively high pressures, i.e. along low geothermal gradients. This facies was originally named glaucophane-schist facies by Eskola. The lawsonite-albite or lawsonite-albite-chlorite facies of some authors are included here in the subgreenschist facies.

4.3.6 Eclogite Facies

Eclogites, metabasites characterized by the occurrence of omphazite + garnet, are the main representatives of a facies that requires unusually high pressures at intermediate to high temperatures. For the most part, eclogites appear as tectonically transported blocks enclosed in rocks of some other facies or in chaotic mélange zones that commonly occur in blueschist terranes. However, some eclogites show a regular zonal distribution suggestive of progressive metamorphism, and isofacial metapelites and metagranitoids with distinct mineral assemblages are increasingly known.

4.3.7 Pressure-Temperature Conditions of Metamorphic Facies

The P-T limits of the various metamorphic facies in Fig. 4.2 should not be taken too literally, and boundaries between facies are gradational because (1) most characterizing mineral assemblages form by continuous reactions over P-T intervals as a consequence of mineral solid solution; (2) the assumption of $P_{H_2O} = P_{total}$ may be incorrect since many metamorphic fluids are compositionally complex solutions. In addition, some data on P-T stabilities of mineral assemblages are still ambiguous.

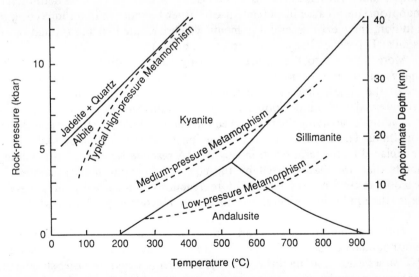

Fig. 4.3. Classification of metamorphic facies series in relation to the Al_2SiO_5 phase diagram (after Bohlen et al. 1991) and the equilibrium curve $Ab = Jd + Qtz$ (after Berman 1988)

Miyashiro (1961) emphasized that certain metamorphic facies are commonly associated to the exclusion of others in different orogenic belts. The sequence of facies characteristic of an individual area or terrain was named a **metamorphic facies series**. Miyashiro recognized three principal types of facies series: (1) a low-P, high-T type, in which andalusite and sillimanite are formed; (2) an intermediate-P, intermediate-T type characterized by kyanite and sillimanite; and (3) a high-P, low-T type, signalled by the occurrence of jadeitic pyroxene and glaucophane. In a P-T diagram, each facies series is represented by a different P-T gradient (Fig. 4.3). Miyashiro also noted the existence of metamorphic facies series intermediate to the three types mentioned above. All intergradations are possible, of course, and result from the fact that each metamorphic belt has been subjected to its own unique set of physical conditions. The merit of the metamorphic facies series concept becomes obvious if regional metamorphism on the broadest possible scale in relation to tectonism is considered.

4.4 Isograds and Reaction-Isograds

The isograd method has been applied by many geologists for several decades, but semantic confusion has arisen because its significance is more complex than was originally assumed. When coining the word isograd, Tilley (1924, p. 169) referred to the boundaries of Barrow's mineral zones as a "… line joining the points of entry of.. (an index mineral) in rocks of the same composition …". Tilley (op.cit.) defined the term **isograd** "as a line joining points of similar P.T. values, under which the rocks as now constituted, originated". This definition is inappropriate because in practice it can never be more than an inference; in addition, there exist several arguments which indicate that the stipulation of "similar P-T values" may be incorrect (see below).

More insight into the nature of an isograd is gained by considering the change of mineral assemblages and a related specific metamorphic reaction across the "line". Since a metamorphic reaction depends on temperature, pressure and fluid composition, an isograd will represent, in general, sets of P-T-X conditions satisfying the reaction equilibrium and not points of equal P-T-X conditions. Therefore, whenever the reaction is known, the term isograd should be replaced by the term reaction-isograd. "A **reaction-isograd** is a line joining points that are characterized by the equilibrium paragenesis of a specific reaction; in certain cases, two (or more) reactions taking place simultaneously define the equilibrium mineral assemblage of a reaction isograd" (Winkler 1979, p. 66).

Tilley's definition of an isograd is identical to what we have called a mineral zone boundary earlier in this chapter, with the additional assumption of similar P-T values existing along this boundary. In order to avoid the latter statement, the terms mineral zone boundary or reaction-isograd will be used in this book instead of the term isograd. The distinction between a mineral zone boundary

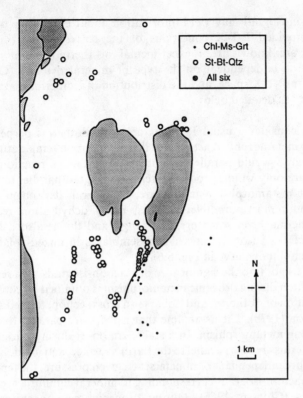

Fig. 4.4. Reaction-isograd based on the reaction Chl + Ms + Grt = St + Bt + Qtz + H₂O from the Whetstone Lake area, Ontario (After Carmichael 1970, Fig. 3)

and a reaction-isograd should be made clear again. Mapping a mineral zone boundary is based on the first appearance of an index mineral, whilst locating a reaction-isograd involves mapping reactants and products of a reaction equation (Fig. 4.4). Such equations may be simple if minerals with fixed compositions are involved (e.g. $Ky = Sil$; $Kln + 2\,Qtz = Prl + H_2O$), but in cases where minerals show extensive solid solution, reactions may be complex and are obtained by applying least-squares regression techniques (e.g. $3.0\,Chl + 1.5\,Grt + 3.3\,Ms + 0.5\,Ilm = 1.0\,St + 3.1\,Bt + 1.5\,Pl + 3.3\,Qtz + 10.3\,H_2O$; Lang and Rice 1985a). The mapping of reaction-isograds in the field is more time-consuming and more ambitious than the mapping of mineral zone boundaries. On the other hand, a reaction-isograd provides more petrogenetic information for two reasons:

1. The position of a reaction-isograd is bracketed by reactants and products, but a mineral zone boundary is limited only towards higher grade and may possibly be displaced towards lower grade as more field data become available. As an example, the chloritoid mineral zone boundary of the Central Alps has been replaced by the reaction-isograd $Prl + Chl \Rightarrow Cld + Qtz + H_2O$, located some 10 km down grade (Frey and Wieland 1975).

2. A reaction-isograd may yield information about conditions of metamorphism provided P-T-X conditions of the corresponding metamorphic
reaction are known from experimental or thermodynamic data. This
information is dependent on the type of mineral reaction (Chap. 3), its
location in P-T-X space, and the distribution of isotherms and isobars in a
rock body as detailed below.

Let us first consider a usual P-T diagram with isotherms perpendicular to
isobars. If a metamorphic reaction was dependent upon temperature only, the
reaction-isograd would parallel an isothermal surface; a reaction dependent
upon pressure only would produce a reaction-isograd parallel to an isobaric
surface; but metamorphic reactions are, in general, dependent upon both
pressure and temperature. Nevertheless, many dehydration reactions are
almost isothermal over some pressure range, and this is also true for some
polybaric traces of isobaric invariant assemblages in mixed-volatile systems
(see, e.g. tremolite-in curve in Fig. 6.6).

   Further insight into the significance of reaction-isograds with respect to the
P-T distribution during orogenic metamorphism is obtained by considering the
relative position of isotherms and isobars in a cross section, termed a **P-T profile**
by Thompson (1978). Let us assume that isobars are parallel to the Earth's
surface during metamorphism. In a stable craton or during burial metamorphism, isotherms are also parallel to the Earth's surface, and in this limiting case
P and T are not independent parameters; at a given pressure, the temperature is
fixed, and reaction-isograds corresponding to mineral equilibria are isothermal
and isobaric (Chinner 1966). During orogenic metamorphism, however,
isotherms are rarely expected to be parallel to isobars. Furthermore, the
temperature gradient–the increase in temperature perpendicular to isotherms,
not to be confused with the geothermal gradient, the increase of temperature
with depth, both expressed in °C/km–is not likely to be constant. Thompson
(1976) and Bhattacharyya (1981) have analyzed the geometrical relations
among reaction-isograds, the angle α between isotherms and isobars, and the
temperature gradient. It was shown that temperature gradient and angle α are
important in determining P-T distributions from reaction-isograd patterns
besides, of course, the slope of the corresponding mineral reaction on a P-T
diagram. As an example, for a given metamorphic reaction with a small P-T
slope of 8°C/kbar, if temperature gradients are low, the angle between such
reaction-isograds and isotherms can be large (Fig. 4.5).

   From the preceding discussion it is clear that along a reaction-isograd, both
P and T will generally vary, and that "similar P-T values", as suggested by Tilley,
either require the special precondition of isotherms being parallel to isobars, or
may be true for nearby localities only. Reaction-isograds intersecting at high
angles as described by Carmichael (1970) provide further convincing evidence
for changeable P-T-X conditions along such metamorphic boundaries.

   Reaction-isograds mentioned so far referred to univariant or discontinuous
mineral reactions. In the field, such reaction-isograds will define relatively
sharp lines which may be mapped within some 10 to 100 m (e.g. Fig. 4.4). Many

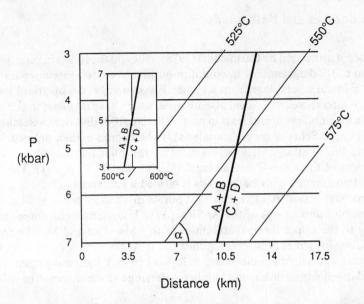

Fig. 4.5. Geometrical relations between a hypothetical mineral reaction $A + B = C + D$, isotherms, and isobars in a P-T profile. The mineral reaction has an assumed dP/dT slope of 8 °C/kbar as shown in the *inset* P-T diagram. Isobars are horizontal with a pressure gradient of 0.286 kbar/km (corresponding to a mean rock density of 2.8 g/cm³). Isotherms are inclined with respect to isobars at an angle α of 60° with a thermal gradient of 5 °C/km. Note the relatively large angle between the mineral reaction and isotherms in the P-T profile

metamorphic reactions are, however, at least divariant or continuous due to extensive solid solution in minerals, e.g. Mg-Fe²⁺ in metapelites. In this case, corresponding reaction-isograds will be smeared out over the erosion surface for several 100 m or km, because of the range of P-T conditions over which the continuous reaction occurs in rocks of different Mg/Fe ratios.

A reaction-isograd line recorded on a map is only a two-dimensional representation of a reaction-isograd, which is a surface when viewed in three dimensions, as already noted by Tilley (1924). The shape of the reaction-isograd surfaces in orogenic belts is of great interest from a geodynamic point of view; but, unfortunately, in many areas such information is difficult to obtain because of poor relief (in addition to poor exposure or scarcity of rock horizons of suitable composition). A few examples are known, however, from the Alps, where the shapes of reaction-isograds have been determined with some confidence (e.g. Bearth 1958; Fox 1975; Thompson 1976). However, even for areas with a long-lasting erosion history and where the vertical dimension is not exposed, Thompson (1978) has described a method to reconstruct the distribution of isotherms in a portion of the Earth's crust at the time of metamorphism from the mapped pattern of reaction-isograds.

4.5 Bathozones and Bathograds

This concept, proposed by Carmichael (1978), relies on the following considerations. On a P-T diagram, any invariant point in any model system separates a lower-P mineral assemblage from a higher-P assemblage. An invariant model reaction relates these two critical assemblages, which have no phase in common (i.e. these assemblages are, at least in part, different from divariant assemblages derived from a Schreinemaker's analysis). A **bathograd** is then defined "as a mappable line that separates occurrences of the higher-P assemblage from occurrences of the lower-P assemblage" (Carmichael 1978, p. 771). The field between two neighbouring bathograds is termed a **bathozone**.

Carmichael considered five invariant points in a model pelitic system: four are based on intersections of the Al_2SiO_5 phase boundaries with three curves referring to the upper thermal stabilities of St + Ms + Qtz and Ms + Qtz ± Na-feldspar; one invariant point corresponds to the Al_2SiO_5 triple point (Fig. 4.6). Carmichael's invariant equilibria are as follows (with P increasing from left to right and in sequential order, and numbers referring to bathozones to be defined below):

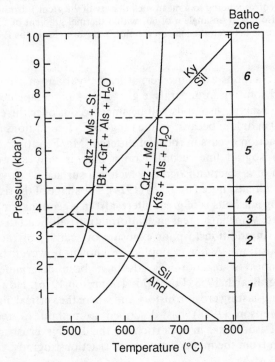

Fig. 4.6. P-T phase diagram for pelitic rocks divided into bathozones (After Carmichael 1978, Fig. 2 and Archibald et al. 1984, Fig 2)

1/2 Kfs + And + H_2O = Qtz + Ms + Sil
2/3 Bt + Grt + And + H_2O = Qtz + Ms + St + Sil
3/4 And = Ky + Sil
4/5 Qtz + Ms + St + Sil = Bt + Grt + Ky + H_2O
5/6 Qtz + Na-feldspar + Ms + Sil = Kfs + Ky + granitic liquid.

Reactants and products of these equilibria define diagnostic mineral assemblages of six bathozones. With the exception of bathozones 1 and 6, which are "open-ended" and therefore characterized by a single diagnostic assemblage, each bathozone is characterized by two diagnostic assemblages, one to constrain its lower-P boundary and the other to constrain its higher-P boundary:

Bathozone 1 : Kfs + And   P < 2.2 kbar
Bathozone 2 : Qtz + Ms + Sil, Bt + Grt + And   P = 2.2–3.5 kbar
Bathozone 3 : Qtz + Ms + St + Sil, And   P = 3.5–3.8 kbar
Bathozone 4 : Ky + Sil, Qtz + Ms + St + Sil   P = 3.8–5.5 kbar
Bathozone 5 : Bt + Grt + Ky, Qtz + Na-feldspar + Ms + Sil   P = 5.5–7.1 kbar
Bathozone 6 : Kfs + Ky (plus granitic material)   P > 7.1 kbar

These bathozones are shown in Fig. 4.6. Note that the apparently precise pressure calibration is somewhat misleading; no errors in the equilibrium curves have been taken into account, and several of the reactions are continuous in natural rocks and also depend on the activity of water. In fact, the equilibria Qtz + Ms + St = Bt + Grt + Als + H_2O were located by Carmichael (1978, p. 793) to be "consistent with a huge body of field data" and later slightly adjusted by Archibald et al. (1984, Fig. 2), but are inconsistent with available experimental or thermodynamic data (see Chap. 7). Nevertheless, Carmichael's scheme provides an elegant and simple way for reconnaissance geobarometry in metapelites of the amphibolite facies, requiring only thin-section observations. Figure 4.7 illustrates such an example for mapping bathozones and bathograds.

4.6 Petrogenetic Grid

P-T curves of different mineral equilibria generally have different slopes and will therefore intersect. Such intersecting curves will thus cut a P-T diagram up into a grid which Bowen (1940, p. 274) called a petrogenetic grid. His idea was to construct a P-T grid with univariant reaction curves bounding all conceivable divariant mineral assemblages for a given bulk chemical composition. Each mineral assemblage would lie in a unique P-T pigeonhole and inform the investigator immediately as to the conditions of metamorphism. When proposing the concept, Bowen regarded it rather as a vision than a tool, realizing that "the determinations necessary for the production of such a grid constitute a task of colossal magnitude.." (p. 274). In general, a **petrogenetic grid**

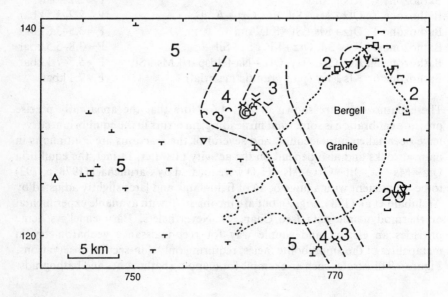

Legend:
- △ Kfs - Ky
- ⩑ Kfs - Qtz - Ab - Ms - Ky - Sil
- ⊤ Qtz - Ab - Ms - Sil
- ⊥ Qtz - Ab - Ms - Bt - Grt - Ky - Sil
- ⊥ Bt - Grt - Ky
- ⋏ Qtz - Ms - Bt - Grt - St - Ky - Sil
- ⌒ Qtz - Ms - St - Sil
- ○ Qtz - Ms - St - Ky - Sil
- ⋃ Ky - Sil
- ✕ Ky - And - Sil
- ⌢ And
- ◇ Qtz - Ms - St - And - Sil
- ⌄ Qtz - Ms - St - Sil
- ⋈ Qtz - Ms - Bt - Grt - St - And - Sil
- ⊓ Bt - Grt - And
- □ Qtz - Ms - Bt - Grt - And - Sil
- ⊔ Qtz - Ms - Sil
- ⩣ Qtz - Kfs - Ms - And - Sil
- △ Kfs - And

Fig. 4.7. Bathozones in metapelites from the Bergell Alps of Switzerland and Italy (After Carmichael 1978, Fig. 6)

is a diagram whose coordinates are intensive parameters characterizing the rock-forming conditions (e.g. P, T) on which may be plotted equilibrium curves delimiting the stability fields of specific minerals and mineral assemblages.

Early petrogenetic grids consisted of a few experimentally determined mineral equilibria belonging to different chemical systems, and thus the geometrical rules of Schreinemakers could not be applied. This resulted in rather vague conclusions with regard to the conditions of formation of associated rocks. Today, univariant curves delineating stability limits of many end-member minerals and mineral assemblages have been determined by experimental studies. Furthermore, internally consistent sets of thermodyn-

Table 4.2. Information on selected petrogenetic grids

System components/ rock type	Phase components, excess phases are in italics	P-T range	Substitutions considered[a]	Reference
CFMASH Ultramafics	Am, Brc, Cal, *Chl*, Cpx, Opx, Ol, Dol, Mgs, Qtz, Srp, Tlc, *Fluid* (H_2O-CO_2)	2 kbar 440–600°C	FeMg$_{-1}$ Tsch	Will et al. (1990a)
CMSH–CO_2 Carbonates	Atg, Ath, Brc, Cal, Di, Dol, En, Fo, Mgs, Per, Qtz, Tlc, Tr, Fluid (H_2O-CO_2)	0–10 kbar 400–800°C	None	Carmichael (1991)
CMSH–CO_2 Carbonates	Atg, Cal, Di, Dol, En, Mgs, Qtz, Tlc, Tr, ± Fluid (H_2O-CO_2)	1.5–14 kbar 465–715°C	None	Conolly and Trommsdorff (1991)
CMASH–CO_2 Carbonates	Cal, Chl, Di, Dol, Fo, Qtz, Spl, Tr, Fluid (H_2O-CO_2)	1 kbar/400–640°C 4 kbar/630–700°C	Fixed Tr activity	Chernosky and Berman (1988)
CFMASH–CO_2 Carbonates and basites	Am, An, And, Cal, Chl, Cld, Czo, Dol, Grt, Mrg, *Qtz*, Tlc, *Fluid* (H_2O-CO_2)	2 and 5 kbar 400–500°C	FeMg$_{-1}$ Tsch	Will et al. (1990b)
NCMASH Basites	Ab, An, Chl, Czo, Di, Gln, Jd, Lws, Pg, Pmp, Prp, Qtz, Tr, Fluid (H_2O)	2–20 kbar 250–600°C	Activity-corrected curves	Evans (1990)
NCMASH Basites	*Ab*, An, *Cln*, Gln, Grs, Hul, Jd, Lmt, Lws, Pg, Prh, Pmp, *Qtz*, Stb, Tr, Wr, Zo, *Fluid* (H_2O)	0.1–10 kbar 0–500°C	Activity-corrected curves	Frey et al. (1991)
FMASH Pelites	Cp, Chl, Cld, Kln, Ky, Prl, Qtz, Sud, Fluid (H_2O)	3–15 kbar 150–500°C	Activity-corrected curves	Vidal et al. (1992)
KFMASH Pelites	And, Bt, Chl, Cld, Crd, Grt, Ky, *Ms, Qtz*, Sil, St, Fluid (H_2O)	0–18 kbar 460–720°C	FeMg$_{-1}$ Tsch	Powell and Holland (1990)
NFMASH Pelites	Ab, Cp, Chl, Cld, Gln, Grt, Jd, Ky, Pg, *Qtz/Cs*, Tcl, *Fluid* (H_2O)	5–50 kbar 300–850°C	FeMg$_{-1}$ Tsch	Guiraud et al. (1990)
MnNKFMASH Pelites	Ab, And, Bt, Chl, Cld, Crd, Grt, Kfs, Ky, Ms, Prl, Qtz, Sil, St, Fluid (H_2O)	3, 5, 7 kbar 400–650°C	FeMg$_{-1}$ FeMn$_{-1}$	Symmes and Ferry (1992)

[a] Tsch = Tschermak exachange, $MgSiAl_{-1}Al_{-1}$.

amic data are now available for many rock-forming minerals, making it possible to calculate petrogenetic grids with all stable reaction curves of some model systems. Such an approach will be extensively used in Part II of this book.

However, even after five decades of extensive experimental and theoretical effort, the goal of a wholly comprehensive petrogenetic grid has still not been achieved. The main limitation is due to widespread solid solution in minerals like chlorite, mica, garnet, pyroxene and amphibole, and one of the main thrusts in future studies will be the evaluation of activity-composition relations for such phases. Nevertheless, available petrogenetic grids for simple end-member systems already provide important constraints on the P-T conditions of metamorphism in chemically complex systems. Some information on selected petrogenetic grids is given in Table 4.2.

4.7 Geothermobarometry

Geologic thermometry and barometry or, in short, **geothermobarometry** refer to the science of inferring the temperatures and pressures at which a rock formed. During the last two decades, laboratory experiments, thermodynamic modelling and calculations, and mineral analyses with the advent of the electron microprobe have led to a much better understanding of P-T conditions prevailing during metamorphism. This is a rapidly evolving field of search and recent general reviews on this topic are provided by Essene (1982), Bohlen and Lindsley (1987) and Essene (1989). More specific reviews refer to the geobarometry of granulites (Bohlen et al. 1983a; Newton 1983) and the geothermobarometry of eclogites (Newton 1986; Carswell and Harley 1989).

4.7.1 Assumptions and Precautions

Successful geothermobarometry relies on several assumptions and many pitfalls have to be avoided; some of these are shortly mentioned below.

1. Question of Equilibrium. A basic assumption for thermobarometry is that a mineral assemblage formed at equilibrium, but it is impossible to prove that minerals in an individual rock ever achieved equilibrium (Chap. 2). It is highly desirable, on the other hand, to find different generations of sub-assemblages in the same rock, each hopefully preserved in a state of local equilibrium. It might then be possible to infer the pressures and temperatures at various stages in the history of a rock, and to reconstruct its pressure-temperature-time path. As an example, St-Onge (1987) found zoned poikiloblastic garnets with inclusions of biotite, plagioclase, quartz and less commonly Al_2SiO_5. This enabled simultaneous determination of successive metamorphic P and T conditions from core to rim of a single garnet grain, using the garnet-biotite (Fe-Mg) exchange reaction and the garnet-Al_2SiO_5-plagioclase-quartz solid-solid reaction.

2. Retrograde Effects. If an equilibrium assemblage has remained unchanged, determined P-T values presumably refer to peak temperature conditions (see Chap. 3). Such situations may be nearly valid for some quickly cooled or low temperature rocks, but are not realized in slowly cooled rocks formed at high temperatures (higher amphibolite facies and granulite facies). In some cases, retrogression is obvious from thin-section examination, e.g. chloritization of biotite and garnet, pinitization of cordierite, or exsolution in feldspars, carbonates, oxides and sulphides. In other cases, however, retrograde diffusion may be detected only by careful microchemical analysis. The following two examples serve to demonstrate how such problems may be overcome. Bohlen and Essene (1977) used feldspar and iron-titanium-oxide equilibria to determine metamorphic temperatures in a granulite facies terrain. Exsolution of albite in alkali feldspar and of ilmenite in magnetite was observed, and geologically reasonable temperatures were obtained only after reintegration of exsolved components. Edwards and Essene (1988) applied two-feldspar and garnet-biotite thermometry to an amphibolite-granulite facies transition zone, whereby biotite-garnet pairs gave erratic temperatures compared to two-feldspar temperatures. In detail, it was found that some garnets were homogeneous in composition, and some were zoned with rapidly changing composition near the rim but with sillimanite included. Biotites had different compositions depending on whether they were included in garnet, touching but not included in garnet, or not touching garnet. In addition, included and touching biotites from the same sample had highly variable Fe/Mg ratios, whereas biotites not touching garnet tended to have the same composition. These observations were explained by the retrograde exchange of Fe and Mg between biotite and garnet and by the retrograde hydration reaction $Grt + Kfs + H_2O = Bt + Sil + Qtz$. It is clear, that the use of biotites not touching garnet and garnet cores should have resulted in peak metamorphic temperatures, but this was not the case with the calibrations applied.

3. Quality of Calibration. The basic reaction or equilibrium must be well calibrated, either through experimental or thermodynamic data. The experiments that were used to calibrate a given system should be well reversed and not rely on synthesis runs only. Furthermore, starting materials and run products should be carefully characterized.

4. Long Extrapolations in P-T. Some geothermobarometers were calibrated at much higher P-T conditions than those of metamorphism, generating large errors if long extrapolations are required. As an example, the reaction $3An = Grs + 2Ky + Qtz$ has been determined experimentally over the temperature range of 900–1600°C, but might be applied at temperatures as low as 500° C. This problem may be diminished if an "accurate" slope calculation is feasible, but calculated uncertainty limits may turn out to be disappointingly large (see, e.g. Hodges and McKenna 1987, Fig. 4).

5. Sensitivity of Thermobarometer. Some systems are sensitive over a restricted P-T range only. Many solvus thermometers, for example, are useful at relatively high temperatures, but less so at relatively low temperatures where mineral compositions are located on the steep limbs of the solvus. In such cases, even small errors involved in the location of the solvus as well as in the mineral analyses will produce large errors in inferred temperatures. In contrast to solvus thermometry, fractionations of oxygen isotopes between coexisting minerals are most sensitive at low temperatures, because fractionations decrease to almost zero at high temperatures (800–1000°C).

6. Variable Structural State. Many minerals have variably ordered cation distributions, but the extent and temperature dependence of the ordering process are presently not well known for most phases. In an ideal case, the synthetic products should have the same structural state as the metamorphic minerals had at the time of metamorphism; of course, the structural state of disordered metamorphic minerals may reset upon cooling. In using feldspar thermometry, for instance, one must decide whether the metamorphic feldspar was ordered, partially disordered or completely disordered.

7. Effect of Other Components. Most geothermobarometers are based on mineral reactions using simple mineral chemistries, whilst most natural minerals represent complex solid solutions. In order to account for these compositional differences, thermodynamic solution modelling must be applied. This involves the determination of activity-composition relations, where activity represents the "thermodynamic concentration" to be determined and mineral composition is analyzed with an electron microprobe. Unfortunately, the relations between activity and composition are often complex, and are only poorly understood for most minerals. As a consequence, there is often disagreement among workers about the most suitable solution model. This has led to the widespread practice of reporting multiple P-T estimates reflecting the effects of different solution models. For the garnet-biotite thermometer, for example, seven calibrations based on the experimental study of Ferry and Spear (1978) are currently available. "In some cases, P-T estimates calculated using different solution models can vary widely, implying that uncertainties in solution behaviour constitute a major source of error in thermobarometry. In general, this error cannot be quantified because the assumptions inherent in different models are often mutually exclusive" (Hodges and McKenna 1987, p. 672).

If metamorphic minerals deviate strongly from the end-member compositions of a calibrated geothermobarometer, then the effect of "other" components will lead to large errors of extrapolation. For instance, the garnet-Al_2SiO_5-quartz-plagioclase thermobarometer is commonly applied to metapelites containing garnets with only 5–10% grossular and plagioclase with only 10–30% anorthite, far away from the end-member reaction 3An = Grs + 2Als + Qtz.

8. Estimation of Fe^{2+}/Fe^{3+} in Mineral Analysis. The electron microprobe renders information on total iron content, and the ferric and ferrous iron contents must be calculated assuming charge balance and some site occupancy model. For relatively simple minerals containing low Fe_2O_3 contents (e.g. garnets, some pyroxenes) such calculations are believed to be reliable in most cases. For other, more complex minerals with vacant cation sites and variable H_2O contents (e.g. amphiboles and micas), such calculations are often questionable or even meaningless. Realizing this shortcoming, many workers either assume all iron to be in the ferrous state or a fixed ferrous-ferric iron ratio; but an independent check by wet-chemical analysis or Mössbauer spectroscopy would be advisable.

4.7.2 Exchange Reactions

Exchange equilibria involve interchange of two similar atoms between different sites in one mineral (intracrystalline exchange) or between two minerals (intercrystalline exchange). The atoms may be elements of similar charge and ionic radius or light stable isotopes. Because the volume changes involved are very small and the entropy changes relatively large, exchange reactions are largely independent of pressure and have a good potential as thermometers. Unfortunately, retrograde exchange by diffusion below peak metamorphic conditions occurs very easily in many cases and generally leaves no textural evidence. This is especially true for intracrystalline cation distributions, as the diffusion paths are but a fraction of a unit cell. Therefore, intracrystalline exchange reactions are unsuitable for peak metamorphic thermometry and will not be considered further.

The most widely applied exchange thermometers in metamorphic rocks involve Fe^{2+} and Mg, including the following minerals: olivine, garnet, clinopyroxene, orthopyroxene, spinel, ilmenite, cordierite, biotite, phengite, chlorite and hornblende. For many pairs of this list there exist calibrations for some end-member mineral compositions, either from experimental work or from thermodynamic calculation. In a few cases, experimental calibrations are also available for complex "dirty" systems. Based on these calibrations, many reformulations exist, using different solution models to account for the effects of the impurities. Furthermore, some exchange thermometers were calibrated empirically against natural parageneses. Data for some 18 systems are available at present, and some of these are listed in Table 4.3. Most of the experiments on the anhydrous pairs were obtained at $T > 800\,°C$ and many were not compositionally reversed. If not reset by retrograde exchange, their best application is in high-temperature rocks from the granulite facies and the mantle, and extrapolation of these data down to relatively low temperatures is often unjustified. The potential problem with Mg-Fe^{2+} thermometry in using microprobe analyses was already mentioned earlier.

In the following, the two most commonly used Fe-Mg exchange thermometers are briefly reviewed.

Garnet-Clinopyroxene. The Grt-Cpx thermometer has been considered by many workers because of its potential importance for garnet-granulites, garnet-

Table 4.3. Some Fe-Mg exchange reactions used for geothermometry

No.	Mineral pair	Exchange reaction	Range of application[b]	Reference[c]
1	Ol-Spl	$\frac{1}{2}$Fo + Hc = $\frac{1}{2}$Fa + Spl	GRE, AMP, GRA	Engi (1983)
2	Ol-Opx	$\frac{1}{2}$Fo + $\frac{1}{2}$Fs = $\frac{1}{2}$Fa + $\frac{1}{2}$En	GRA, ECL	Docka et al. (1986), Carswell and Harley (1989)
3	Ilm-Opx	Gei + $\frac{1}{2}$Fs = Ilm + $\frac{1}{2}$En	GRA	Docka et al. (1986)
4	Ilm-Cpx	Gei + Hd = Ilm + Di	GRA, AMP	Docka et al. (1986)
5	Ilm-Ol	Gei + $\frac{1}{2}$Fa = Ilm + $\frac{1}{2}$Fo	GRA	Docka et al. (1986)
6	Opx-Cpx	$\frac{1}{2}$En + Hd = $\frac{1}{2}$Fs + Di	GRA	Stephenson (1984), Docka et al. (1986)
7	Opx-Bt	$\frac{1}{2}$En + $\frac{1}{3}$Ann = $\frac{1}{2}$Fs + $\frac{1}{3}$Phl	GRA	Fonarev and Konilov (1986), Sengupta et al. (1990)
8	Crd-Spl	$\frac{1}{2}$Mg-Crd + Hc = $\frac{1}{2}$Fe-Crd + Spl	GRA	Vielzeuf (1983)
9	Grt-Ol	$\frac{1}{3}$Prp + $\frac{1}{2}$Fa = $\frac{1}{3}$Alm + $\frac{1}{2}$Fo	GRA, ECL	Carswell and Harley (1989)
10	Grt-Opx	$\frac{1}{3}$Prp + $\frac{1}{2}$Fs = $\frac{1}{3}$Alm + $\frac{1}{2}$En	GRA, ECL	Carswell and Harley (1989), Bhattacharya et al. (1991)
11	Grt-Cpx	$\frac{1}{3}$Prp + Hd = $\frac{1}{3}$Alm + Di	GRA, ECL, ± AMP	Pattison and Newton (1989), Carswell and Harley (1989)
12	Grt-Crd	$\frac{1}{3}$Prp + $\frac{1}{2}$Fe-Crd = $\frac{1}{3}$Alm + $\frac{1}{2}$Mg-Crd	GRA	Bhattacharya et al. (1988)
13	Grt-Bt	$\frac{1}{3}$Prp + $\frac{1}{3}$Ann = $\frac{1}{3}$Alm + $\frac{1}{3}$Phl	AMP, GRE, ± GRA	Thoenen (1989)
14	Grt-Phe	$\frac{1}{3}$Prp + Fe-Cel = $\frac{1}{3}$Alm + Mg-Cel	ECL, BLU	Carswell and Harley (1989)
15	Grt-Hbl[a]	$\frac{1}{3}$Prp + $\frac{1}{4}$Fprg = $\frac{1}{3}$Alm + $\frac{1}{4}$Prg	AMP, ± GRA	Graham and Powell (1984)
16	Grt-Chl[a]	$\frac{1}{3}$Prp + $\frac{1}{5}$Fe-Chl = $\frac{1}{3}$Alm + $\frac{1}{5}$Mg-Chl	GRE, BLU, ± AMP	Laird (1988), Grambling (1990)

[a] Empirical calibrations; all other calibrations were determined by experiment or by calculation.
[b] Abbreviations of metamorphic facies: AMP Amphibolite, BLU Blueschist, ECL Eclogite, GRA Granulite, GRE Greenschist.
[c] One or two recent references for each reaction are given only.

amphibolites, garnet-peridotites and eclogites. There exist at least ten calibrations of this thermometer, relying on a large body of experimental work. Experiments were conducted in the temperature range of 600–1500°C, but predominantly at T > 900°C. Only three calibrations will be mentioned here, whilst references to other studies can be found in Pattison and Newton (1989).

The most often used calibration of the Grt-Cpx thermometer is that derived by Ellis and Green (1979). These authors first experimentally evaluated the effect of grossular substitution (X_{Ca}^{Grt}) on the exchange reaction (11) (Table 4.3) and derived the following equation:

$$T(°C) = (3030 + 10.86 \ P(kbar) + 3104 \ X_{Ca}^{Grt})/(\ln K_D + 1.9034) - 273,$$

where K_D is the distribution ratio $(Fe^{2+}/Mg)^{Grt}/(Fe^{2+}/Mg)^{Cpx}$. More recent experimental studies have shown that the Ellis and Green geothermometer overestimates temperatures when applied to granulites formed at ca. 10 kbar. According to Green and Adams (1991, p. 347) "this overestimate could be of the order of 50–150°C". However, no new thermometric equation was presented.

Krogh (1988), using earlier experimental data, derived a new expression for this thermometer with a curvilinear relationship between $\ln K_D$ and X_{Ca}^{Grt}

$$T(°C) = [1879 + 10 \ P(kbar) - 6173(X_{Ca}^{Grt})^2 + 6731 \ X_{Ca}^{Grt}]/(\ln K_D + 1.393) - 273.$$

Most recently, Pattison and Newton (1989) presented a large set of new experimental data. They found that the Grt-Cpx Fe/Mg distribution curves were asymmetric, implying non-ideal solid solution behaviour, in contrast to those of previous experimental studies. The geothermometer expression of Pattison and Newton is as follows:

$$T(°C) = (a'X^3 + b'X^2 + c'X + d)/(\ln K_D + a_o X^3 + b_o X^2 + c_o X + d_o)$$
$$+ 5.5[P(kbar) - 15] - 273,$$

where a', b', c', d' and a_o, b_o, c_o, d_o are constants for a given X_{Ca}^{Grt} (to be found in Pattison and Newton 1989, p. 99), and where X stands for the Mg number (Mg/Mg + Fe) of the garnet. Note that this expression is valid only within the experimental Mg number range of 0.125–0.600.

The Grt-Cpx thermometer as formulated by Pattison and Newton (1989) is recommended for upper amphibolite facies, granulite facies and high-T eclogite facies rocks. Application to low- and medium-T crustal eclogites is uncertain because their clinopyroxenes are ordered omphacites of high jadeite content, much different from those of the simpler system investigated by Pattison and Newton. In almost all recent studies, the Ellis and Green (1979) calibration has been applied to eclogites. In many cases, a long extrapolation in temperature is involved, and in this respect the lowest T estimates obtained are critical, i.e. in the range of 400–500°C. As an example, Schliestedt (1986) determined Grt-Cpx temperatures of 471 ± 31°C (range = 392–512°C, 16 samples) on eclogites and blueschists from Sifnos, based on a preliminary version of the Krogh (1988) calibration. Schliestedt concluded that these temperatures were in accordance with those estimated by other methods (calcite-dolomite solvus and oxygen isotope fractionation). Note that the Ellis and Green (1979) thermometer gave too high equilibrium temperatures by ca. 50°C for these rocks (Krogh 1988). These limited data seem to indicate that reasonable results can be obtained with the Krogh formulation, even in low-T eclogites.

The Grt-Cpx thermometer calibrations mentioned above do not include any possible effects of chemical variations in Cpx. Such effects were noted for variations in the jadeite content, with K_D tending to decrease with increasing of X_{Jd}^{Cpx} of the sodic omphacites, and especially at $X_{Jd}^{Cpx} > 0.6$ (e.g. Koons 1984; Heinrich 1986; Benciolini et al. 1988). However, X_{Jd}^{Cpx} will in most cases be lower than 0.6 and will exert only a minor effect on K_D. On the other hand, no definite compositional dependence of K_DS on the Fe^{3+} content in clinopyroxene (aegirine component) was observed by Benciolini et al. (1988), but it should be remembered that the problem of the determination of ferrous/ferric iron in the clinopyroxenes from microprobe data is not solved. Additional chemical effects on the Fe-Mg partition coefficient K_D are discussed by Krogh (1988).

In conclusion, the present status of Grt-Cpx Fe-Mg exchange geothermometry is certainly far from being ideal, but is nevertheless a valuable thermometer for granulites and eclogites. The calibrations by Krogh (1988) and Pattison and Newton (1989) are recommended for low-T plus medium-T eclogites and for high-T eclogites plus granulites, respectively.

Garnet-Biotite. The garnet-biotite exchange thermometer is the most popular of all geothermometers because of its wide application in a large variety of rocks covering a broad range of metamorphic grade. At least 18 versions of this thermometer are available: 3 have been calibrated by field observations, 2 by laboratory experimentation and 13 were derived from the experimental calibrations by using different non-ideal mixing models for garnet and biotite. A characterization of some of the different versions and a comparison of the thermometric equations is provided by Thoenen (1989).

Ferry and Spear (1978) published experimental data on the Fe-Mg exchange between synthetic annite-phlogopite and almandine-pyrope in systems with Fe/ (Fe + Mg) held at 0.9. The following equation was derived:

$$T(°C) = [2089 + 9.56\ P\ (kbar)]/(0.782 - \ln K_D) - 273,$$

with $K_D = (Fe/Mg)^{Bt}/(Fe/Mg)^{Grt}$. As it is likely that solid solution of additional components affects this thermometer, Ferry and Spear (1978) suggested that it should be restricted for usage with garnets low in Ca and Mn, with (Ca + Mn)/ (Ca + Mn + Fe + Mg) ≤ 0.2, and with biotites low in Al^{vi} and Ti, with $(Al^{vi} + Ti)/$ $(Al^{vi} + Ti + Fe + Mg) \leq 0.15$.

Perchuk and Lavrent'eva (1983) performed the second experimental study on the Fe-Mg exchange between garnet and biotite. For most experiments, natural minerals served as starting materials in systems that covered a range of 0.3–0.7 Fe/(Fe + Mg). As pointed out by Thoenen (1989), the thermometric equation given by Perchuk and Lavrent'eva is unlikely because it gives negative Clausius-Clapeyron slopes. A corrected version of the equation is

$$T(°C) = [3890 + 9.56 \ P(kbar)]/(2.868 - lnK_D) - 273.$$

Temperatures obtained from this calibration are $\approx 30\,°C$ higher in the $500\,°C$ temperature range, but $\approx 60\,°C$ lower in the $700\,°C$ temperature range compared to the Ferry and Spear calibration.

Several modifications of the Ferry and Spear calibration have been made, based upon the effects of Ca and Mn in garnet and of Al and Ti in biotite. Despite strong efforts of research, the magnitude of the effects of impurities on the Fe-Mg exchange between garnet and biotite is not sufficiently well known at present (for a recent discussion, see Bhattacharya et al. 1992). This makes it difficult to choose between the many versions of this geothermometer. Some authors have tried to evaluate the quality of different calibrations by comparing the calculated temperatures with T estimates based on some other mineral equilibria or by comparing the scatter of data within a given metamorphic zone. Still another approach was taken by Chipera and Perkins (1988), who applied a trend surface analysis to determine how well the temperatures fitted a regional temperature surface across their study area. Eight calibrations of the garnet-biotite thermometer were compared and it was found that the Perchuk and Lavrent'eva (1983) thermometer yielded the most precise results.

In general, the Grt-Bt thermometer seems to work fairly well in the high-T greenschist facies and the amphibolite facies. For several areas the Grt-Bt temperatures indicate a regular increase in temperature with increasing metamorphic grade over a range of some $100–150\,°C$ (e.g. Ferry 1980; Lang and Rice 1985b; Holdaway et al. 1988). In the uppermost amphibolite facies and granulite facies, however, retrograde Fe-Mg exchange generally occurs, resulting in anomalously low Grt-Bt temperatures if garnet rims and biotites in contact with garnets are analyzed. Nevertheless, reasonable prograde temperatures can still be obtained from garnets that have undergone retrograde reactions. As suggested by Tracy et al. (1976), in biotite-rich rocks retrograde reactions will cause negligible changes in matrix biotite composition while the garnet-rim composition changes greatly. In biotite-rich rocks, therefore, reasonable estimates of maximum prograde temperatures can be made using garnet-core composition and matrix biotite composition (provided that these two mineral compositions once were in chemical equilibrium).

Isotopic Thermometry. Most exchange thermometers involve cations. Isotopic thermometers are an important exception that are based on the temperature dependence of equilibrium partition of light stable isotopes (typically of the elements C, O, S) between two coexisting phases. Stable isotopes are chosen because their ratios will not vary with time due to radioactive decay. Heavy isotopes are not useful for thermometry because they do not fractionate as much as light isotopes. The fractionations between light isotopes are dependent on temperature, mineral chemical composition and crystal structure. Fractionations between two coexisting minerals are largest at low temperatures and decrease to almost zero at high temperatures (800–1000°C). Fractionations tend to be largest between phases of widely different compositions and structures, so pairs like magnetite-quartz, ilmenite-quartz, rutile-quartz and calcite-quartz are most useful. Isotopic exchange has virtually no volume change, which means that this thermometer is effectively independent of pressure.

Isotopic thermometry has potentially a great power and it would appear to be the ideal thermometer for metamorphic rocks. Several problems exist, however. Many different experimentally determined calibration curves are in use which are generally not internally consistent. Retrograde isotopic exchange will occur during slow cooling from high temperatures, yielding discordant temperature values for different mineral pairs from the same rock. The set of paleotemperatures obtained for a rock will, in general, give neither the mineral closure temperatures nor the formation or crystallization temperatures. On the other hand, the cooling rate of the rock may be derived from the data (Giletti 1986). Lack of equilibrium may be a problem, especially during contact metamorphism and at low temperatures. For further information the reader is referred to Valley et al. (1986), Hoefs (1987) and Savin and Lee (1988).

4.7.3 Solvus Relations

Some mineral pairs with similar structures exhibit continuous solid solution at high temperatures, but exhibit an expanding miscibility gap to lower temperatures. The curved P-T-X line or surface that separates the field of homogeneous solid solution from the field of limited mutual solid solution will here be loosely termed a solvus. Miscibility gaps are strongly temperature-dependent and thus serve mainly as thermometers, but very often there will be a pressure dependence as well which then cannot be neglected. Because of the general shape of any solvus, small changes in mineral compositions correspond to large temperature differences along the steep limbs, but correspond to moderate temperature differences near the crest of a solvus. Thus, solvus geothermobarometers become more sensitive temperature indicators as temperature increases. If disorder occurs in one or both minerals, then there will be an infinity of solvi, one for each degree of partial disorder.

The following mineral pairs are often used for solvus geothermometry: orthopyroxene-clinopyroxene, plagioclase-alkali feldspar, calcite-dolomite and muscovite-paragonite; these will be discussed shortly below.

Orthopyroxene-Clinopyroxene. A large body of experimental and theoretical data exists on the enstatite-diopside solvus, and it has been shown that at high temperatures, above about 800°C, the diopside limb of the solvus is temperature-dependent. Because natural pyroxenes rarely belong to the binary join $Mg_2Si_2O_6$-$CaMgSi_2O_6$, the major problem with this thermometer lies in correction for other components. This has led to the formulation of several multicomponent two-pyroxene thermometers. If applied to ultramafic rocks, where pyroxenes closely approach the enstatite-diopside binary join, satisfactory results have been obtained (e.g. Carswell and Gibb 1987), but much less so for iron-rich pyroxenes from granulites.

A different, and apparently more successful approach to Opx-Cpx thermometry was taken by Lindsley (1983). This author determined phase equilibria for pure Ca-Mg-Fe pyroxenes over the temperature range of 800–1200°C and for pressures up to 15 kbar. Solvus isotherms for coexisting Opx-Cpx were displayed graphically in the quadrilateral pyroxene system composed of Di-Hd-En-Fs. The effect of pressure was found to be ≤ 8°C/kbar, and graphs are presented for solvus relations at 1 atm and at 5, 10 and 15 kbar, along with approximate formulas for interpolating between these pressures. Pyroxenes with appreciable contents of non-quadrilateral components require special projections onto the Di-Hd-En-Fs pyroxene quadrilateral, and a corresponding method has been presented by Lindsley and Anderson (1983) and Lindsley (1983). This correction scheme is largely empirical, and Lindsley (1983) suggests that application be limited to pyroxenes with relatively low (< 10%) amounts of non-quadrilateral components.

Slowly cooled pyroxenes often show exsolution lamellae, and reintegration of exsolved material in the microprobe analyses is necessary to obtain meaningful thermometric results. Also, the incorrect calculation of ferric iron will affect this geothermometer.

Plagioclase-Alkali Feldspar. Most feldspars are well represented by the three components Ab, Or, An, with extensive solid solution between Ab and An (the plagioclase series) and between Ab and Or (the alkali feldspar series). The compositions of coexisting feldspars contain important thermometric information, as recognized by T.F.W. Barth as early as 1936. The majority of formulations of the two-feldspar thermometer concentrate on two feldspar binaries, Ab-An and Ab-Or, and obtain temperatures from the partitioning of albite component between coexisting plagioclase and alkali feldspar (e.g. Stormer 1975; Haselton et al. 1983). This approach may be valid at lower temperatures where feldspars have only trivial amounts of the third component. At higher temperatures, however, the effects of ternary solution become increasingly important. Thus, most recently developed feldspar thermometers are based on ternary solution models that account for the effects of K in plagioclase and Ca in alkali feldspar (e.g. Fuhrman and Lindsley 1988; Elkins and Grove 1990). A great advantage of the ternary approach is that it allows three temperatures to be calculated, one for each component, providing a valuable test of equilibrium. A drawback of the more sophisticated thermometers is that simple thermometric equations are no longer included in a publication, but a thermometer program has to be ordered from the authors.

Two-feldspar thermometry has been applied with some success to high-grade gneisses from the granulite facies (e.g. Bohlen et al. 1985; see also Fuhrman and Lindsley 1988). In such rocks exsolution of Ab-rich plagioclase from alkali feldspar and of K-rich feldspar from plagioclase upon cooling is a common phenomena, and it is essential that exsolved grains are reintegrated to obtain the compositions of the original feldspars (see Bohlen et al. 1985, for analytical details). Applications of feldspar geothermometry to rocks from the amphibolite and greenschist facies have been less successful, often yielding unreasonable low temperature values. Evidentally, alkali exchange with a fluid phase below peak metamorphic conditions must have taken place.

Calcite-Dolomite. In the system $CaCO_3$-$MgCO_3$, a solvus exists between calcite and dolomite, and the temperature dependence of the amount of $MgCO_3$ in calcite in equilibrium with dolomite can be used for estimating metamorphic temperatures. This solvus appears to be well determined and the effect of pressure on temperature estimates is small. Most natural carbonates, however, contain additional components like $FeCO_3$ or $MnCO_3$, and the effect of $FeCO_3$ on calcite-dolomite thermometry has been evaluated, e.g. by Powell et al. (1984) and Anovitz and Essene (1987a).

Application of the two-carbonate thermometer to metamorphic rocks is complicated by retrograde resetting. Magnesian calcite may exsolve dolomite which must be reintegrated during microprobe analysis. More disadvantageous is the case where the original high magnesian content is obliterated through diffusion, and temperatures are on the order of 300–400°C for high-grade marbles (e.g. Essene 1983). Therefore, this thermometer is more useful in low-grade orogenic metamorphic rocks and in contact metamorphic rocks where cooling times are reduced.

Muscovite-Paragonite. The binary solvus between coexisting muscovite and paragonite has been repeatedly investigated experimentally (see Essene 1989, p. 8, for references), the most recent and most detailed study being that of Flux and Chatterjee (1986). However, even their solvus is not yet well constrained, relying on three experimental brackets only, and with one bracket determined with 1M micas run products (instead of the stable $2M_1$ polytype). The solvus calculated by Chatterjee and Flux (1986) shows a considerable pressure dependence.

The use of the muscovite-paragonite solvus geothermometer has not been successful in the past (see, e.g. Guidotti 1984, p. 409). Nevertheless, the muscovite limb of the solvus determined by Chatterjee and Flux (1986, Fig. 6) may be provisionally used to obtain geothermometric data for coexisting muscovite-paragonite pairs, provided (1) the equilibrium pressure is independently known and (2) the chemical composition of the micas is close to the ideal muscovite-paragonite join. The last-mentioned requirement will be rarely fulfilled for phengitic muscovites formed at high pressures and/or low temperatures, and use of the solvus should be restricted to muscovites with Si < 3.05 atoms per formula unit.

4.7.4 Polymorphic Transitions

Examples of polymorphic transitions in metamorphic rocks include andalusite-kyanite-sillimanite, calcite-aragonite, quartz-coesite and diamond-graphite. Theoretically these simple conversions correspond to univariant reactions which are relatively well known in P-T space (Fig. 4.8). Due to the small energy changes involved in these reactions, however, such equilibria may become divariant through the preferential incorporation of minor element contents, by order-disorder relations, by kinetic factors, or by the influence of strain effects. Neglecting these complications for the moment, polymorphic transitions are easy to apply to rocks because the mere presence of an appropriate phase (or pseudomorphs thereafter) is sufficient to yield some P-T information. A disadvantage is that only P-T limits are usually provided, and absolute P-T

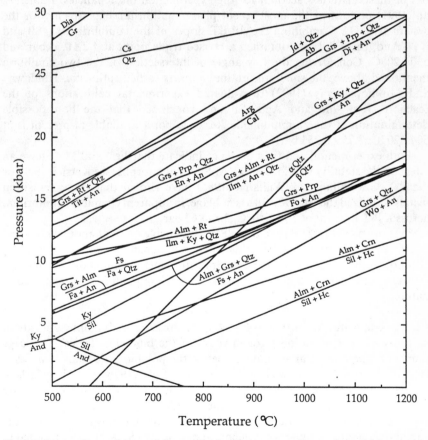

Fig. 4.8. Some polymorphic transitions and univariant net-transfer reactions which are useful thermobarometers. The Dia = Gr boundary is after Bundy et al. (1961). All other equilibria were calculated using thermodynamic data of Berman (1988)

values can be determined only in rare cases where univariant assemblages are preserved.

Andalusite-Kyanite-Sillimanite. Because of their common occurrence in peraluminous rocks, the three Al_2SiO_5 polymorphs have been widely used as index minerals, for the definition of facies series (Sect. 4.3), and for estimating P-T in metamorphic rocks (e.g. Sect. 4.5). The system Al_2SiO_5 has an eventful history of experimentation (see, e.g. Kerrick 1990, for a review). During the last two decades, most petrologists referred to the Al_2SiO_5 triple point of Richardson et al. (1969), 5.5 kbar and 620°C, or that of Holdaway (1971), 3.8 kbar and 500°C. Both determinations were obtained in a hydrostatic pressure apparatus. The main difference between the Richardson et al. and Holdaway studies concerns the And = Sil equilibrium. Richardson et al. used fibrolitic sillimanite whilst Holdaway experimented with coarse-grained sillimanite, and Salje (1986) suggested that different thermodynamic properties of these materials account for the experimental discrepancies. Bohlen et al. (1991) performed the most recent phase equilibrium experiments for the system Al_2SiO_5. Combining the dP/dT slopes of the equilibria Ky = Sil and Ky = And, these authors obtained a revised triple point at 4.2 ± 0.3 kbar and 530 ± 20°C. Considering the low angle of intersection of the two equilibria mentioned above, the small uncertainty appears rather optimistic. Holdaway and Mukhopadhyay (1993) re-evaluated experimental calibrations of the reactions Ky = And and And = Sil and concluded that the best possible determination of the aluminum silicate triple point available at present is at 504 ± 20°C, 3.75 ± 0.25 kbar.

Both experimental and field evidence indicate that Fe^{3+} and Mn^{3+} have an effect on the stability relations of the Al_2SiO_5 polymorphs (see Kerrick 1990 for details). Grambling and Williams (1985) have shown that these transition metals stabilized the And-Ky-Sil assemblage, in apparent chemical equilibrium, across a P-T interval of 500–540°C, 3.8–4.6 kbar.

In summary, it is clear that a single Al_2SiO_5 triple point at about 4 kbar and 500°C exists on petrogenetic grids only, but depending on the degree of "fibrolitization" of sillimanite and the transition metal content of the aluminosilicates, some variation in the location of the triple point is to be expected in nature.

Calcite-Aragonite. According to Fig. 4.8, the presence of aragonite appears to be a valuable pressure indicator in rocks of the blueschist facies. The large number of experimental attempts to define the polymorphic transition Cal = Arg are thoroughly reviewed by Newton and Fyfe (1976); see also Carlson (1983). Only small discrepancies exist between results obtained in the pure $CaCO_3$ system by different workers in the late 1960s and early 1970s. The effects of solid solution were evaluated for $MgCO_3$ and $SrCO_3$. It was found that the common contents of these components in metamorphic carbonates resulted in small displacements of the equilibrium boundary, not exceeding a few hundreds of bars.

Application of the simple equilibrium reaction Cal = Arg to metamorphic rocks is, however, complicated by several factors. First, aragonite is a widely distributed phase in some high-pressure metamorphic rocks of the Franciscan formation in California, but is uncommon in other blueschist terrains. These field observations have been explained by high reaction rates, meaning that aragonite-bearing rocks en route to the surface must enter the calcite stability field between 125 and 175°C (Carlson and Rosenfeld 1981), and that aragonite will be transformed into calcite if it crosses the equilibrium curve at higher temperatures. Second, the metastable growth of aragonite in modern and ancient oceans and its subsequent conversion to calcite is a well-known phenomena, and one wonders whether metamorphic aragonite might form outside its stability field as well. It has been shown that aragonite may form instead of calcite at much lower pressures if it is precipitated from fluids with other dissolved ions, rather than from solution in pure water. The frequent occurrence of aragonite in the Franciscan formation as coarse crystals in late-stage veins suggests that it may often form in a solution-precipitation process. The question remains open whether such aragonite grows metastably at a lower pressure than its stability field. Third, severely deformed calcite will accumulate strain energy, and Newton et al. (1969) were able to grow aragonite at the expense of calcite strained by prolonged mortar grinding at pressures several kilobars below the aragonite stability limit. This suggests that deformational energy can be a significant parameter in Cal-Arg stability considerations.

In summary, more experimental and field work will be needed to establish definitely the circumstances under which aragonite forms stably in nature. Where associated with other high-pressure minerals such as jadeite, it is likely that aragonite occurring in metamorphic rocks is an indicator of high pressure.

Quartz-Coesite. The recent discoveries of coesite in eclogites and subducted supracrustal rocks (see Reinecke 1991 and Schmädicke 1991, for references) have generated great interest in the quartz-coesite transition. This equilibrium has been well located as a function of P and T (Bohlen and Lindsley 1987, Fig. 1), because of the rapid conversion of one polymorph to the other. It is thus astonishing that coesite survived the long return trip to the surface. Under static conditions, the experimental results imply unusually high metamorphic pressures of 25–30 kbar, depending on temperature, corresponding to depths in excess of 75 km. However, it is crucial to know whether coesite formed under static conditions or not, because it has been shown experimentally that coesite can grow from highly strained quartz as much as 10 kbar below the transition determined under static conditions. From the description of several coesite localities, it appears that strain energy was considered to be a negligible factor, and that the presence of coesite actually indicated very high pressures of formation.

4.7.5 Net-Transfer Reactions

Net-transfer reactions in which one or more of the phases has appreciable solid solution are of the greatest use for geobarometry. Due to their generally small positive slope in P-T space, solid-solid reactions are primarily used as geobarometers, with metamorphic temperatures determined by exchange or other thermometers. These continuous or sliding or displaced equilibria are multivariant in P-T-X space and, as a result, the entire thermobarometric assemblage can coexist over a wide range of P and T. Application of these reactions requires that (1) the appropriate univariant end-member equilibrium has been well calibrated, (2) the chemical compositions of the phases are known, and (3) activity-composition relations are available for those minerals with appreciable solid solution.

As an example, consider the metapelite assemblage Grt-Rt-Als-Ilm-Qtz. The following end-member reaction can be formulated:

$$Fe_3Al_2Si_3O_{12} + 3\ TiO_2 = 3\ FeTiO_3 + Al_2SiO_5 + 2\ SiO_2.$$

   Alm   Rt   Ilm   Als   Qtz

This univariant reaction is displayed in Fig. 4.8 whilst Fig. 4.9 shows the effect of solid solution in Alm and Ilm, with isopleths of $\log_{10}K = \log_{10}(a_{Ilm}^3/a_{Alm})$ contoured on a P-T diagram. If one is able to convert chemical analyses of Ilm and Grt to the activities of the respective Fe end-members by application of appropriate solution models, then $\log_{10}K$ may be calculated and a P-T line located for a given K.

As mentioned above, thermodynamic models of activity-composition relations are a prerequisite for application of thermobarometers involving solid solutions. Unfortunately, solution models for the minerals most commonly employed in thermobarometry are still poorly constrained (see, e.g. Essene 1989, p. 2), and this represents one of the main obstacles to accurate geothermobarometry. For this reason, the use of any thermodynamic model at large dilutions, i.e. where the actual mineral compositions deviate largely from those of the calibrated end-member reaction, should be avoided or at least regarded with scepticism, except where experimental a-X data are available.

Some of the more commonly used together with some potentially useful solid-solid end-member reactions in thermobarometry are listed in Table 4.4, together with their range of applicability in terms of metamorphic facies. Note that these reactions are listed with increasing number of phases involved, ranging from two up to five. Before some specific reactions are considered, a few general comments concerning Table 4.4 are given. (1) No equilibria involving cordierite or sapphirine have been considered because the effects of various fluid species (contained in cordierite), order/disorder, and non-ideal a-X relations are not known well enough (Essene, 1989). (2) For six of the mineral assemblages listed there are two reactions available, one each for the Fe and Mg end member of appropriate phases involved. (3) Most of these thermobarometers have a rather restricted range of applicability, often limited to one or two

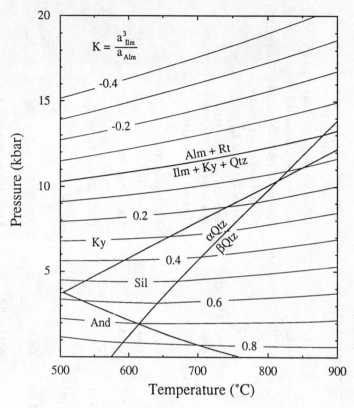

Fig. 4.9. $Log_{10}K$ contoured on a P-T diagram for the reaction $Alm + Rt = Ilm + Als + Qtz$, using thermodynamic data of Berman (1988)

metamorphic facies. (4) There are at least 19 thermobarometers based on solid-solid reactions available for the granulite facies and still 9 for the amphibolite facies, but the situation is much less favourable for the other facies.

Clinopyroxene-Plagioclase-Quartz. Reaction 4 (Table 4.4) provides the basis of an important thermobarometer in many assemblages of the blueschist and the eclogite facies. Because of its petrological significance, this reaction has been evaluated by several experimentalists (for references see, e.g. Essene 1982, p. 176; Berman 1988, Fig. 38). All these experiments have been performed at $T > 500 °C$, using high albite for starting material. Many petrologists have used the latest determination by Holland (1980) for their P-T estimates, and this procedure is also recommended by Carswell and Harley (1989, p. 93) for the eclogite facies. However, Liou et al. (1987, p. 91) found that Holland's data are not consistent with some observed mineral assemblages and mineral chemical data from the blueschist facies. For this reason, at low temperatures, Liou et al. recommended using the Jd-Ab-Qz stability relations as determined by Popp and Gilbert (1972).

Table 4.4. Some net-transfer reactions used in geothermobarometry

No.	Mineral assemblage	Net-transfer reaction	Range of application[c]	Reference[d]
1	Grt-Opx	1 Prp = 1 En + 1 MgTs	ECL	Carswell and Harley (1989)
2	Opx-Ol-Qtz	1 Fs = 1 Fa + 1 Qtz	GRA	Bohlen and Boettcher (1981), Newton (1983)
3	Cpx-Pl-Qtz	1 CaTs + 1 Qtz = 1 An	GRA	Newton (1983), Gasparik (1984)
4	Cpx-Pl-Qtz	1 Jd + 1 Qtz = 1 Ab	BLU, ECL, ±GRA	Liou et al. (1987), Carswell and Harley (1989)
5	Grt-Pl-Ol	1 Grs + 2 Prp = 3 An + 3 Fo	GRA	Johnson and Essene (1982)
6	Grt-Pl-Ol (GAF)[a]	1 Grs + 2 Alm = 3 An + 3 Fa	GRA	Bohlen et al. (1983a)
7	Sp-Po-Py	–	GRE, AMP, GRA ±BLU	Jamieson and Craw (1987), Bryndzia et al. (1988, 1990)
8	Wo-Pl-Grt-Qtz (WAGS)[a]	1 Grs + 1 Qtz = 1 An + 2 Wo	AMP, GRA	Huckenholz et al. (1981)
9	Grt-Spl-Sil-Qtz	1 Alm + 2 Sil = 3 Hc + 5 Qtz	GRA	Bohlen et al. (1986)
10	Grt-Spl-Sil-Crn	1 Alm + 5 Crn = 3 Hc + 3 Sil	GRA	Shulters and Bohlen (1989)
11	Grt-Pl-Opx-Qtz (GAES)[a]	1 Grs + 2 Prp + 3 Qtz = 3 An + 3 En	GRA	Eckert et al. (1991)
12	Grt-Pl-Opx-Qtz (GAFS)[a]	1 Grs + 2 Alm + 3 Qtz = 3 An + 3 Fs	GRA	Faulhaber and Raith (1991)
13	Grt-Pl-Cpx-Qtz (GADS)[a]	2 Grs + 1 Prp + 3 Qtz = 3 An + 3 Di	GRA	Eckert et al. (1991)
14	Grt-Pl-Cpx-Qtz (GAHS)[a]	2 Grs + 1 Alm + 3 Qtz = 3 An + 3 Hd	GRA	Moecher et al. (1988)
15	Grt-Als-Qtz-Pl (GASP)[a]	1 Grs + 2 Als + 1 Qtz = 3 An	AMP, GRA	Koziol and Newton (1988), McKenna and Hodges (1988)

No.	Assemblage	Reaction	Facies[c]	Reference[d]
16	Grt-Hbl-Pl-Qtz[b]	2 Grs + 1 Prp + 3 Prg + 18 Qtz = 6 An + 3 Ab + 3 Tr	AMP, GRA	Kohn and Spear (1989)
17	Grt-Hbl-Pl-Qtz[b]	2 Grs + 1 Alm + 3 Fprg + 18 Qtz = 6 An + 3 Ab + 3 Fac	AMP, GRA	Kohn and Spear (1989)
18	Gr-Ms-Pl-Bt[b]	1 Grs + 1 Prp + 1 Ms = 3 An + 1 Phl	AMP	Powell and Holland (1988)
19	Grt-Ms-Pl-Bt[b]	1 Grs + 1 Alm + 1 Ms = 3 An + 1 Ann	AMP	Powell and Holland (1988)
20	Ttn-Ky-Pl-Rt	1 Ttn + 1 Ky = 1 An + 1 Rt	ECL	Manning and Bohlen (1991)
21	Grt-Opx-Cpx-Pl-Qtz	1 Prp + 1 Di + 1 Qtz = 2 En + 1 An	GRA	Paria et al. (1988)
22	Grt-Opx-Cpx-Pl-Qtz	1 Alm + 1 Hd + 1 Qtz = 2 Fs + 1 An	GRA	Paria et al. (1988)
23	Grt-Rt-Als-Ilm-Qtz (GRAIL)[a]	1 Alm + 3 Rt = 3 Ilm + 1 Als + 2 Qtz	AMP, GRA	Bohlen et al. (1983b)
24	Grt-Rt-Pl-Ilm-Qtz (GRIPS)[a]	1 Grs + 2 Alm + 6 Rt = 3 An + 6 Ilm + 3 Qtz	AMP, GRA	Bohlen and Liotta (1986), Anovitz and Essene (1987b)

[a] Abbreviations according to Essene (1989).
[b] Empirical calibrations; all other calibrations were determined by experiment or by calculation.
[c] Abbreviations of metamorphic facies: AMP Amphibolite, BLU Blueschist, ECL Eclogite, GRA Granulite, GRE Greenschist.
[d] One or two recent references for each reaction are given only.

Application of the jadeitic pyroxene equilibria to natural parageneses is complicated by the accurate analysis of Na_2O in clinopyroxene, by the order-disorder transitions in albite and clinopyroxene, and by the limited a-X data among the Ca-Na pyroxenes. For a recent discussion of some of these questions, the reader is referred to Carswell and Harley (1989). Also note that many eclogite facies assemblages are devoid of albitic plagioclase, and in this case the jadeite content of the clinopyroxene will yield only a minimum pressure of formation.

Sphalerite-Pyrrhotite-Pyrite. The iron content of sphalerite in equilibrium with hexagonal pyrrhotite and pyrite continuously decreases as a function of pressure, and provides a barometer in the temperature range 300–750°C (e.g. Bryndzia et al. 1988, 1990). Applications of this barometer to metamorphic rocks have yielded excellent to poor results. At T < 300 °C, extensive resetting of sulphide compositions and transformation of hexagonal to monoclinic pyrrhotite occurs, resulting in decreased FeS contents in sphalerite and high apparent pressures. It is important to use unexsolved, coexisting three-phase assemblages following the textural criteria proposed by Hutchison and Scott (1980), and we caution against the indiscriminant use of this sphalerite geobarometer.

Garnet-Aluminosilicate-Quartz-Plagioclase (GASP). Reaction 15 (Table 4.4) has become the most widely used barometer for amphibolite and granulite facies rocks. The advantage of this system is the widespread occurrence of GASP assemblages in metapelites, due to extensive solid solution in garnet and plagioclase. An expression of the popularity of this geobarometer is the existence of at least eight different calibrations. However, the large uncertainties involved (see below) are cause for concern in the successful application of this barometer.

The end-member reaction has been calibrated by several workers (see e.g. Koziol and Newton 1988, for references) in the temperature range 900–1400°C. Application between 500 and 800°C, therefore, requires a rather long extrapolation and contributes a large proportion of the uncertainty in any P estimate from this barometer. Based on 27 brackets from 5 different experimental studies, McKenna and Hodges (1988) give the following equation for use in thermobarometry and with kyanite as the aluminosilicate phase:

$$P(bar) = (22.0 \pm 1.5)\ T\ (K) - (6200 \pm 3000),$$

where the uncertainties are at the 95% confidence level. According to McKenna and Hodges (1988), these inaccuries propagate into paleopressure uncertainties of ± 2.5 kbar.

An additional problem with this barometer is the generally small grossular contents of garnet in natural assemblages. On one hand, this leads to relatively large errors in analyzing a small percentage of Ca in garnet. On the other hand, this requires long extrapolations in composition, and because the activity of grossular at such extreme dilution is unknown, large additional uncertainties are involved in applying solution models. Given these problems, the GASP barometer is not recommended.

Garnet-Rutile-Aluminosilicate-Ilmenite-Quartz (GRAIL). Reaction 23 (Table 4.4) has been tightly reversed by Bohlen et al. (1983b) in the temperature range 750-1100°C. This system has many prerequisites for a valuable barometer for metapelites from the amphibolite and the granulite facies: (1) it was calibrated within or in the vicinity of the P-T range for which it will be used and, therefore, long extrapolations are unnecessary; (2) its location is quite insensitive to temperature; (3) only two minerals in the requisite assemblage, garnet and ilmenite, have appreciable solid solution in ordinary metamorphic rocks; (4) analyses of garnet and ilmenite are for their major components, and long compositional extrapolations are not required. Furthermore, in the absence of rutile, the GRAIL barometer may be used to estimate an upper pressure limit. Application of this barometer in numerous terrains yields geologically reasonable pressures that agree well with other barometers.

4.7.6 Reactions Involving Fluid Species

Univariant equilibria involving a fluid are less directly applicable for thermobarometry because their position in P-T space is dependent on the compositions of the fluid species involved in the reaction (Chap. 3). Dehydration reactions are often used in constructing petrogenetic grids assuming $P_{H_2O} = P_s$ or $a_{H_2O} = 1$, but it is now well known that CH_4 is a significant fluid species in many low-T rocks whilst CO_2 is abundant in many high-T rocks. For this reason, dehydration reactions are, with one exception, not further considered here.

Phengite-Biotite-K-Feldspar-Quartz. Phengites are intermediate members of the muscovite-celadonite solid solution series. The two end members are related by the Tschermak exchange vector, $(Mg,Fe^{2+})Si(AlAl)_{-1}$, and silica content is the preferred measure of this substitution. In the KMASH paragenesis phengite + phlogopite + K-feldspar + quartz + H_2O, the composition of phengite is invariant at fixed P, T and a_{H_2O}. The reaction

$$3KMgAlSi_4O_{10}(OH)_2 = KMg_3AlSi_3O_{10}(OH)_2 + 2KAlSi_3O_8 + 3SiO_2 + 2H_2O$$

  celadonite phlogopite K-feldspar quartz

has been investigated experimentally by Massonne and Schreyer (1987) for varying Si content in phengite. In a P-T diagram, Si isopleths were found to be rectilinear with a small positive slope and with a strong, almost linear increase of the Si content with pressure. In practice, from the assemblage Phe-Bt-Kfs-Qtz only phengite is chemically analyzed, and if the temperature is known, then the pressure can be read from the corresponding Si isopleth. If K-feldspar and/ or biotite is lacking from the above assemblage, a minimum pressure can be derived.

This is a potentially powerful geobarometer, especially in low- to medium-grade rocks, where calibrated barometers are scarce. However, several draw-backs exist: (1) the isopleths determined by Massonne and Schreyer (1987) are based on synthesis experiments that produced mainly 1M and Md K-white mica polytypes. (2) Since the above reaction involves dehydration, estimates of a_{H_2O} are crucial for correct application; water activities below unity shift the Si isopleths of phengite towards higher pressure, but Massonne and Schreyer argued that the effects are relatively small. (3) Experiments were conducted in an iron-free system whilst natural micas may contain considerable amounts of Fe^{2+} and Fe^{3+}. Since the ratio Fe^{2+}/Mg is generally higher in biotite than coexisting muscovite, the introduction of Fe will stabilize biotite + K-feldspar and thus reduce the celadonite content of the K-white mica, displacing the Si-isopleths to higher pressure (Evans and Patrick 1987). Application of the Massonne and Schreyer phengite geobarometer yielded reasonable results for metagranites from the eclogite facies (Massonne and Chopin 1989), but gave pressures up to 4 kbar too high for greenschist and amphibolite facies metamorphic terrains (Patrick and Ghent 1988).

The system Fe-Ti-O offers a mineral pair which may be useful as a thermometer and oxygen barometer as mentioned below.

Magnetite-Ilmenite. These two Fe-Ti oxides coexist in some metamorphic rocks. Magnetite shows solid solution with ulvöspinel (Fe_2TiO_4) whilst ilmenite shows solid solution with hematite. Chemical equilibrium between titanian magnetite and ferrian ilmenite can be described by a temperature-dependent Fe-Ti exchange reaction:

$$Fe_3O_4 + FeTiO_3 = Fe_2TiO_4 + Fe_2O_3$$

and an oxidation reaction:

$$4\ Fe_3O_4 + O_2 = 6\ Fe_2O_3.$$

Buddington and Lindsley (1964) presented a graphical form of this ther-mometer and oxybarometer. They contoured isopleths of ilmenite (X_{FeTiO_3}) and ulvöspinel ($X_{Fe_2TiO_4}$) on a log f_{O2} – T diagram and showed that intersection of these isopleths yields f_{O2} and T. Since then, there have been a number of thermodynamic treatments. The best of the published versions is that of Andersen and Lindsley (1988), and programs to calculate T and f_{O2} for coexisting Mag_{ss}-Ilm_{ss} are available from these authors.

Application of the Mag-Ilm thermometer is limited because these oxides are easily reset during retrogression. Upon cooling, Ti-bearing magnetite will be oxidized with the appearance of ilmenite lamellae within magnetite; conversely, Fe_2O_3-rich ilmenite will be reduced, yielding lamellae of Ti-magnetite within ilmenite. In order to obtain peak metamorphic temperatures, these "exsolution" lamellae must be reintegrated (Bohlen and Essene 1977).

4.7.7 Thermobarometry Using Multi-Equilibrium Calculations

The most common method of thermobarometric analysis consists of two intersecting mineral equilibria with different P-T slopes. These equilibria are calibrated directly using experimental results or indirectly from empirical observations of lnK data in natural assemblages. Recently, thermodynamic databases have become available from which any mineral equilibrium reaction may be calculated, as long as the dataset encompasses reliable data for all phases in the desired reaction. Using an internally consistent database (e.g. Berman 1988; Holland and Powell 1990), all possible equilibria among minerals of a given metamorphic assemblage are then computed in P-T space, plotted for visual inspection, and utilized to compute an average pressure and temperature simultaneously from all intersections. Corresponding computer programs

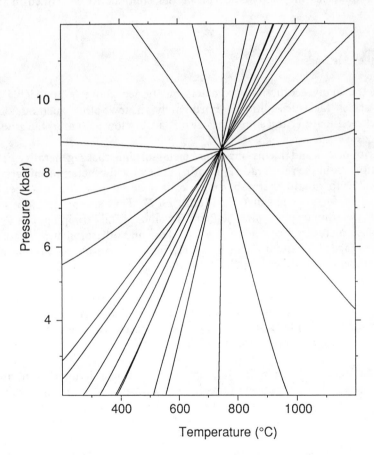

Fig. 4.10. P-T diagram showing the results of a TWEEQU calculation for an amphibolite sample containing the mineral assemblage Grt-Cpx-Hbl-Pl-Qtz-Ilm-Rt. 14 equilibria between (Grs-Alm-Prp)-(Di-Hd)-An-Qtz-Ilm-Rt, excluding hornblende, were considered. Note that only 3 of the 14 equilibria are independent. (After Lieberman and Petrakakis 1991, Fig. 8b)

most widely used in metamorphic petrology are THERMOCALC (Powell and Holland 1988) and TWEEQU (Berman 1991; Lieberman and Petrakakis 1991). Such calculations are performed with a set of internally consistent thermodynamic data for end members and activity models of those phases showing solid solution. Ideally, if thermodynamic and compositional data are perfect, and if all mineral last equilibrated at the same pressure and temperature, all equilibria will intersect at a single P-T point. In practice, significant deviations from this ideal case may occur. When results for several rocks are compared to each other and to petrographic observations, it may become possible to conclude whether inaccurate thermodynamic data, compositional data or disequilibrium may have caused discrepant results. In Fig. 4.10 the result of such a calculation is shown, and further examples may be found, e.g. in Berman (1991) and Lieberman and Petrakakis (1991). This new thermobarometric technique may become popular as thermodynamic properties continue to be refined in the future.

4.7.8 Other Methods

None of the many geothermobarometers mentioned above is applicable to subgreenschist facies conditions. Fortunately, a few other methods are available to estimate T, and sometimes also P, at very low metamorphic grade.

Fluid Inclusions. Fluid inclusions from metamorphic rocks generally yield information about the retrograde P-T path, but at very low metamorphic grade it is possible to obtain thermobarometric data about peak (or near-peak) conditions of metamorphism (e.g. Mullis 1987). This author developed a method to determine approximative P and T values of fluid trapping based on fluid inclusion studies from fissure quartz. This microthermometric method takes advantage of immiscibility in H_2O-CH_4 fluids. As a result, the homogenization temperature of water-rich inclusions saturated with CH_4 can be interpreted as temperature of formation. Methane-rich inclusions, on the other hand, allow the measurement of the density of CH_4. With these data the pressure of inclusion formation can be determined from known experimental P-V-T-X properties. Application of this method recorded temperatures of up to 270 °C and pressures of up to about 3 kbar.

Vitrinite Reflectance. Vitrinite is a predominant constituent in many coals and in finely dispersed organic matter in sedimentary rocks. As organic material is converted from complex hydrocarbons to graphite, the optical reflectance of vitrinite increases and indicates the degree of transformation. Various models exist relating vitrinite reflectance and physical variables. While temperature is regarded uniformly as the main controlling variable, the role played by the time factor is controversial and ranges from "important" to "negligible". Presumably the best model available at present is that by Sweeney and Burnham (1990), called EASY%R_o, based on a chemical kinetic model of vitrinite maturation by

Burnham and Sweeney (1989). In this Arrhenius reaction model, equations were integrated over temperature and time to account for the elimination of water, carbon dioxide, methane and higher hydrocarbons from vitrinite. The model can be implemented on a spreadsheet or in a small computer program on a personal computer, but results are also shown graphically in nomograms of vitrinite reflectance vs. exposure time and maximum temperature (Sweeney and Burnham 1990, Fig. 5). EASY%R_o can be used for vitrinite reflectance values of 0.3 to 4.5%, and for heating rates ranging from igneous intrusions (1°C/day), and geothermal systems (10°C/100 year) to burial diagenesis (1°C/10 Ma).

Conodont Colour Alteration Index (CAI). Conodonts are apatitic marine microfossils that contain trace amounts of organic material that colours them, occurring from the Cambrian to the Triassic, mainly in carbonate rocks. In the range of diagenesis and incipient metamorphism five stages of CAI are distinguished according to colour changes from pale yellow to brown to black. Epstein et al. (1977) and Rejebian et al. (1987) reproduced these colour changes by heating in the laboratory and found a time and temperature dependence. Experimental data obtained between 300 and 950°C and measured up to 10^3 h were extrapolated to geologic times. A graph of reciprocal absolute temperature versus a logarithmic time scale forms then the basis of this geothermometer, which has a resolution of possibly 50°C. An advantage of the conodont method is that it can be used in carbonate rocks in which vitrinite is rarely present.

4.7.9 Uncertainties in Thermobarometry

The development of geothermobarometry has been one of the most exciting advances in metamorphic petrology during the past 20 years. It has allowed petrologists to quantify the conditions of metamorphism; but how accurate are such data? Ideally, application of geothermobarometry should be accompanied by a statement regarding the uncertainties of T and P. In practice, however, uncertainties are rarely reported because it is difficult to quantify some sources of errors.

Kohn and Spear (1991) performed an evaluation of the uncertainty in pressure when a barometer is applied to an assemblage in a rock. Four barometers were considered, including GASP and GRAIL discussed in Section 4.7.5. Sources of error considered included (with estimated 1s uncertainties in pressure given in brackets): accuracy of the experimentally determined, barometric end-member reaction (± 300 to ± 400 bar); volume measurement errors (± 2.5 to ± 10 bar); analytical imprecision of mineral analyses using an electron microprobe (± 55 to ± 185 bar); thermometer calibration errors (± 250 to ± 1000 bar); variation in garnet and plagioclase activity models (± 60 to ± 1500 bar); and compositional heterogeneity of natural minerals (± 150 to ± 500 bar). Collectively, the accuracy of a barometer may typically range from ± 600 to ± 3250 bar (1s), with the most significant sources of error being uncertainty in thermometer calibration and poorly constrained activity models. However, continued experimental and empirical work should substantially reduce these uncertainties in the future.

The situation is much better, however, if we are interested in comparative thermobarometry, which involves applying a single pair of thermobarometers to different samples in order to calculate differences in P-T conditions. In this case, the systematic errors associated with experimental calibrations and solution modelling are eliminated, leaving only the effects of analytical uncertainties. It is then possible to confidently resolve P-T differences of as little as a few hundred bars and a few tens of degrees.

References

Anderson DJ, Lindsley DL (1988) Internally consistent solution models for Fe-Mg-Mn-Ti oxides: Fe-Ti oxides. Am Mineral 73:714-726

Anovitz IM, Essene EJ (1987a) Phase equilibria in the system $CaCO_3$-$MgCO_3$-$FeCO_3$. J Petrol 28:389-414

Anovitz IM, Essene EJ (1987b) Compatibility of geobarometers in the system CaO-FeO-Al_2O_3-SiO_2-TiO_2 (CFAST): implications for garnet mixing models. J Geol 95:633-645

Archibald DA, Krogh TE, Armstrong RL, Farrar E (1984) Geochronology and tectonic implications of magmatism and metamorphism, southern Kootenay Arc and neighbouring regions, southeastern British Columbia. Part II: Mid-Cretaceous to Eocene. Can J Earth Sci 21:567-583

Barrow G (1893) On an intrusion of muscovite biotite gneiss in the S.E. Highlands of Scotland and its accompanying metamorphism. Q J Geol Soc Lond 49:330-358

Barrow G (1912) On the geology of lower Deeside and the southern highland border. Proc Geol Assoc 23:268-284

Bearth P (1958) Ueber einen Wechsel der Mineralfazies in der Wurzelzone des Penninikums. Schweiz Miner Petr Mitt 38:363-373

Bégin NJ (1992) Contrasting mineral isograd sequences in metabasites of the Cape Smith Belt, northern Québec, Canada: three new bathograds for mafic rocks. J Metamorph Geol 10:685-704

Benciolini L, Lombardo B, Martin S (1988) Mineral chemistry and Fe/Mg exchange geothermometry of ferrogabbro-derived eclogites from the Northwestern Alps. N Jb Mineral Abh 159:199-222

Berman RG (1988) Internally-consistent thermodynamic data for minerals in the system K_2O-Na_2O-CaO-MgO-FeO-Fe_2O_3-Al_2O_3-SiO_2-TiO_2-H_2O-CO_2. J Petrol 29:445-522

Berman RG (1991) Thermobarometry using multi-equilibrium calculations: a new technique, with petrological applications.Can Mineral 29:833-855

Bhattacharya A, Mazumdar AC, Sen SK (1988) Fe-Mg mixing in cordierite: constraints from natural data and implications for cordierite-garnet geothermometry in granulites. Am Mineral 73:338-344

Bhattacharya A, Krishnakumar KR, Raith M, Sen SK (1991) An improved set of a-X parameters   for Fe-Mg-Ca garnets and refinements of the orthopyroxene-garnet thermometer and the orthopyroxene-garnet-plagioclase-quartz barometer. J Petrol 32:629-656

Bhattacharya A, Mohanty L, Maji A, Sen SK, Raith M (1992) Non-ideal mixing in the phlogopite-annite binary: constraints from experimental data on Mg-Fe partitioning and a reformulation of the biotite-garnet geothermometer. Contrib Mineral Petrol 111:87-93

Bhattacharyya DS (1981) Geometry of isograds in metamorphic terrains. Tectonophysics 73:385-395

Bohlen SR, Boettcher AL (1981) Experimental investigations and geological applications of orthopyroxene geobarometry. Am Mineral 66:951–964

Bohlen SR, Essene EJ (1977) Feldspar and oxide thermometry of granulites in the Adirondack Highlands. Contrib Mineral Petrol 62:153–169

Bohlen SR, Lindsley DH (1987) Thermometry and barometry of igneous and metamorphic rocks. Annu Rev Earth Planet Sci 15:397–420

Bohlen SR, Liotta JJ (1986) A barometer for garnet amphibolites and garnet granulites. J Petrol 27:1025–1056

Bohlen SR, Wall VJ, Boettcher AL (1983a) Experimental investigation and application of garnet granulite equilibria. Contrib Mineral Petrol 83:52–61

Bohlen SR, Wall VJ, Boettcher AL (1983b) Experimental investigations and geologic applications of equilibria in the system $FeO-TiO_2-Al_2O_3-SiO_2-H_2O$. Am Mineral 68:1049–1058

Bohlen SR, Valley JW, Essene EJ (1985) Metamorphism in the Adirondacks. 1. Pressure and temperature. J Petrol 26:971–992

Bohlen SR, Dollase WA, Wall VJ (1986) Calibration and application of spinel equilibria in the system $FeO-Al_2O_3-SiO_2$. J Petrol 27:1143–1156

Bohlen SR, Montana A, Kerrick DM (1991) Precise determinations of the equilibria kyanite = sillimanite and kyanite = andalusite and a revised triple point for Al_2SiO_5 polymorphs. Am Mineral 76:677–680

Bowen NL (1940) Progressive metamorphism of siliceous limestone and dolomites. J Geol 48:225–274

Bryndzia LT, Scott SD, Spry PG (1988) Sphalerite and hexagonal pyrrhotite geobarometer: Experimental calibration and application to the metamorphosed sulfide ores of Broken Hill, Australia. Econ Geol 83:1193–1204

Bryndzia LT, Scott SD, Spry PG (1990) Sphalerite and hexagonal pyrrhotite geobarometer: Correction in calibration and application. Econ Geol 85:408–411

Buddington AF, Lindsley DH (1964) Iron-titanium oxide minerals and their synthetic equivalents. J Petrol 5:310–357

Bundy FR, Bovenkerk HP, Strong HM, Wentorf RH (1961) Diamond-graphite equilibrium line from growth and graphitisation of diamond. J Chem Phys 35:383–391

Burnham AK, Sweeney JJ (1989) A chemical kinetic model of vitrinite maturation and reflectance. Geochim Cosmochim Acta 53:2649–2657

Carlson WD (1983) The polymorphs of $CaCO_3$ and the aragonite-calcite transformation. In: Reeder RJ (ed) Carbonates: mineralogy and chemistry. Reviews in mineralogy 11. Mineralogical Society of America, Washington DC, pp 191–225

Carlson WD, Rosenfeld JL (1981) Optical determination of topotactic aragonite-calcite growth kinetics: metamorphic implications. J Geol 89:615–638

Carmichael DM (1970) Intersecting isograds in the Whetstone Lake area, Ontario. J Petrol 11:147–181

Carmichael DM (1978) Metamorphic bathozones and bathograds: a measure of the depth of post-metamorphic uplift and erosion on a regional scale. Am J Sci 278:769–797

Carmichael DM (1991) Univariant mixed-volatile reactions: pressure-temperature phase diagrams and reaction isograds. Can Mineral 29:741–754

Carswell DA, Gibb FGF (1987) Evaluation of mineral thermometers and barometers applicable to garnet lherzolite assemblages. Contrib Mineral Petrol 95:499–511

Carswell DA, Harley SL (1989) Mineral barometry and thermometry. In: Carswell DA (ed) Eclogites and related rocks. Blackie, Glasgow, pp 83–110

Chatterjee ND, Flux S (1986) Thermodynamic mixing properties of muscovite-paragonite crystalline solutions at high temperatures and pressures, and their geological applications. J Petrol 27:677–693

Chernosky JV, Berman RG (1988) The stability of Mg-chlorite in supercritical H_2O-CO_2 fluids. Am J Sci 288A:393–420

Chinner GA (1966) The distribution of pressure and temperature during Dalradian metamorphism. Q J Geol Soc Lond 122:159–186

Chipera SJ, Perkins D (1988) Evaluation of biotite-garnet geothermometers: application to the English River subprovince, Ontario. Contrib Mineral Petrol 98:40–48

Connolly JAD, Trommsdorff V (1991) Petrogenetic grids for metacarbonate rocks: pressure-temperature phase-diagram projection for mixed-volatile systems. Contrib Mineral Petrol 108:93–105

Docka JA, Berg JH, Klewin K (1986) Geothermometry in the Kiglapait aureole. II. Evaluation of exchange thermometry in a well-constrained setting. J Petrol 27:605–626

Eckert JO, Newton RC, Kleppa OJ (1991) The ΔH of reaction and recalibration of garnet-pyroxene-plagioclase-quartz geobarometers in the CMAS system by solution calorimetry. Am Mineral 76:148–160

Edwards RL, Essene EJ (1988) Pressure, temperature and C-O-H fluid fugacities across the amphibolite-granulite facies transition, NW Adirondack Mtns., NY. J Petrol 29:39–72

Elkins LT, Grove TL (1990) Ternary feldspar experiments and thermodynamic models. Am Mineral 75:544–559

Ellis DJ, Green DH (1979) An experimental study of the effect of Ca upon garnet-clinopyroxene Fe-Mg exchange equilibria. Contrib Mineral Petrol 71:13–22

El-Shazly AK, Liou JG (1991) Glaucophane chloritoid-bearing assemblages from NE Oman: petrologic significance and a petrogenetic grid for high P metapelites. Contrib Mineral Petrol 107:180–201

Engi M (1983) Equilibria involving Al-Cr spinel: I. Mg-Fe exchange with olivine. Experiments, thermodynamic analysis, and consequences for geothermometry. Am J Sci 283A:29–71

Epstein AG, Epstein JB, Harris LD (1977) Conodont color alteration – an index to organic metamorphism. US Geol Surv Prof Pap 955, US Government Printing Office, Washington DC, 27 pp

Eskola P (1915) On the relations between the chemical and mineralogical composition in the metamorphic rocks of the Orijarvi region. Bull Comm Geol Finlande 44

Essene EJ (1982) Geologic thermometry and barometry. In: Ferry JM (ed) Characterization of metamorphism through mineral equilibria. Reviews in mineralogy 10. Mineralogical Society of America, Washington DC, pp 153–205

Essene EJ (1983) Solid solutions and solvi among metamorphic carbonates with applications to geologic thermometry. In: Reeder RJ (ed) Carbonates: mineralogy and chemistry. Reviews in mineralogy 11. Mineralogical Society of America, Washington DC, pp 77–96

Essene EJ (1989) The current status of thermobarometry in metamorphic rocks. In: Daly JS, Cliff RA, Yardley, BWD (eds) Evolution of metamorphic belts. Geol Soc Spec Publ 43, Blackwell, Oxford, pp 1–44

Evans BW (1990) Phase relations of epidote-blueschists. Lithos 25:3–23

Evans BW, Patrick BE (1987) Phengite-3T in high-pressure metamorphosed granitic orthogneisses, Seward Peninsula, Alaska. Can Mineral 25:141–158

Faulhaber S, Raith M (1991) Geothermometry and geobarometry of high-grade rocks: a case study on garnet-pyroxene granulites in southern Sri Lanka. Mineral Mag 55:33–56

Ferry JM (1980) A comparative study of geothermometers and geobarometers in pelitic schists from southern-central Maine. Am Mineral 65:720–732

Ferry JM, Spear FS (1978) Experimental calibration of the partitioning of Fe and Mg between biotite and garnet. Contrib Mineral Petrol 66:113–117

Flux S, Chatterjee ND (1986) Experimental reversal of the Na-K exchange reaction between muscovite-paragonite crystalline solutions and a 2 molal aqueous (Na,K)Cl fluid. J Petrol 27:665–676

Fonarev VI, Konilov AN (1986) Experimental study of Fe-Mg distribution between biotite and orthopyroxene at P=490 MPa. Contrib Mineral Petrol 93:227–235

Fox JS (1975) Three-dimensional isograds from the Lukmanier Pass, Switzerland, and their tectonic significance. Geol Mag 112:547–564

Frey M, de Capitani C, Liou JG (1991) A new petrogenetic grid for low-grade metabasites. J Metamorph Geol 9:497–509

Frey M, Wieland B (1975) Chloritoid in autochthon-parautochthonen Sedimenten des Aarmassivs. Schweiz Miner Petr Mitt 55:407–418

Fuhrman ML, Lindsley DH (1988) Ternary feldspar modeling and thermometry. Am Mineral 73:201–215

Gasparik T (1984) Experimental study of subsolidus phase relations and mixing properties of pyroxene in the system $CaO-Al_2O_3-SiO_2$. Geochim Cosmochim Acta 48:2537–2546

Giletti BJ (1986) Diffusion effects on oxygen isotope temperatures of slowly cooled igneous and metamorphic rocks. Earth Planet Sci Lett 77:218–228

Graham CM, Powell R (1984) A garnet-hornblende geothermometer and application to the Peloma Schist, southern California. J Metamorph Geol 2:13–32

Grambling JA (1990) Internally-consistent geothermometry and H_2O barometry in metamorphic rocks: the example garnet-chlorite-quartz. Contrib Mineral Petrol 105:617–628

Grambling JA, Williams ML (1985) The effects of Fe^{3+} and Mn^{3+} on aluminum silicate phase relations in north-central New Mexico, USA J Petrol 26:324–354

Green TH, Adam J (1991) Assessment of the garnet-clinopyroxene Fe-Mg exchange thermometer using new experimental data. J Metamorph Geol 9:341–347

Guidotti CV (1984) Micas in metamorphic rocks. In: Bailey SW (ed) Micas. Reviews in mineralogy 13. Mineralogical Society of America, Washington DC, pp 357–467

Guidotti CV, Dyar MB (1991) Ferric iron in metamorphic biotite and its petrologic and crystallochemical implications. Amer Mineral 76:161–175

Guiraud M, Holland T, Powell R (1990) Calculated mineral equilibria in the greenschist-blueschist-eclogite facies in $Na_2O-FeO-MgO-Al_2O_3-SiO_2-H_2O$. Contrib Mineral Petrol 104:85–98

Haselton HT, Hovis GL, Hemingway BS, Robie RA (1983) Calorimetric investigation of the excess entropy of mixing in analbite-sanidine solid solutions: lack of evidence for Na, K short range order and implications for two-feldspar thermometry. Am Mineral 68:398–413

Heinrich CA (1986) Eclogite facies regional metamorphism of hydrous mafic rocks in the Central Alpine Adula Nappe. J Petrol 27:123–154

Hodges KV, McKenna LW (1987) Realistic propagation of uncertainties in geologic thermobarometry. Am Mineral 72:671–680

Hodges KV, Spear FS (1982) Geothermometry, geobarometry and the Al_2SiO_5 triple point at Mt. Moosilauke, New Hampshire. Am Mineral 67:1118–1134

Hoefs J (1987) Stable isotope geochemistry. 3rd edn. Springer, Berlin Heidelberg New York, 241 pp

Holdaway MJ (1971) Stability of andalusite and the aluminum silicate phase diagram. Am J Sci 271:97–131

Holdaway MJ, Mukhopadhyay B (1993) A reevaluation of the stability relations of andalusite: thermochemical data and phase diagram for the aluminum silicates. Am Mineral 78:298–315

Holdaway MJ, Dutrow BL, Hinton RW (1988) Devonian and Carboniferous metamorphism in west-central Maine: the muscovite-almandine geobarometer and the staurolite problem revisited. Am Mineral 73:20–47

Holland TJB (1980) The reaction albite = jadeite + quartz determined experimentally in the range 600–1200 °C. Am Mineral 65:129–134

Holland TJB, Powell R (1990) An enlarged and updated internally consistent thermodynamic dataset with uncertainties and correlations: the system $K_2O-Na_2O-CaO-MgO-MnO-FeO-Fe_2O_3-Al_2O_3-TiO_2-SiO_2-C-H_2-O_2$. J Metamorph Geol 8:89–124

Huckenholz HG, Lindhuber W, Fehr KT (1981) Stability of grossular + quartz + wollastonite + anorthite: the effect of andradite and albite. N Jb Mineral Abh 142:223–247

Hutchison MN, Scott SD (1980) Sphalerite geobarometry applied to metamorphosed sulfide ores of the Swedish Caledonides and U.S. Appalachians. Nor Geol Unders 360:59–71

Jamieson RA, Craw D (1987) Sphalerite geobarometry in metamorphic terranes: an appraisal with implications for metamorphic pressure in the Otago Schist. J Metamorph Geol 5:87–99

Johnson CA, Essene EJ (1982) The formation of garnet in olivine-bearing metagabbros from the Adirondacks. Contrib Mineral Petrol 81:240–251

Kerrick DM (1990) The Al_2SiO_5 polymorphs. Reviews in mineralogy 22. Mineralogical Society of America, Washington DC, 406 pp

Kohn MJ, Spear FS (1989) Empirical calibration of geobarometers for the assemblage garnet + hornblende + plagioclase + quartz. Am Mineral 74:77–84

Kohn MJ, Spear FS (1991) Error propagation for barometers: 2. Application to rocks. Am Mineral 76:138–147

Koons PO (1984) Implications to garnet-clinopyroxene geothermometry of non-ideal solid solution in jadeitic pyroxenes. Contrib Mineral Petrol 88:340–347

Koziol AM, Newton RC (1988) Redetermination of the garnet breakdown reaction and improvement of the plagioclase-garnet-Al_2SiO_5-quartz geobarometer. Am Mineral 73:216–223

Kretz R (1983) Symbols for rock-forming minerals. Am Mineral 68:277–279

Krogh EJ (1988) The garnet-clinopyroxene Fe-Mg geothermometer–a reinterpretation of existing experimental data. Contrib Mineral Petrol 99:44–48

Laird J (1988) Chlorites: metamorphic petrology. In: Bailey SW (ed) Hydrous phyllosilicates. Reviews in mineralogy 19. Mineralogical Society of America, Washington DC, pp 404–454

Lang HM, Rice JM (1985a) Regression modelling of metamorphic reactions in metapelites, Snow Peak, northern Idaho. J Petrol 26:857–887

Lang HM, Rice JM (1985b) Geothermometry, geobarometry and T-X (Fe-Mg) relations in metapelites, Snow Peak, northern Idaho. J Petrol 26:889–924

Lieberman J, Petrakakis K (1991) TWEEQU thermobarometry: analysis of uncertainties and applications to granulites from western Alaska and Austria. Can Mineral 29:857–887

Lindsley DH (1983) Pyroxene thermometry. Am Mineral 68:477–493

Lindsley DH, Andersen DJ (1983) A two-pyroxene thermometer. Proceedings of the Thirteenth Lunar and Planetary Science Conference, Part 2. J Geophys Res 88, Suppl A887-A906

Liou JG, Maruyama S, Cho M (1987) Very low-grade metamorphism of volcanic and volcaniclastic rocks – mineral assemblages and mineral facies. In: Frey M (ed) Low temperature metamorphism. Blackie, Glasgow, pp 59–114

Manning CE, Bohlen SR (1991) The reaction titanite + kyanite = anorthite + rutile and titanite-rutile barometry in eclogites. Contrib Mineral Petrol 109:1–9

Massonne HJ, Chopin C (1989) P-T history of the Gran Paradiso (western Alps) metagranites based on phengite geobarometry. In: Daly JS, Cliff RA, Yardley BWD (eds) Evolution of metamorphic belts. Geol Soc Spec Publ 43, Blackwell, Oxford, pp 545–549

Massonne HJ, Schreyer W (1987) Phengite geobarometry based on the limiting assemblage with K-feldspar, phlogopite, and quartz. Contrib Mineral Petrol 96:212–224

McKenna LW, Hodges KV (1988) Accuracy versus precision in locating reaction boundaries: Implications for the garnet-plagioclase-aluminum silicate-quartz geobarometer. Am Mineral 73:1205–1208

Miyashiro A (1961) Evolution of metamorphic belts. J Petrol 2:277–318

Moecher DP, Essene EJ, Anovitz LM (1988) Calculation of clinopyroxene-garnet-plagioclase-quartz geobarometers and application to high grade metamorphic rocks. Contrib Mineral Petrol 100:92–106

Mullis J (1987) Fluid inclusion studies during very low-grade metamorphism. In: Frey M (ed) Low temperature metamorphism. Blackie, Glasgow, pp 162–199

Newton RC (1983) Geobarometry of high-grade metamorphic rocks. Am J Sci 283-A:1–28

Newton RC (1986) Metamorphic temperatures and pressures of Group B and C eclogites. Geol Soc Am Spec Pap 164:17–30

Newton RC, Fyfe WS (1976) High pressure metamorphism. In: Bailey DK, MacDonald R (eds) The evolution of the crystalline rocks. Academic Press, London, pp 100–186

Newton RC, Smith JV (1967) Investigations concerning the breakdown of albite at depth in the earth. J Geol 75:268–286

Newton RC, Goldsmith JR, Smith JV (1969) Aragonite crystallization from strained calcite at reduced pressures and its bearing on aragonite in low-grade metamorphism. Contrib Mineral Petrol 22:335–348

Paria P, Bhattacharya A, Sen A (1988) The reaction garnet + clinopyroxene + quartz = 2 orthopyroxene + anorthite: a potential geobarometer for granulites. Contrib Mineral Petrol 99:126–133

Patrick BE, Ghent ED (1988) Empirical calibration of the K-white mica + feldspar + biotite geobarometer. Program with Abstracts 13. Mineralogical Society of Canada, St. John's, p A95

Pattison DRM, Newton RC (1989) Reversed experimental calibration of the garnet-clinopyroxene Fe-Mg exchange thermometer. Contrib Mineral Petrol 101:87–103

Perchuk LL, Lavrent'eva IV (1983) Experimental investigation of exchange equilibria in the system cordierite-garnet-biotite. In: Saxena SK (ed) Kinetics and equilibrium in mineral reactions. Advances in physical geochemistry 3. Springer, Berlin Heidelberg New York, pp 199–239

Popp RK, Gilbert MC (1972) Stability of acmite-jadeite pyroxenes at low pressure. Am Mineral 57:1210–1231

Powell R, Holland TJB (1988) An internally consistent dataset with uncertainties and correlations: 3. Applications to geobarometry, worked examples and a computer program. J. metamorphic Geol. 6:173–204

Powell R, Holland T (1990) Calculated mineral equilibria in the pelite system, KFMASH (K_2O-FeO-MgO-Al_2O_3-SiO_2-H_2O). Amer Mineral 75:367–380

Powell R, Condliffe DM, Condliffe E (1984) Calcite-dolomite geothermometry in the system $CaCO_3$-$MgCO_3$-$FeCO_3$: an experimental study. J Metamorph Geol 2:33–41

Reinecke T (1991) Very-high-pressure metamorphism and uplift of coesite-bearing metasediments from the Zermatt-Saas zone, Western Alps. Eur J Mineral 3:7–17

Rejebian VA, Harris AG, Huebner JS (1987) Conodont color and textural alteration: An index to regional metamorphism, contact metamorphism, and hydrothermal alteration. Geol Soc Am Bull 99:471–479

Richardson SW, Gilbert MC, Bell PM (1969) Experimental determination of kyanite-andalusite and andalusite-sillimanite equilibrium; the aluminium silicate triple point. Am J Sci 267:259–272

Salje E (1986) Heat capacities and entropies of andalusite and sillimanite: the influence of fibrolization on the phase diagram of the Al_2SiO_5 polymorphs. Am Mineral 71:1366–1371

Savin SM, Lee M (1988) Isotopic studies of phyllosilicates. In: Bailey SW (ed) Hydrous phyllosilicates. Reviews in mineralogy 19. Mineralogical Society of America, Washington DC, pp 189–223

Schliestedt M (1986) Eclogite-blueschist relationships as evidenced by mineral equilibria in the high-pressure metabasic rocks of Sifnos (Cycladic Islands), Greece. J Petrol 27:1437–1459

Schmädicke E (1991) Quartz pseudomorphs after coesite in eclogites from the Saxonian Erzgebirge. Eur J Mineral 3:231–238

Sengupta P, Dasgupta S, Bhattacharya PK, Mukherjee M (1990) An orthopyroxene-biotite geothermometer and its application in crustal granulites and mantle-derived rocks. J Metamorph Geol 8:191–197

Shulters JC, Bohlen SR (1989) The stability of hercynite and hercynite-gahnite spinels in corundum- or quartz-bearing assemblages. J Petrol 30:1017–1031

Stephenson NCN (1984) Two-pyroxene thermometry of Precambrian granulites from Cape Riche, Albany-Fraser Province, Western Australia. J Metamorph Geol 2:297–314

St-Onge MR (1987) Zoned poikiloblastic garnets: P-T paths and syn-metamorphic uplift through 30 km of structural depth, Wopmay Orogen, Canada. J Petrol 28:1–21

Stormer JC (1975) A practical two-feldspar geothermometer. Am Mineral 60:667–674

Sweeney JJ, Burnham AK (1990) Evaluation of a simple model of vitrinite reflectance based on chemical kinetics. Am Assoc Pet Geol Bull 74:1559–1570

Symmes GH, Ferry JM (1992) The effect of whole-rock MnO content on the stability of garnet in pelitic schists during metamorphism. J Metamorph Geol 10:221–237

Thoenen T (1989) A comparative study of garnet-biotite geothermometers. Unpubl Doctoral Dissertation, University of Basel, 118 pp

Thompson PH (1976) Isograd patterns and pressure-temperature distributions during regional metamorphism. Contrib Mineral Petrol 57:277–295

Thompson P (1978) Archean regional metamorphism in the Slave structural province – a new perspective on some old rocks. In: Fraser JA, Heywood WW (eds) Metamorphism in the Canadian Shield. Geol Surv Can Pap 78–10:85–102

Tilley CE (1924) The facies classification of metamorphic rocks. Geol Mag 61:167–171

Tracy RJ, Robinson P, Thompson AB (1976) Garnet composition and zoning in the determination of temperature and pressure of metamorphism, Central Massachusetts. Am Mineral 61:762–775

Turner FJ (1981) Metamorphic petrology–mineralogical, field and tectonic aspects. 2nd edn. McGraw-Hill, New York, 524 pp

Valley JW, Taylor HP, O'Neil JR (eds) (1986) Stable isotopes in high temperature geological processes. Reviews in mineralogy 16. Mineralogical Society of America, Washington DC, 570 pp

Vidal O, Goffé B, Theye T (1992) Experimental study of the stability of sudoite and magnesiocarpholite and calculation of a new petrogenetic grid for the system FeO-MgO-Al$_2$O$_3$-SiO$_2$-H$_2$O. J Metamorph Geol 10:603–614

Vielzeuf D (1983) The spinel and quartz associations in high grade xenoliths from Tallante (S.E. Spain) and their potential use in geothermometry and barometry. Contrib Mineral Petrol 82:301–311

Will TM, Powell R, Holland TJB (1990a) A calculated petrogenetic grid for ultramafic rocks in the system CaO-FeO-MgO-Al$_2$O$_3$-SiO$_2$-CO$_2$-H$_2$O at low pressures. Contrib Mineral Petrol 105:347–358

Will TM, Powell R, Holland T, Guiraud M (1990b) Calculated greenschist facies mineral equilibria in the system CaO-FeO-MgO-Al$_2$O$_3$-SiO$_2$-CO$_2$-H$_2$O. Contrib Mineral Petrol 104:353–368

Winkler HGF (1979) Petrogenesis of metamorphic rocks. 5th edn, Springer, Berlin Heidelberg New York

Zwart HJ (1973) Metamorphic map of Europe. Leiden/UNESCO, Paris

Part II Metamorphism of Different Rock Compositions

5 Metamorphism of Ultramafic Rocks

5.1 Introduction

The Earth's mantle consists predominantly of ultramafic rocks. The mantle is, with the exception of some small anomalous regions, in the solid state. The ultramafics undergo continuous recrystallization due to large-scale convection in sub-lithosphere mantle and tectonic processes in the lithosphere and any other processes causing pressure and temperature variations in a given volume of mantle rocks (magmatism, cooling and hydration). The majority of mantle rocks, therefore, qualify as metamorphic rocks. Metamorphic ultramafic rocks build up the largest volume of rocks on a global scale.

Most of the ultramafic rocks found in the Earth's crust and consequently found in surface outcrops have been tectonically implanted from the mantle during orogenesis. Such ultramafic mantle fragments in the crust are often referred to as Alpine-type peridotites. Ultramafic rocks of magmatic origin do occur in the crust but are relatively rare. Most of the magmatic ultramafic rocks in the crust formed from fractional crystallization from basic (gabbroic, basaltic) magmas in crustal magma chambers (olivine-saturated cumulate layers).

There are two basic types of mantle fragments in the crust: (1) mantle fragments from beneath oceanic crust. They constitute integral members of ophiolite sequences and are often lherzolitic in bulk composition (see below); (2) mantle fragments from sub-continental mantle occur in rock associations typical for continental crust. Such ultramafitites are usually of harzburgitic (or dunitic) composition.

5.2 Metamorphic Ultramafic Rocks

Ultramafic mantle rock fragments experience strong mineralogical and structural modifications during emplacement in the crust and subsequent crustal deformation and metamorphism. Two completely different situations can be conceived.

1. The mantle fragments retain parts of the original mineralogy and structure. Equilibration is incomplete, because of limited access of water or slow

reaction kinetics at low temperature. The ultramafics do not (or only partially) display mineral assemblages that equilibrated at the same conditions as the surrounding crustal rocks. Such ultramafic rocks may be termed **allofacial**. Allofacial ultramafics are particularly typical for ophiolite complexes, metamorphosed under relatively low grade conditions.

2. The ultramafic rocks completely equilibrate at the same pT conditions as the surrounding crustal rocks. They only show rare relics of original mineralogy and structure (e.g. cores of Cr-spinel in metamorphic magnetite). The ultramafic rocks register identical pT histories as their crustal envelope. Ultramafic rocks showing such characteristics may be designated **isofacial**. Most ultramafics in higher grade terrains conserve very little of their mantle heritage. On the other hand, transitions between the two extremes, as usual, do occur (or are probably the rule). For example, the assemblage Fo + Atg + Di (see below) found in antigorite schists may be isofacial but structures and relict minerals often permit a fairly confident reconstruction of the pre-crustal state of the ultramafics (chlorite formation from high-Al mantle CPX, Cr-spinel, polygonization of mantle olivine megacrysts etc.).

5.2.1 Rock Types

A few examples of metamorphic ultramafics are listed below. **Serpentinites** are massive or schistose rocks containing abundant minerals of the serpentine group. Serpentinites represent hydrated low-temperature versions of lherzolites or harzburgites. **Peridotite** is often used for olivine-bearing ultramafic rocks. Rocks containing metamorphic olivine and enstatite are consequently named **En + Fo felses**. Carbonate-bearing serpentinites are often designated **ophicarbonates** (ophicalcites, etc). Carbonate-bearing talc schists (talc felses) are known as **soapstones**. The name **sagvandite** is often used for carbonate-bearing enstatite felses.

5.2.2 Chemical Composition

Ultramafic rocks consist predominantly of ferro-magnesian silicates. Anhydrous ultramafics contain the three minerals olivine, orthopyroxene and calcic clinopyroxene in various proportions. Olivine, OPX and CPX together dominate the modal composition of anhydrous ultramafitites. Consequently, the system components SiO_2, Fe_2O_3, MgO, and CaO constitute > 95 wt% of almost all anhydrous ultramafitites. Fe_2O_3 represents an important component in most ultramafic rocks. It is most often bound to a spinel phase (magnetite, spinel, chromite). Fe-Mg substitution in olivine and enstatite in such rocks seldom exceeds 5–10 mol% ($X_{Mg} \sim 0.9$ to 0.95). Thus, phase relationships in ultramafic rocks can be discussed in Fe-free systems as a first approximation. We will discuss possible effects of additional components such as iron where necessary and appropriate. Most ultramafic rocks contain hydrates (amphi-

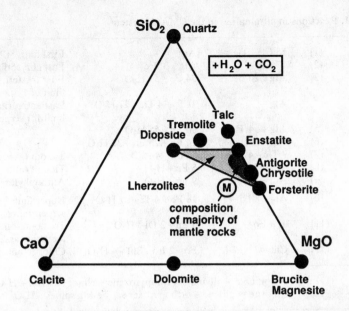

Fig. 5.1. Chemography of the ultramafic CMS-HC system projected from CO_2 and H_2O onto the plane CaO-MgO-SiO$_2$, showing some rock and mineral compositions in ultramafic rocks

boles, sheet silicates etc.) and very often also carbonates. Therefore, H_2O and CO_2 must be added to the set of components in order to describe phase relationships in partially or fully hydrated (and/or carbonated) versions of Ol + OPX + CPX rocks.

The system SiO_2-MgO-CaO-H_2O-CO_2 (CMS-HC system) or subsystems thereof is adequate for a discussion of metamorphism of ultramafic rocks (note: the same system will be used to describe phase relationships in meta-sedimentary carbonate rocks, Chap. 6). Figure 5.1 shows a chemographic projection of the CMS-HC system from H_2O and CO_2 onto the SiO_2-CaO-MgO plane. The composition of mantle rocks with the anhydrous mineralogy Ol + OPX + CPX is restricted to the shaded area defined by forsterite (Ol), enstatite (OPX) and diopside (CPX). Rocks with compositions inside the triangle Fo-En-Di are usually termed lherzolites (undepleted mantle peridotite). Rocks falling on the Fo-En join are harzburgites (depleted mantle peridotites) and rocks falling close to the Fo-corner in Fig. 5.1 are referred to as dunites. Pyroxenites are rocks with compositions along the Di-En join (including the Di- and En-corners).

Most of the accessible mantle material (tectonic fragments in the crust, xenoliths in basalts and other mantle-derived volcanic rocks) and of the ultramafic cumulate material is olivine normative. In this chapter, we will not consider metamorphism of pyroxenites and their hydrated equivalents. In addition, mantle lherzolites are often rich in Ol + OPX, whereas CPX (Di) occurs in modal amounts below 20–30%. The composition of the majority of

Table 5.1. Reactions in ultramafites in the CMASH system

MSH	(1)	$15 \, Ctl + Tlc \Rightarrow$	Atg	First antigorite
	(2)	$17 \, Ctl \Rightarrow$	$Atg + 3 \, Brc$	Last chrysotile
	(3)	$Atg + 20 \, Brc \Rightarrow$	$34 \, Fo + 51 \, H_2O$	First forsterite (low-T limit of Fo)
	(4)	$Atg \Rightarrow$	$18 \, fo + 4 \, Tlc + 27 \, H_2O$	Last antigorite (high-T limit for serpentinites)
	(5)	$Atg + 14 \, Tlc \Rightarrow$	$90 \, En + 55 \, H_2O$	
	(6)	$Atg \Rightarrow$	$14 \, Fo + 20 \, En + 31 \, H_2O$	
	(7)	$9 \, Tlc + 4 \, Fo \Rightarrow$	$5 \, Ath + 4 \, H_2O$	Tlc-out (at lower P)
	(8)	$Tlc + Fo \Rightarrow$	$5 \, En + H_2O$	Tlc-out (at higher P)
	(9)	$Ath + Fo \Rightarrow$	$9 \, En + H_2O$	Anthophyllite-out
CMSH	(10)	$Atg + 8 \, Di \Rightarrow$	$18 \, Fo + 4 \, Tr + 27 \, H_2O$	Upper limit for Di in serpentinites
	(11)	$Tr + Fo \Rightarrow$	$5 \, En + 2 \, Di + H_2O$	Tremolite-out, lherzolite assemblage
MASH	(12)	$Chl \Rightarrow$	$Fo + 2 \, En + Spl + 4 \, H_2O$	Chlorite-out

(11a)	Amphibole + olivine $\Rightarrow$ orthopyroxene + clinopyroxene + H_2O	
(12a)	Chlorite $\Rightarrow$ olivine + orthopyroxene + Fe-Mg spinel + H_2O	

mantle rocks is restricted to the heavy shaded range (labelled M) in Fig. 5.1. The present chapter is dealing with such rock compositions.

All important minerals occurring in ultramafic rocks are shown in Fig. 5.1. The serpentine minerals chrysotile and lizardite are structural polymorphs of very similar composition (Lz is more aluminous), whereas antigorite is slightly less magnesian than Ctl/Lz. The minerals of the serpentine group are related by Eq. (2) in Table 5.1. Ctl/Lz represent the low temperature serpentine minerals and typically occur at metamorphic grades below middle greenschist facies. Antigorite is the typical greenschist/blueschist and lower amphibolite facies serpentine mineral. All three carbonate minerals of Fig. 5.1 occur in ultramafites. It is apparent from the figure that a large number of potential mineral assemblages in this system are excluded on rock compositional grounds. For example, the assemblage $En + Tr + Tlc$ is outside the composition range of "normal" ultramafic rocks, it will not occur in such rocks and all mineral reactions leading to this assemblage are irrelevant for meta-lherzolites and have not to be considered on the subsequent phase diagrams. Other examples of assemblages outside the lherzolite composition (with reference to Fig. 5.1) are $Brc + Fo$, $Di + En + Tlc$. We also will not consider assemblages of the type $Dol + Fo + Brc$ for the same reason.

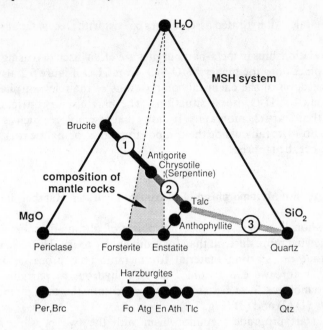

Fig. 5.2. Chemography of the ultramafic MSH system showing the relevant mineral compositions in ultramafic rocks of harzburgitic composition (depleted subcontinental upper mantle, *shaded area*). Maximum hydrated ultramafics contain the assemblages labelled *1, 2* and *3*. The projection of the MSH system from H_2O onto the MgO-SiO$_2$ binary is shown in the *lower half of the figure* together with the composition range for harzburgites

5.3 Metamorphism in the MSH System

5.3.1 Chemographic Relations in the MSH System

Figure 5.2 shows the chemographic relationships in the three-component MSH system. The base line is anhydrous, and mantle rocks are normally restricted to the composition range between Fo and En (= harzburgites). Hydrated versions of harzburgites occupy the shaded area in Fig. 5.2. The H_2O in the system may be stored in solid hydrates such as amphiboles (Oam, anthophyllite), sheet silicates (talc, antigorite), hydroxide (brucite) or as a free fluid phase, depending on pT conditions and hydrologic situation.

Mantle harzburgites may, during weathering or other near-surface hydration processes, reach a state of maximum hydration. Three maximum hydrated assemblages are possible in the MSH system. The assemblages are shown in Fig. 5.2: (1) brucite + chrysotile, (2) chrysotile + talc, (3) talc + quartz. The latter assemblage is outside the meta-harzburgite composition. However, a "super" hydrated assemblage chrysotile + quartz, (tie line chrysotile-quartz in Fig. 5.2) often occurs in near-surface modifications of harzburgites. This assemblage is probably metastable under all geological conditions relative to chrysotile + talc. All hydrated assemblages coexist with free water at subcritical conditions.

Phase relationships in meta-harzburgites are often discussed using projected chemographies onto the binary $MgO-SiO_2$ from H_2O. Figure 5.2 also shows a binary projection of the chemical compositions of the relevant phases in the MSH system for H_2O excess situations. The harzburgite restriction is also shown on the binary chemography. It shows that assemblages such as En + Ath, Tlc + Ath, Fo + Brc fall outside the composition range of mantle rocks and they will, therefore, be ignored.

5.3.2 Progressive Metamorphism of Maximum Hydrated Harzburgite

The discussion of progressive metamorphism of ultramafic rocks is conceptually somewhat more difficult than for sedimentary rocks. The difficulty lies in the fact that the starting material for metamorphic processes has to be "prepared" before we can go on. That is, anhydrous or partially hydrated mantle assemblages have first to be converted into the maximum hydrated equivalents (1),(2) and (3) (Fig. 5.2).

Let us start prograde metamorphism with the two possible maximum hydrated versions of harzburgite in structural and chemical equilibrium in a very low-grade terrain. The two assemblages are: Brc + Ctl and Ctl + Tlc (or metastable Ctl + Qtz) respectively. The phase relationships for the system are shown in Fig. 5.3 and the reactions are listed in Table 5.1. In Fig. 5.3, three geological situations are superimposed on the phase relationships in the form of two different orogenic geotherms and a contact metamorphic situation at 2 kbar.

Chrysotile + talc is the first assemblage that undergoes prograde metamorphic transformation when heat is added to it (reaction 1, Table 5.1). However, the equilibrium condition for the reaction is outside the pT window of Fig. 5.3. Antigorite forms already at temperatures below 200°C. The maximum temperature for chrysotile is given by reaction (2) and the pT coordinates for reaction (2) are at about 250°C irrespective of the position of the geotherms. Chrysotile represents a very low temperature weathering or alteration product of ultramafic rocks and the stable serpentine mineral under metamorphic conditions is antigorite. From Fig. 5.3 it follows that antigorite serpentinites cover a wide temperature range [between reactions (2) and (4), that is about 300°C along the Ky-geotherm].

Above about 250°C, MSH rocks may contain two different assemblages depending on bulk composition: Brc + Atg and Atg + Tlc. The bulk chemistry control on the assemblages can be seen in Fig. 5.2. Harzburgites with Fo/En ratios > 0.7 will contain Brc + Atg, harzburgites with Fo/En < 0.7 consist of Atg + Tlc, and finally harzburgites with Fo/En = 0.7 will be present as monomineralic antigorite serpentinites.

Brc + Atg is the next assemblage that becomes unstable with increasing temperature. At about 400°C the assemblage is replaced by Fo + Atg [reaction

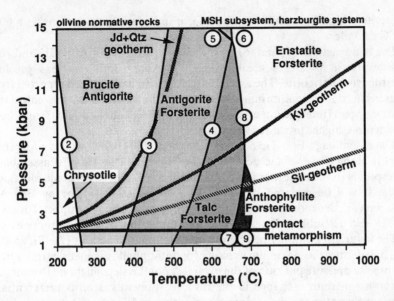

Fig. 5.3. Pressure temperature diagram for equilibria in the MSH (harzburgite) system, [reactions (1) through (9); Table 5.1]. The figure is valid if pure H_2O is present as a fluid phase. *Superimposed:* four geological geotherms

(3), Table 5.1]. Brucite-antigorite schists are diagnostic rocks for lower to middle greenschist facies conditions. The product assemblage Fo + Atg is diagnostic for upper greenschist to lower amphibolite facies conditions. The upper T limit of the assemblage along the Ky geotherm is near 570°C (see below). Antigorite serpentinites are therefore stable in the middle amphibolite facies, but also under blueschist and low-T eclogite conditions (Fig. 5.3). Reaction (3) also represents the low-temperature limit for forsterite (olivine) in the presence of an aqueous fluid phase. It is important to remember that the isograd boundary represented by reaction (3) only affects rocks with normative Fo/En > 0.7. The widespread assemblage Atg + Tlc is unaffected by reaction (3); such rocks are, in contrast to Brc + Atg rocks, inappropriate for locating the 400°C isograd in terrains with abundant ultramafic rocks.

The upper thermal stability for antigorite is given by reaction (4) (Table 5.1). The high-temperature limit for serpentinites is at about 570°C along the Ky-geotherm, ~550°C along the Sil geotherm and near 510°C in a contact aureole at 2 kbar. Generally, the last serpentinites disappear approximately with the beginning of middle amphibolite facies conditions in orogenic belts. However, because of the positive slope of equilibrium (3) in Fig. 5.3, serpentinites are stable at fairly high temperatures if pressure is high, for instance in connection with subduction and crustal thickening processes (the absolute maximum temperature for antigorite is near 650°C at 14 kbar). In high-pressure low-temperature ophiolite complexes produced by subduction of oceanic crust, antigorite schists (serpentinites) often occur in isofacial associations with 600–650 °C eclogites.

The two reactions (5) and (6) limit the upper stability of Atg + Tlc and Atg, respectively, at pressures greater than ~14 kbar. Both reactions produce enstatite from antigorite. The assemblage Atg + En has not been reported from rocks as a stable equilibrium assemblage. The position of the calculated invariant point [intersection of reactions (4), (5), (6), and (8); Fig. 5.3] may be at considerable higher pressures in nature.

The assemblage Fo + Tlc occupies an about 100–150 °C-wide temperature interval along three of the geotherms of Fig. 5.3. The Tlc + Fo assemblage corresponds to the middle amphibolite facies conditions in orogenic metamorphism. It will be replaced by either En + Fo [reaction (8)] or by Fo + Ath [reaction (7)], depending on the position of the geotherm. The upper T limit of the Tlc + Fo assemblage is near 670 °C and rather independent of pressure.

The invariant point defined by the intersection of reactions (7), (8), and (9) (Fig. 5.3) defines the maximum pressure for anthophyllite in meta-harzburgites. The precise pressure position of the invariant point is dependent on the amount of iron-magnesium exchange in the involved minerals. Iron is preferentially taken up by anthophyllite. Consequently, the field for Fo + Ath expands for relatively Fe-rich ultramafic rocks. The effect is very small on the temperature position of the equilibria 7, 8, and 9. However, because of the low-angle intersection of equilibrium 7 and 9, the invariant point is displaced by several kbars along the slightly displaced equilibrium 8.

Nevertheless, anthophyllite + forsterite schists and felses occur in contact aureoles and Sil-type metamorphism. The assemblage is often related to partial hydration (retrogression) of En + Fo rocks during decompression and uplift in Ky-type metamorphism. The precise location of the invariant point must be calculated for each particular field case separately, by using mineral composition data and appropriate solution models (which is outside the scope of this book). Reactions (7) and (8) mark the approximate beginning of upper amphibolite facies conditions.

Anthophyllite decomposes in the presence of forsterite according to reaction (9) at temperatures near 700 °C. The harzburgite assemblage enstatite + forsterite has its low temperature limit (~670 °C) in the presence of an aqueous fluid. The limit is defined by reactions (6), (8), and (9), respectively. Towards higher temperatures, the assemblage forsterite + enstatite remains stable at all geologically possible temperatures in the crust.

5.4 Metamorphism in the CMASH System

5.4.1 Progressive Metamorphism of Hydrated Al-Bearing Lherzolites

The phase relationships for ultramafic rocks in the CMASH system are presented in Fig. 5.4. Calcium, if present, is stored in ultramafic rocks in two

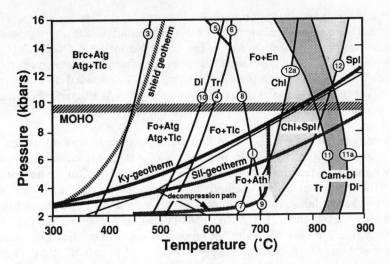

Fig. 5.4. Equilibrium position of all reactions of Table 5.1 in pressure temperature space for conditions of H₂O pressure = total pressure. *Superimposed:* three geotherms and a path of rapid isothermal decompression prior to cooling; also shown is the position of the MOHO under normal stable continents. Note that the upper mantle beneath the large continental shield areas is near the Brc + Atg field under water-present conditions. *Shaded areas between curve 11 and 11a.* Divariant field for simultaneous presence of amphibole and Cpx; *between curves 12 and 12a,* divariant field for coexistence of chlorite and spinel in ultramafic rocks. The alumosilicate phase diagram is shown for comparison

minerals; calcic clinoamphibole (Cam) and/or calcic clinopyroxene (CPX). Diopside is, somewhat surprising, the stable calcic phase at low temperatures (at extremely low temperatures, tremolite is probably replacing diopside). Diopside occurs as stable Ca-phase in Atg + Brc, Atg + Tlc and Atg + Fo schists (serpentinites). However, it is consumed by reaction (10) about 20–30 °C below the upper limit of antigorite [reaction (4)]. This means that rocks containing the product assemblage Atg + Fo + Tr are diagnostic for a narrow range of temperatures depending on the geologic situation. The temperature for the assemblage is near 540 °C along the Ky geotherm, 520 °C along the Sil-geotherm and 500 °C in the 2 kbar contact aureole.

Tremolite is the only calcic phase in isofacial ultramafics over the entire amphibolite facies. The calcic amphibole is removed from En + Fo rocks at the onset of granulite facies conditions. Reaction (11) replaces tremolite by diopside in lherzolitic rock compositions. The product assemblage Fo + En + Di is equivalent to the mantle lherzolite mineralogy. The onset of CPX production in the pure CMSH system is at about 800 °C along the Ky geotherm. However, pure CMSH ultramafics are very rare. The presence of Al (Chl, Spl), Na (fluid), Ti (Ilm) and other elements in natural lherzolites has the consequence that calcic amphiboles become increasingly aluminous, sodic and titaniferous (etc) at high temperatures. The product CPX of reaction (11) will also not be pure diopside. However, the partitioning of the "extra components" between the two minerals (CPX and Cam) results in a substantial increase of the amphibole field towards higher temperatures. The position of equilibrium (11) is therefore composition dependent. Reaction (11a) expresses the general amphibolite facies to granulite facies transition in ultramafic rocks. The curve 11a in Fig. 5.4 represents the upper limit of amphibole for moderate departures from tremolite endmember compositions ($\sim 30\%$ tremolite + 70% pargasite). The shaded area between curves (11) and (11a) denotes the amphibolite/granulite facies transition zone in H_2O-saturated regimes. Fo + En + Tr + Di (or Ol + OPX + Cam + CPX) is the characteristic assemblage in this zone.

Aluminium, if present, is stored in low-temperature hydrated ultramafics almost exclusively in chlorite. In the absence of quartz, Mg-chlorite is a very stable mineral. Its maximum temperature limit exceeds 800 °C at the base of the crust! It decomposes to Fo + En and spinel according to reaction (12) (if the metamorphic evolution follows typical Ky- or Sil-geotherms). The equilibrium conditions for reaction (12) in the MASH system are shown in Fig. 5.4. However, the presence of "extra" components has a dramatic effect on the thermal stability of chlorite. In this case, iron strongly partitions into the spinel phase (hercynite, magnetite component). This significantly modifies the position of equilibrium 12. The general breakdown of chlorite is described by [Eq. (12a)]. The onset of this spinel-producing reaction is 100–150 °C lower than in the pure MASH system. Curve (12a) in Fig. 5.4 marks the first appearance of Mg-rich spinel from the prograde decomposition of chlorite (spinel-hercynite overgrowth on low-grade magnetite, the latter originally produced by the serpentinization process; see below). In the shaded pT range between curves (12a) and (12), both chlorite and spinel may occur simultaneously. Note, by reference to Fig. 5.4, Chl + Spl may occur at low pressures together with Fo + En + Tr, with Fo + En + Cam + CPX in the pressure range 8–17 kbar and with the assemblage Fo + En + CPX at relatively high pressures and high temperatures.

Along the Ky-geotherm, chlorite begins to decompose near 740 °C, and amphibole is not removed from the ultramafics before a temperature of about 850 °C is reached. "Extra" components in ultramafic rocks have opposite effects on the tremolite- and chlorite-out reactions, respectively. They tend to increase the upper limit for calcic clinoamphibole, but decrease the equivalent limit for chlorite. Particularly dramatic are the effects along the Ky-geotherm. Pure tremolite decomposes at a temperature that is about 50 °C lower than the maximum temperature for pure Mg clinochlore. In the "dirty" systems, chlorite begins to decompose 150 °C below the last amphibole disappears.

The use of Fig. 5.4 for the more pragmatically oriented field geologist can be illustrated by making a specific example. Suppose that a rock contains Fo + En and CPX overgrowth on pargasitic amphibole in addition to structures that indicate chlorite replacement by Mg-rich spinel (e.g. Mg-spinel/hercynite overgrowth on magnetite together with resorbed chlorite). If this rock has been collected in a typical Ky-type terrain, Fig. 5.4 suggests metamorphic conditions in the range 800–840 °C and 10–11 kbar.

5.4.2 Effects of Rapid Decompression and Uplift Prior to Cooling

In many orogenic belts metamorphic rocks experience a period of rapid decompression and uplift after equilibration at the maximum temperature of the reaction history. Possible effects related to rapid decompression must always be considered. Suppose an isofacial ultramafic rock body has equilibrated along the Ky-geotherm (Fig. 5.4) at 710°C, 8.2 kbar. It contains the assemblage Fo + En + Tr + Chl. The rock may subsequently follow the decompression path indicated in Fig. 5.4. In that case the chlorite breakdown reaction (12a) may begin to produce some Mg-Fe spinel at pressures below about 6 kbar. The decompression and cooling path may also cross the stability field of the assemblage anthophyllite + forsterite. Production of the assemblage, however, requires access of H_2O during cooling. Late Ath + Fo rocks often develop along fractures and vein systems that provide access for late aqueous fluids during the decompression and cooling history of a Ky-type terrain.

5.5 Isograds in Ultramafic Rocks

The successive sequence of diagnostic mineral assemblages in isofacial ultramafic rocks that have been metamorphosed along a Ky geotherm is shown in Fig. 5.5. The assemblage boundaries (zone boundaries) shown on the figure are all very well suited for isograd mapping and isograd definitions. They are equivalent to reaction isograds because all boundaries are related to well-defined reactions (Table 5.1) in relatively simple chemical systems. As discussed above, there is some overlap of the chlorite and spinel fields, as well as of the tremolite and diopside fields respectively. Along the Sil-geotherm, the reaction isograds at lower grades are displaced to somewhat lower temperatures, an Ath + Fo field may appear near 700°C and amphibole decomposes at higher temperatures compared with the Ky geotherm.

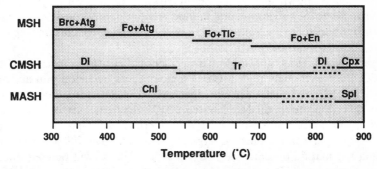

Fig. 5.5. Compilation of mineral assemblages and position of isograds in ultramafic rocks along the Ky geotherm

The stable sequence of mineral assemblages and the positions of reaction isograds for ultramafic rocks heated by a magmatic intrusion at 2 kbar can be seen along the bottom axis of Fig. 5.4. The reactions (3), (10) and (4) occur in the stability field for andalusite. The latter mineral may occur in meta-pelites, meta-sandstones and metamorphic quartzo-feldspathic rocks associated with the ultramafics. The first occurrence of sillimanite in the felsic rocks is in the middle of the Tlc + Fo field in ultramafic rocks. Chlorite will decompose to spinel assemblages at temperatures in the vicinity of 650°C, a temperature usually realized at the immediate contact to granitoid intrusions. Figure 5.4 suggests that the assemblages Ath + Fo and En + Fo should normally not be observed at the contacts to shallow level granitoid intrusions, a conclusion that is, unfortunately, in conflict with field data.

A little warning must be added at this point: since all phase relationships presented in this book are based on thermodynamic databases that in turn are based on experimental data, it is inevitable that some discrepancies and inconsistencies with nature and natural assemblages may occur. For example, in contact metamorphism caused by the intrusion of Qtz-diorite into a large body of antigorite schist (170 km^2) in the Central Alps produced locally En + Fo at the immediate contact, and a well-developed anthophyllite + forsterite zone is present along the contact to the quartz-diorite (Trommsdorff and Evans, 1972). At the immediate contact, 620°C and 3 kbar have been estimated from a variety of different independent geological thermobarometers. This suggests that the positions for reactions (7), (8), and (9) in Figs. 5.3 and 5.4 may be 50–100°C lower in natural systems with well-crystallized, stable and ordered minerals.

5.6 Mineral Assemblages in the Uppermost Mantle

The boundary between continental crust and the upper mantle (MOHO) is at a depth of about 30–40 km in periods without active tectonic processes (subduction, collision, rifting). The precise location of the MOHO depends on the state and geological history of the continent. MOHO depth has been converted into pressure and is shown in Fig. 5.4 below 10 kbar along with a typical shield geotherm. Since the subcontinental mantle predominantly consists of harzburgites with subordinate lherzolite, a number of interesting conclusions regarding mantle petrology can be drawn from Fig. 5.4.

Spinel lherzolites are not stable in the presence of H$_2$O at temperatures below ~850°C. Such extremely high MOHO temperatures occur only in active rift zones or in collision belts undergoing thermal relaxation or in other anomalous regions. Harzburgites in the MSH system (OPX-Ol) are not stable below ~670° C if an H$_2$O-rich fluid is present at MOHO depth. The MOHO temperature in Central Europe, for instance, is typically between 600–700°C. Any water reaching the mantle beneath the MOHO at T < 670 °C will be consumed by hydration reaction (8) [reactions (11a), (12) consume H$_2$O at even much higher temperatures but affect only lherzolites and/or Al-phases in the rocks]. The assemblage Talc + Fo + Tr + Chl represents the stable mantle assemblage under fluid-present conditions (H_2O fluid) beneath most of the young continental areas. Under the Precambrian shields the MOHO temperature ranges usually between 350 and 450°C (blueschist facies conditions). The upper mantle ultramafics will, therefore, be transformed into stable serpentinites (Brc + Atg; Atg + Tlc; Atg + Fo) if water crosses the MOHO along deep faults and shear zones. Very prominent seismic reflectors observed in the mantle beneath the Celtic sea and Great Britain (Warner and McGeary 1987) are probably caused by partially hydrated mantle rocks (serpentinites).

Beneath most continental areas there are two types of mantle rocks present: (1) weakly hydrated or completely anhydrous peridotites may occur under fluid-absent conditions or in areas with absolutely abnormally high geothermal gradients; (2) partially or fully hydrated ultramafics (serpentinites and talc-schists) under fluid-present conditions.

5.7 Serpentinization of Peridotites

Serpentinization of mantle rocks occurs normally in three different environments: (1) in the mantle itself (see above), (2) in oceanic ophiolite complexes where serpentinization may be related to oceanic metamorphism, (3) in the crust during collision belt formation. Some aspects of low temperature alteration of ultramafics in ophiolite and other associations will be discussed below.

Generally, if an ultramafic rock of the MSH system (e.g. mantle harzburgite) is subjected to conditions to the left of curve (4) in Figs. 5.3 and 5.4 it will be partially or fully serpentinized in the presence of an aqueous fluid. The reversed reaction (4) (Table 5.1) describes the equilibrium serpentinization process. Metastable serpentinization (or stable at P > 14 kbar) occurs through the reversed reactions (5) and (6).

The presence of iron in ultramafic rocks complicates the description of the general process somewhat. Both chrysotile and antigorite are extremely iron-pure minerals unable to accommodate iron present in olivine and orthopyroxene. Although the metastable reaction (6) [or reactions (3) and (4)] describes serpentinization of En + Fo, it does not describe the decomposition of the fayalite and ferrosilite components in olivine and orthopyroxene, respectively. The following reaction scheme illustrates the serpentinization of Fe-bearing olivine (e.g. typical $Fo_{90}Fa_{10}$ olivine).

$$3\ Fe_2SiO_4 + O_2 \Rightarrow 2\ Fe_3O_4 + 3\ SiO_{2aq} \tag{13}$$

Fayalite   Magnetite

$$3\ Mg_2SiO_4 + SiO_{2aq} + 2\ H_2O \Rightarrow 2\ Mg_3Si_2O_5(OH)_4. \tag{14}$$

Forsterite   chrysotile

Reaction (13) oxidizes the fayalite component to magnetite and releases SiO_2 component that in turn is consumed by reaction (14) forming serpentine from forsterite component. Because olivine in mantle rocks always contains much more Fo than Fa, all SiO_2 released by reaction (13) will be consumed by reaction (14). Equilibrium of reaction (13) is dependent of the oxygen pressure in the fluid. However, the overall serpentinization process is dominated by reactions in the MSH system that are independent of the oxygen pressure. The resulting assemblage of the total process is chrysotile (serpentine) + magnetite, unreacted olivine (with resorption structures) may remain present in the rock. Five to ten vol% magnetite typically occurs in many serpentinites. Magnetite often nucleates on primary chromites (Cr-spinel).

In the progression of metamorphism of such a Ctl + Mag rock the iron remains fixed in magnetite and does not dissolve in the product phases of the reactions in Table 5.1. As a result, olivine and orthopyroxene in rocks produced by prograde metamorphism of serpentinites are usually very Mg-rich in contrast to primary mantle peridotites that may have lower X_{Mg} in Ol and OPX.

Low-temperature alteration of high-T clinopyroxene (Al, Ti-rich CPX) results in the decomposition of the pyroxene and the formation of chlorite and a pure diopside (the latter is stable in very low-temperature environments as discussed above).

5.8 Reactions in Ultramafic Rocks at High Temperatures

With high temperatures we understand temperatures above the stability limit for chlorite and amphibole ($> 850\,°C$). Such high temperatures occur in deeper parts of the mantle or at lower pressures in thermally very abnormal regions.

Three reactions are important in such environments:

$$2\,Mg_2SiO_4 + CaAl_2Si_2O_8 \Rightarrow 2\,MgSiO_3 + CaMgSi_2O_8 + MgAl_2O_4 \tag{15}$$
forsterite   anorthite   enstatite   diopside   spinel

$$MgSiO_3 + MgAl_2O_4 \Rightarrow Mg_2SiO_4 + Mg_3Al_2Si_3O_{12} \tag{16}$$
enstatite   spinel   forsterite   pyrope

$$Orthopyroxene + spinel \Rightarrow olivine + Ca\text{-}Mg\text{-}Fe\text{-}garnet \tag{16a}$$

$$CaMg_5Si_8O_{22}(OH)_2 + 2\,MgAl_2O_4 \Rightarrow$$
   tremolite   spinel

$$3\,Mg_2SiO_4 + MgSiO_3 + 2\,CaAl_2Si_2O_8 + H_2O. \tag{17}$$
forsterite   enstatite   anorthite

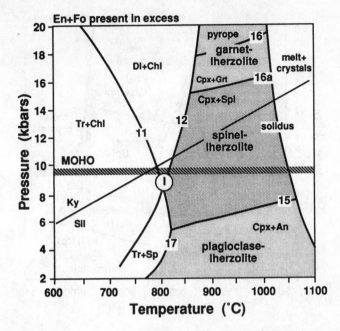

Fig. 5.6. Mineral reactions in iron-free H_2O-saturated ultramafic rocks (aluminous lherzolites) at high temperatures. Point *(I)* represents indifferent crossing of reaction (11) and (12). Its position is very sensitive to FM substitution in the phases involved. It is displaced towards higher pressures with increasing Fe content of the rocks (Fig. 5.7). *Shaded fields* represent stability fields for three important anhydrous assemblages

The equilibrium conditions for the three reactions are shown in Fig. 5.6 (Fe-free system) and Fig. 5.7 (Fe-bearing system) together with reactions shown on previous Figs. Ultramafic rocks with anhydrous assemblages occupy three characteristic PT fields (shaded) in Fig. 5.6. The stability fields for the assemblages are bounded towards higher temperatures by melt-involving reactions (solidus). Towards lower temperatures the three anhydrous assemblages are limited by the hydration reactions (11) and (12), respectively, and by the reaction (17) above.

The low-pressure assemblage is CPX + An (+ En + Fo) and rocks containing it are referred to as plagioclase lherzolites. Rocks containing forsterite + anorthite are diagnostic for low pressure of formation (< 6 kbar). Because the assemblage also requires very high temperatures (> 800 °C) olivine + anorthite is most often found in shallow level ultramafic or gabbroic intrusions (olivine gabbros, troctolites).

Spinel-lherzolites contain Spl + CPX in addition to Fo + En. Spinel-lherzolites are related to plagioclase-lherzolites by reaction (15). Spinel-lherzolites occur in granulite facies metamorphism and in the uppermost part of the mantle (provided that the MOHO temperature is sufficiently high and/or H_2O is not available).

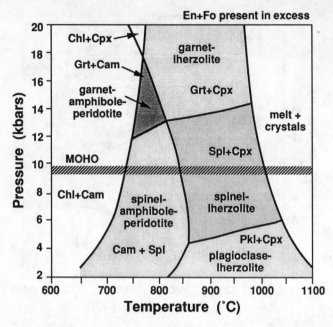

Fig. 5.7. Mineral reactions in ultramafic rocks of the CFMASH system at high temperatures. *Shaded fields* represent stability field for three important anhydrous assemblages

Garnet replaces spinel as the Al-bearing phase in anhydrous ultramafic rocks at higher pressures. Reaction (16) limits the stability field for pyrope + forsterite in Al-bearing harzburgites towards lower pressure. The presence of calcium in lherzolites has the consequence that product garnet dissolves Ca in the form of grossular component, thus expanding the stability field for Ca-Mg-garnet + forsterite [reaction (16a)].

The strong fractionation of iron into spinel, the melt phase and garnet has rather significant effects on the high-T phase relationships in ultramafic rocks as shown in Fig. 5.7. The solidus is displaced to lower temperatures. The garnet-lherzolite field is considerably enlarged towards lower T and p. The pT field for spinel-bearing amphibole peridotite is much larger than in Fig. 5.6. In contrast to the Fe-free system, Fig. 5.7 also displays a stability field for garnet-amphibole-peridotite. All phase boundaries are calculated (or estimated) for typical iron contents found in Alpine peridotites ($X_{Mg} \sim 0.9$). For more iron-rich rock compositions, garnet-amphibole- and spinel-amphibole-peridotites may occur at even lower temperatures than indicated in Fig. 5.7. Garnet-amphibole-peridotites occasionally occur in association with high-T eclogites (see Chap. 9).

5.9 Thermometry and Geobarometry in Ultramafic Rocks

Geothermobarometers applicable to ultramafic rocks that are independent of the activity of volatile species include: intercrystalline FM exchange between any pair of anhydrous FM phases in ultramafic rocks (e.g. Ol-Spl, Ol-Grt, OPX-CPX), intracrystalline FM exchange in CPX, temperature dependence of miscibility gap between En-Di, Mg-Tschermak content in OPX in various defining assemblages (e.g. OPX-Grt), Ca-Tschermak content in CPX in a number of assemblages (e.g. CPX-Ol-Spl).

Most of the listed geothermobarometers obviously can only be used for estimating equilibration p and T for ultramafic rocks metamorphosed in eclogite- and granulite-facies (also pyroxene-hornfelses). Reliable pT estimates can be calculated for rocks that equilibrated within the shaded areas of Figs. 5.6 and 5.7. In greenschist- and amphibolite-facies ultramafic rocks, thermobarometers are lacking. pT estimates from such rock types depend entirely on phase relationships and reactions involving volatile species. This requires some assumptions or estimates on the activities of volatile species (H_2O). In progressive metamorphism it is often justified to assume H_2O-saturated conditions (as in Figs. 5.3, 5.4, 5.6, and 5.7). In terrains with a polymetamorphic (polyphase) evolution, it may be difficult to estimate the activity of H_2O during equilibration of the ultramafic rocks. In many ultramafic rocks the presence of carbonate minerals (magnesite, dolomite) confirms that the fluid cannot be treated as a pure H_2O fluid but rather as a mixed volatile fluid (e.g. binary CO_2-H_2O fluid).

The problem appears severe; however, it is probably not. Let us consider as a specific example: a Fo-En-fels contains scattered flakes of talc apparently replacing enstatite but no carbonates. The rock also contains occasional talc veins. Is it possible to estimate the temperature at which talc formation occurred? The presence of vein structures can be used as evidence that the talc formation is related to the infiltration of an aqueous (H_2O-rich) fluid. By reference to Fig. 5.3, direct talc formation from enstatite by reaction (8) requires pressures in excess of 6.5 kbar and, provided that metamorphism occurred near the Ky-geotherm, temperatures below 670°C. The minimum temperature is given by the position of reaction (4) in Fig. 5.3 because the alteration did not produce antigorite.

5.10 Carbonate-Bearing Ultramafic Rocks

Carbonate minerals (magnesite, calcite and dolomite) are found in many ultramafic rocks. In the Scandinavian Caledonides, for example, almost all Alpine-type ultramafics contain carbonate minerals in an amazing variety of associations with silicates and oxides. The carbonates are in some instances certainly of mantle origin. However, most ultramafics in the crust are carbonated in the same way as they are hydrated. Addition of external CO_2 to ultramafic rocks leads to saturation of the rocks with carbonate phases. Maximum hydrated and carbonated low temperature varieties of ultramafic rocks such as talc + magnesite + dolomite rocks or antigorite + talc + dolomite rocks encompass the composition range of typical mantle rocks (see chemographical relationships shown in Fig. 5.1).

Phase relationships in carbonate-bearing ultramafics are relatively complex and can provide excellent information about pT regimes in a particular metamorphic terrain. Two specific geologic environments will be used to demonstrate the usefulness of analyzing phase relationships in carbonate-bearing ultramafic rocks. It is obvious that the two examples cannot cover all possible geological situations. This strongly suggests that our presented phase diagrams must be modified or recalculated in order to use them in other environments.

5.10.1 Metamorphism of Ophicarbonate Rocks

Serpentinites containing carbonate minerals in appreciable amounts are referred to as ophicarbonates (ophicalcite, ophimagnesite ...). Ophicarbonates form from carbonate-free serpentinites by reaction with crustal CO_2. Carbon dioxide evolves from progressing decarbonation reactions in carbonate-bearing metasediments. Serpentinites are very efficient CO_2 buffers. Small amounts of CO_2 in the fluid are sufficient to convert serpentine assemblages into carbonate-bearing assemblages.

Figure 5.8 shows an example of phase relationships in ophicarbonates in terms of an isobaric TX section at 2 kbar. All reaction stoichiometries are given in Table 5.2, except for the two dehydration reactions (4) and (10) that have been discussed previously and are listed in Table 5.1. The diagram can be used to analyze assemblages in contact metamorphic terrains. The positions of the equilibria are displaced to higher T with increasing pressure (it is therefore necessary to recalculate the figure for higher or lower pressure terrains).

The figure shows that antigorite-bearing rocks do not tolerate much CO_2 in the fluid phase. Reaction (18) limits the stability field for antigorite in the presence of CO_2-bearing fluids. Under the conditions of Fig. 5.8, 13 mol% CO_2 in the fluid is the maximum CO_2 content of a fluid coexisting with antigorite. Therefore, serpentinites emplaced in the crust may behave as very efficient CO_2 absorbers. The stability fields for the three assemblages antigorite + calcite (ophicalcites, light shaded), antigorite + dolomite (ophidolomites, patterned frame) and antigorite + magnesite (ophimagnesite, dark shaded) are also shown in Fig. 5.8. The assemblage antigorite + calcite is diagnostic for extremely H_2O-rich fluids.

Five low-variant assemblages define a sequence of five potential isograds in the field. The five assemblages are (Fig. 5.8): (1) Atg + Fo + Di + Tr + Cal (Atg + Di out), (2) Atg + Fo + Tr + Dol + Cal (Atg + Cal out), (3) Atg + Fo + Tr + Tlc + Mgs (Atg + Mgs out), (4) Atg + Fo + Tr + Tlc + Dol (Atg + Dol out),

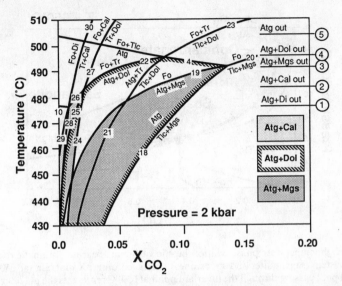

Fig. 5.8. Low temperature phase relationships in carbonate-bearing ultramafic rocks (ophicarbonate rocks) depicted on a TX diagram (at 2 kbar)

Table 5.2. Reactions in ophicarbonates in the CMS-HC system

MS-HC	(18)	17 Tlc + 45 Mgs + 45 H_2O ⇒	2 Atg + 45 CO_2
	(19)	Atg + 20 Mgs ⇒	34 Fo + 20 CO_2 + 31 H_2O
	(20)	Tlc + 5 Mgs ⇒	4 Fo + 5 CO_2 + 1 H_2O
CMS-HC	(21)	47 Tlc + 30 Do + 30 H_2O ⇒	15 Tr + 2 Arg + 60 CO_2
	(22)	40 Dol + 13 Atg ⇒	20 Tr + 282 Fo + 383 H_2O + 80 CO_2
	(23)	13 Tlc + 10 Dol ⇒	12 Fo + 5 Tr + 8 H_2O + 20 CO_2
	(24)	107 Dol + 17 Tr + 107 H_2O ⇒	4 Atg + 141 Cal + 73 CO_2
	(25)	20 Dol + 1 Atg ⇒	34 Fo + 20 Cal + 31 H_2O + 20 CO_2
	(26)	40 Cal + 11 Atg ⇒	20 Tr + 214 Fo + 40 CO_2 + 321 H_2O
	(27)	Tr + 11 Dol ⇒	8 Fo + 13 Cal + 1 H_2O + 9 CO_2
	(28)	31 Tr + 45 Cal ⇒	1 Atg + 107 Di + 45 CO_2
	(29)	20 Cal + 3 Atg ⇒	62 Fo + 20 Di + 93 H_2O + 20 CO_2
	(30)	3 Tr + 5 Cal ⇒	11 Di + 2 Fo + 5 CO_2 + 3 H_2O

(5) Atg + Fo + Tlc (Atg out). The position of these isograds has been successfully mapped by Trommsdorff and Evans (1977) in a contact aureole of the Central Alps and the temperature resolution of the isograds is within a few °C. Such a high thermal resolution is very unusual and it is strongly recommended to study and map ophicarbonate assemblages for investigating the thermal fine structure of a metamorphic terrain.

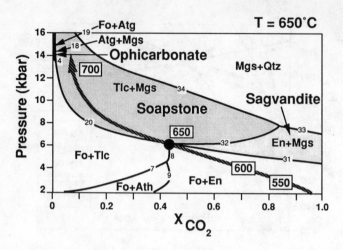

Fig. 5.9. High temperature phase relationships in carbonate-bearing ultramafic rocks of the MS-HC system (sagvandites and soapstones) portrayed on a pX diagram (at 650°C). The patterned path passing through the invariant point at 650°C represents the polythermal trace of the assemblage Fo + En + Tlc + Mgs. The temperature in °C indicates the position of the invariant point at these temperatures. The *arrow* at the high-PT end of the curve indicates that the invariant point is at P > 16 kbar at 700°C. *Shaded fields* symbolize magnesite-bearing varieties of common ultramafic rocks

5.10.2 Soapstones and Sagvandites

The term soapstone is used for ultramafic rocks containing predominantly talc + magnesite (or dolomite). Sagvandites are ultramafic rocks that are modally dominated by the assemblage enstatite + magnesite. Both rock types usually develop from ordinary carbonate-free ultramafic rocks by interaction with an externally derived CO_2-H_2O fluid phase. Ultramafic rocks emplaced in the crust are always accompanied by partly carbonated versions of these rocks.

Phase relationships in the MS-HC system are depicted on an isothermal pX diagram at 650°C (Fig. 5.9). The figure shows the distribution of diagnostic assemblages relevant for harzburgitic (Fo + En) total rock compositions at middle amphibolite facies temperatures as a function of pressure (depth). Reactions (19), (20) and (31) separate magnesite-bearing assemblages from carbonate-free assemblages. The three major types of carbonate-bearing ultramafic rocks are shown as shaded areas in Fig. 5.9. At low pressures carbonate saturation is not reached even in pure CO_2 fluids. At 3 kbar fluid pressure, for instance, the stable assemblage in pure H_2O fluids is Fo + Tlc (compare with Fig. 5.3). In the fluid composition range 0.3–0.45 Fo + Ath replaces Fo + Tlc. Fo + En represents the stable assemblage in CO_2-rich fluids (X_{CO_2} up to 1.0).

In addition to the reactions listed in Tables 5.1 and 5.2, there are four CO_2-involving reactions affecting harzburgitic bulk compositions:

$$Mgs + En \Rightarrow Fo + CO_2 \tag{31}$$

$$Tlc + Mgs \Rightarrow 4\ En + CO_2 + H_2O \tag{32}$$

$$Qtz + Mgs \Rightarrow En + CO_2 \tag{33}$$

$$4\ Qtz + 3\ Mgs + H_2O \Rightarrow Tlc + 3\ CO_2. \tag{34}$$

In most carbonate-bearing ultramafitites the composition of the fluid phase is buffered by the reactions (19), (20), and (31). Magnesite + quartz assemblages are extremely rare, which suggests that reactions (33) and (34) are not important for ultramafic rocks. Reaction (32) relates the sagvandite assemblage (En + Mgs) to the soapstone assemblage (Tlc + Mgs). Reactions (20), (32) and (31) terminate in an isothermal invariant point (6.5 kbar and 0.42 at T = 650 °C). The invariant assemblage Fo + En + Tlc + Mgs is very widespread in Alpine-type ultramafic rocks from many orogenic belts. Different processes lead to that assemblage: in a first step harzburgites may undergo partial carbonation by the inverse of reaction (31) that produces En + Mgs from Fo. The progressing reaction buffers the fluid composition to the invariant point and, finally, the invariant reaction produces Tlc. The presence of the stable assemblage Fo + En + Tlc + Mgs is restricted to the heavy patterned line passing through the invariant point at 650 °C (polythermal trace of the invariant point). The curve represents the position of the invariant point as a function of temperature that is labelled along the curve. For instance, if the assemblage Fo + En + Tlc + Mgs is found at a 550 °C locality, the corresponding equilibrium pressure is near 3 kbar and the fluid composition is around 0.85. The invariant point occurs at rapidly increasing pressure above 650 °C (at 700 °C the pressure exceeds 16 kbar). Consequently, the assemblage represents a fairly good temperature indicator (680 ± 20 °C) in high pressure rocks (> 10 kbar). It also indicates H_2O-rich fluids in such rocks. In rocks that have equilibrated at pressures below 10 kbar the assemblage supplies the field geologist with useful pressure estimates.

Reaction (18) relates the soapstone assemblage (Tlc + Mgs) to the ophicarbonate assemblage (Atg + Mgs). Reactions (4), (18), (19) and (20) terminate in an invariant point characterized by the assemblage Atg + Tlc + Fo + Mgs. The assemblage is also relatively common in CO_2-infiltrated serpentinites (ophicarbonates, see Section 5.10.1). The coordinates for the assemblage are at 650 °C, 13.5 kbar and $X_{CO_2} = 0.02$ (Fig. 5.9). A second coordinate point for this important assemblage is given in Fig. 5.8: 490 °C, 2 kbar and $X_{CO_2} = 0.13$. The polythermal trace for this assemblage in Fig. 5.9 can be represented by a steep curve with a pressure difference of 12 kbar within the temperature interval 490–650 °C. The assemblage can be used for pressure estimates at temperatures up to about 550 °C. Such pressure estimates, however, require rather precise independent T information. At pressures above about 5 kbar the assemblage is diagnostic for very H_2O-rich fluids and temperatures near 600–650 °C.

The discussion of mineral reactions and phase relations in soapstones and sagvandites above is restricted to the MS-HC system. Ca-bearing rocks give rise to a number of additional reactions and diagnostic assemblages. Particularly important are reactions involving Dol, Tr, and Di in addition to the phases already considered. The relationships, reactions and assemblages presented above for the MS-HC are, of course, also valid in a more complex system (e.g. CMAS-HC). Phase relationships for carbonate-bearing ultramafic rock of spinel-lherzolithic bulk compositions are very useful for deciphering fine details of the metamorphic evolution of a terrain.

References and Further Reading

Berman RG, Engi M, Greenwood HJ, Brown TH (1986) Derivation of internally-consistent thermodynamic data by the technique of mathematical programming: a review with application to the system MgO- SiO_2-H_2O. J Petrol 27:1331–1364

Brey G, Brice WR, Ellis DJ, Green DH, Haris KL, Ryabchikov ID (1983) Pyroxene-carbonate reactions in the upper mantle. Earth and Planet Sci Lett 62:63–74

Bucher-Nurminen K (1991) Mantle fragments in the Scandinavian Caledonides. Tectonophysics 190:173–192

Carswell DA (1981) Clarification of the petrology and occurrence of garnet lherzolites and eclogite in the vicinity of Rödhaugen, Almklovdalen, West Norway. Nor Geol Tidsskr 61:249–260

Carswell DA (1985) The metamorpic evolution of Norwegian garnet peridotites. Terra Cognita 5:439

Carswell DA, Gibb FGF (1980) The equilibrium conditions and petrogenesis of European crustal garnet lherzolites. Lithos 13:19–29

Chidester AH, Cady WM (1972) Origin and emplacement of alpine-type ultramafic rocks. Nat Phys Sci 240:27–31

Dawson JB (1981) The nature of the upper mantle. Mineral Mag 44:1–18

Dymek RF, Boak JL, Brothers SC (1988) Titanic chondrodite- and titanian clinohumite-bearing metadunite from the 3800 Ma Isua supracrustal belt, West Greenland. Chemistry, petrology, and origin. Am Mineral 73, 5–6:547–558

Eckstrand OR (1975) The Dumont Serpentinite: a model for control of nickeliferous opaque mineral assemblages by alteration reactions in ultramafic rocks. Econ Geol 70:183–201

Evans BW (1977) Metamorphism of Alpine peridotite and serpentine. Ann Rev Earth Planet Sci 5:397–447

Evans BW, Trommsdorff V (1970) Regional metamorphism of ultramafic rocks in the central Alps: parageneses in the system CaO- MgO- SiO_2- H_2O. Schweiz Mineral Petrogr Mitt 50:481–492

Evans BW, Trommsdorff V (1974) Stability of enstatite + talc, and CO_2-metasomatism of metaperidotite, Val d'Efra, Lepontine Alps. Am J Sci 274:274–296

Evans BW, Trommsdorff V (1978) Petrogenesis of garnet lherzolite, Cima di Gagnone, Lepontine Alps. Earth Planet Sci Lett 40:333–348

Evans BW, Johannes W, Oterdoom H, Trommsdorff V (1976) Stability of chrysolite and antigorite in the serpentinite multisystem. Schweiz Mineral Petrogr Mitt 56:79–93

Frost BR (1975) Contact metamorphism of serpentinite chlorite blackwall and rodingite at Paddy–Go–Easy pass, central Cascades, Washington. J Petrol 16:272–313

Frost BR (1985) On the stability of sulfides, oxides and nativ metals in serpentinite. J Petrol 26:31–63

Lieberman JE, Rice JM (1986) Petrology of marble and peridotite in the Seiad ultramafic complex, northern California, USA. J Metamorph Geol 4:179–199

Naldrett AJ, Cabri LJ (1976) Ultramafic and related mafic rocks: their classification and genesis with special references to the concentrations of nickel sulfides and platinium-group elements. Econ Geol 71:1131–1158

Nishiyama T (1990) CO_2-metasomatism of a metabasite block in a serpentine melange from the Nishisonogi metamorphic rocks, southwest Japan. Contrib Mineral Petrol 104:35–46

O'Hanley DS, Dyar MD (1993) The composition of lizardite 1T and the formation of magnetite in serpentinites. Am Mineral 78:391–404

Schreyer W, Ohnmacht W, Mannchen J (1972) Carbonate-orthopyroxenites (sagvandites) from Troms, northern Norway. Lithos 5:345–363

Soto JI (1993) PTMAFIC: software for thermobarometry and activity calculations with mafic and ultramafic assemblages. Am Mineral 78:840–844

Trommsdorff V (1983) Metamorphose magnesiumreicher Gesteine: Kritischer Vergleich von Natur, Experiment und thermodynamischer Datenbasis. Fortschr Mineral 61:283–308

Trommsdorff V, Connolly JAD (1990) Constraints on phase diagram topology for the system CaO- MgO- SiO_2- CO_2- H_2O. Contrib Mineral Petrol 104:1–7

Trommsdorff V, Evans BW (1972) Progressive metamorphism of antigorite schist in the Bergell tonalite aureole (Italy). Am J Sci 272:487–509

Trommsdorff V, Evans BW (1977) Antigorite–ophicarbonates: contact metamorphism in Valmalenco, Italy. Contrib Mineral Petrol 62:301–312

Warner M, McGeary S (1987) Seismic reflection coefficients from mantle fault zones. Geophys J R Astronom Soc 89:223–230.

Wyllie PJ, Huang W.-L., Otto J, Byrnes AP (1983) Carbonation of peridotites and decarbonation of siliceous dolomites represented in the system CaO- MgO- SiO_2- CO_2 to 30 kbar. Tectonophysics 100:359–388

6 Metamorphism of Dolomites and Limestones

6.1 Introduction

6.1.1 General

Sedimentary carbonate rocks consist predominantly of carbonate minerals (as the name suggests). There are two main classes of carbonate rocks, dolomites and limestones. The first one is modally dominated by dolomite $[CaMg(CO_3)_2]$, the second by calcite $(CaCO_3)$. The rocks often also contain variable amounts of quartz (SiO_2) in addition to the two carbonates minerals (siliceous dolomites, siliceous limestones).

Magnesite-bearing sedimentary carbonate rocks occur occasionally in evaporitic environments. They are extremely rare, however, and will not be considered further here. In contrast, siliceous dolomites and limestones represent a large portion of the sedimentary record, particularly in Phanerozoic sediments. Their metamorphic equivalents are designated **marbles** (dolomitic marbles, dolomite marbles; calcitic marbles, calcite marbles). Marbles are very widespread in metamorphic terrains associated with orogenic belts. Marbles are also found in contact aureoles around shallow level magmatic intrusions into dolomite and limestone sequences.

6.1.2 Chemical Composition

The dominant mineralogy of sedimentary carbonate rocks (dolomite, calcite, quartz), defines a very simple chemical composition space. H_2O-bearing minerals are often absent in sedimentary carbonate rocks but do occur in metamorphic equivalents (e.g. talc, tremolite). Therefore, H_2O was either present as water in the pore space at low grades or it was introduced to the marbles under metamorphic conditions. In order to discuss the phase relationships in marbles, H_2O must be added to the four oxide components of Cal, Dol and Qtz. The five components of the siliceous dolomite system used in the discussion below are $CaO\text{-}MgO\text{-}SiO_2\text{-}H_2O\text{-}CO_2$ (CMS-HC system).

Dolomites are very iron-poor rocks and metamorphic ferro-magnesian minerals in dolomitic marbles commonly show $X_{Mg} > 0.95$ (often 0.99). Iron can be ignored in the discussion of phase relationships of most marbles. On the other hand, this compositional restriction also precludes the use of Fe-Mg exchange thermometry in marble assemblages. Other chemical impurities and their effect on the phase relationships in marbles will be discussed at the end of this chapter.

6.1.3 Chemographic Relationships

Most of the metamorphic transformations in marbles can be discussed in the five component CMS-HC system. A chemographic representation of the siliceous dolomite system is shown in Fig. 6.1. The figure is a projection of the five-component system onto the $CaO-MgO-SiO_2$ plane from CO_2 and H_2O. It shows the composition of all relevant minerals found in marbles that can be described in this system. Other minerals of this system, such as enstatite and anthophyllite, do not occur in marbles. Some minerals of the CMS-HC system (e.g. monticellite, Åkermanite) are found in very high temperature low pressure contact aureoles around gabbro intrusions (hydrous granite is too cold). Their phase relationships will not be considered in this book.

The shaded areas in Fig. 6.1 represent the compositional range of sedimentary carbonate rocks within the triangle Dol-Cal-Qtz. Table 6.1 lists all stable reactions among the minerals of Fig. 6.1 that affect rocks of the shaded composition range. The composition space can be subdivided further, by considering dolomite-rich rocks (1) and quartz-rich rocks (2) separately.

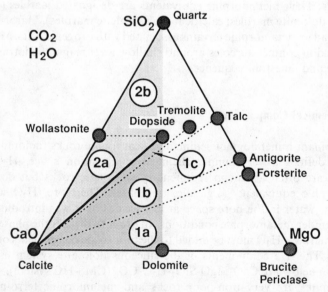

Fig. 6.1. Chemography of the siliceous dolomite system projected from CO_2 and H_2O onto the $CaO-MgO-SiO_2$ plane. Dolomite marbles are represented by the composition fields *1a, 1b* and *1c,* calcsilicate marbles by the fields *2a* and *2b*

Table 6.1. Reactions in siliceous dolomites

---

1. Dolomite–rich rock compositions

Upper limit of quartz

(1)	$3\,Dol + 4\,Qtz + H_2O \Rightarrow$	$Tlc + 3\,Cal + 3\,CO_2$	Talc
(2)	$5\,Dol + 8\,Qtz + H_2O \Rightarrow$	$Tr + 3\,Cal + 7\,CO_2$	Tremolite
(3)	$Dol + 2\,Qtz \Rightarrow$	$Di + 2\,CO_2$	Diopside

Upper limit of talc

| (4) | $2\,Tlc + 3\,Cal \Rightarrow$ | $Tr + Dol + CO_2 + H_2O$ | Tremolite |

Upper limit of tremolite

(5)	$3\,Cal + Tr \Rightarrow$	$Dol + 4\,Di + H_2O + CO_2$	Diopside
(6)	$11\,Dol + Tr \Rightarrow$	$13\,Cal + 8\,Fo + 9\,CO_2 + H_2O$	Forsterite
(7)	$5\,Cal + 3\,Tr \Rightarrow$	$11\,Di + 2\,Fo + 5\,CO_2 + 3\,H_2O$	Fo + Di
(8)	$107\,Dol + 17\,Tr + 107\,H_2O \Rightarrow$	$141\,Cal + 4\,Atg + 73\,CO_2$	Antigorite

Upper limit of diopside

| (9) | $3\,Dol + Di \Rightarrow$ | $4\,Cal + 2\,Fo + 2\,CO_2$ | Forsterite |

Upper limit of antigorite

| (10) | $20\,Dol + Atg \Rightarrow$ | $20\,Cal + 34\,Fo + 20\,CO_2 + 31\,H_2O$ | Forsterite |

Upper limit of dolomite

| (11) | $Dol \Rightarrow$ | $Cal + Per + CO_2$ | Periclase |
| (12) | $Dol + H_2O \Rightarrow$ | $Cal + Brc + CO_2$ | Brucite |

Upper limite of brucite

| (13) | $Brc \Rightarrow$ | $Per + H_2O$ | Periclase |

---

2. Quartz–rich rock compositions

Upper limit of dolomite

(1)	$3\,Dol + 4\,Qtz + H_2O \Rightarrow$	$Tlc + 3\,Cal + 3\,CO_2$	Talc
(2)	$5\,Dol + 8\,Qtz + H_2O \Rightarrow$	$Tr + 3\,Cal + 7\,CO_2$	Tremolite
(3)	$Dol + 2\,Qtz \Rightarrow$	$Di + 2\,CO_2$	Diopside

Upper limit of talc

| (14) | $5\,Tlc + 6\,Cal + 4\,Qtz \Rightarrow$ | $3\,Tr + 6\,CO_2 + 2\,H_2O$ | Tremolite |

Upper limit of tremolite

| (15) | $3\,Cal + 2\,Qtz + Tr \Rightarrow$ | $5\,Di + H_2O + 3\,CO_2$ | Diopside |

Upper limit of quartz + calcite

| (16) | $Cal + Qtz \Rightarrow$ | $Wo + CO_2$ | Wollastonite |

Dolomites and dolomitic limestones will fall into category (1), limestones in either category (1) or (2) depending on the Dol/Qtz ratio of the rock. The boundary that separates the two compositional categories is the tie line between calcite and diopside. Metamorphic rocks of the composition range (2) are often referred to as calcsilicate marbles or calcsilicate rocks at higher grades (particularly if they contain aluminum as an additional component, see Chap. 8).

The reason for a separate discussion of the two groups of rocks is given by different reactions affecting them at higher metamorphic grades. Reaction (3) of Table 6.1 ($Qtz + Dol \Rightarrow Di + CO_2$) describes the separation of the composition space. A sedimentary carbonate rock consisting of $Cal + Dol + Qtz$ will, after completion of reaction (3) consist of either $Cal + Dol + Di$ [rock run out of Qtz, Dol-rich composition group(1)], or $Cal + Qtz + Di$ [rock run out of Dol, Qtz-rich composition group(2)]. The Cal-Dol-Di assemblage of the Dol-rich marble will undergo transformations described by reaction (9), whereas the Cal-Qtz-Di assemblage of the Qtz-rich composition will experience reaction (6) at high temperatures. Note that the first three reactions of Table 6.1 describe the upper limit of the sedimentary Dol-Cal-Qtz assemblage and are identical for the two composition groups (1) and (2). Specifically, reactions (1), (2) and (3) all involve $Dol + Qtz$ as reactants and if the reactions run to completion the rock will contain the product assemblage plus unused reactants which can be either dolomite **or** quartz.

The equilibrium conditions of the reactions listed in Table 6.1 depend on p, T and, in the presence of a binary CO_2-H_2O fluid phase, on the composition of this fluid. The fluid composition can be expressed as X_{CO_2} (see Chap. 3). At a given pressure and temperature (e.g. along an orogenic geotherm), the assemblages of the reactions in Table 6.1 coexist with a fluid of a fixed given composition.

6.2 Orogenic Metamorphism of Dolomites

The phase relationships of dolomite-rich marbles in orogenic belts are shown in Fig. 6.2 along a kyanite-type geotherm as defined earlier. The figure is valid for assemblages that contain dolomite + calcite throughout. The sedimentary assemblage is therefore represented by the quartz field. $Dol + Qtz$ will react to form talc, tremolite or diopside according to reactions (1), (2) and (3), depending on the composition of the fluid phase initially present in the pore space or introduced to the marble along the geotherm. Note that both the talc- and tremolite-forming reactions consume H_2O.

Talc will form in H_2O-rich fluids at $T < 500\,°C$. Talc-bearing marbles are restricted to $T < 500\,°C$ at $p < 5$ kbar. However, talc in marbles of orogenic belts is usually associated with late hydrothermal activity or is of retrograde origin. Talc is the only metamorphic mineral in dolomites at lower metamorphic grades. The steep equilibrium position of reaction (1) in pTX space (Fig. 6.2) has the consequence that only little modal talc is produced by this reaction under closed system conditions. Tremolite is therefore usually the first recognizable metamorphic mineral in marbles. Talc is removed from the marbles by reaction (4).

Tremolite first occurs at the invariant point involving $Qtz + Tlc + Tr$ ($+ Dol + Cal$) at about $500\,°C$ and 5 kbar. The first occurrence of tremolite represents an excellent mappable isograd in the field and approximately coincides with the beginning of the amphibolite facies ($=$ tremolite-in isograd).

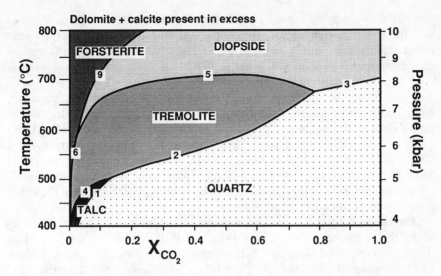

Fig. 6.2. PTX phase relationships in dolomite marbles containing excess dolomite and calcite. The *vertical axis* represents an orogenic geotherm characteristic for kyanite-type terrains

The temperature of the isograd is rather independent of the precise position of the geotherm and close to 500°C. Reaction (2) will continue to produce tremolite as the temperature increases along the geotherm and the rock will effectively control the composition of the fluid phase along the boundary of the quartz and tremolite fields of Fig. 6.2 (the rock will contain the assemblage Dol + Cal + Qtz + Tr). Finally, the rock will run out of reactant quartz and will be able to enter the divariant tremolite field (Dol + Cal + Tr). This assemblage is **the** characteristic assemblage in lower to middle amphibolite facies dolomite marbles. At a typical middle amphibolite facies temperature (say 600°C, 6.5 kbar, Fig. 6.2) dolomite marbles may contain three assemblages: Dol-Cal-Tr (common), Dol-Cal-Qtz-Tr [reaction (2)], Dol-Cal-Qtz (rare).

Diopside forms in dolomite marbles from reactions (3) and (5), and from the reaction in invariant point Qtz + Tr + Di (+ Dol + Cal). The position of this invariant point defines the lowest possible occurrence of diopside in dolomite marbles (closed system). In other words, the first occurrence of diopside represents a sharp mappable boundary in the field and coincides with a temperature of about 670°C along the geotherm of Fig. 6.2 (= diopside-in isograd).

If the rocks do not reach the invariant point along a prograde path, that is if reaction (2) has consumed all quartz at some lower temperature, then diopside will form from reaction (5). This reaction replaces the divariant assemblage Dol + Cal + Tr by Dol + Cal + Di. Reaction (5) separates the tremolite from the diopside field in Fig. 6.2 but also in the field. The maximum temperature for the assemblage Dol + Cal + Tr is given by the T maximum of reaction (5). Along the Ky-type geotherm no tremolite-bearing dolomite marbles can be stable at

T > 720 °C (= tremolite-out isograd). In the pT interval between the diopside-in and the tremolite-out isograd, tremolite- and diopside-bearing marbles may occur side by side, that is, the four-phase assemblage of reaction (5) may be found in rocks. The T interval is about 40 °C.

Diopside may also form from reaction (3) if the rock is sufficiently quartz-rich to leave the invariant point along the quartz-diopside boundary [= equilibrium of reaction (3)] in the course of prograde metamorphism or if the initial fluid contained more than about 80 mol% CO_2. Reaction (3) will also produce diopside in rocks devoid of a fluid phase. Reaction (3) will create a pure CO_2 fluid phase in such rocks and impose the absolute maximum limit for the Dol + Qtz assemblage (~700 °C at 8 kbar). Under such conditions, no mineralogical changes will occur in siliceous dolomites until pT conditions of the upper amphibolite facies are reached.

Forsterite will not form in orogenic metamorphism of siliceous dolomites along the geotherm of Fig. 6.2. In principle, diopside will be removed from dolomite marbles by reaction (9), that replaces the Dol + Cal + Di assemblage of the diopside field by the Dol + Cal + Fo assemblage of the forsterite field. Reaction (9) represents the boundary between the two fields in Fig. 6.2. At temperatures < 800 °C forsterite can only be produced by interaction of the marble with an externally derived H_2O-rich fluid phase (e.g. along fractures or shear zones, veins). Closed system marbles will be free of forsterite, provided that the area has been metamorphosed along a Ky-type geotherm. At pressures in the range < 8 to 5 kbar (Sil type geotherm), forsterite may form in dolomites under granulite facies conditions [T > 800 °C (8 kbar) or T > 700 °C (5 kbar)].

Summary. In regional metamorphic terrains of orogenic belts, dolomite-rich marbles (Fig. 6.1) may contain small amounts of talc in addition to Dol + Cal + Qtz in the upper greenschist facies. Tremolite-bearing marbles are characteristic for the lower to middle amphibolite facies. Diopside-bearing marbles are stable from the middle amphibolite facies onwards. Forsterite marbles are diagnostic for granulite facies conditions or for infiltration by H_2O-rich fluids.

6.3 Orogenic Metamorphism of Limestones

Phase relationships in quartz-rich marble compositions or calcsilicate marbles (composition range 2 in Fig. 6.1) in orogenic metamorphism along the Ky-geotherm are given in Fig. 6.3. The figure is valid for rocks with excess calcite + quartz. Because dolomite is the limiting mineral in the reactions (1), (2) and (3), for these compositions the field for the sedimentary assemblage (Dol + Cal + Qtz) is labelled *dolomite* in Fig. 6.3. The production of Tlc, Tr and Di by reactions (1), (2) and (3) is identical to the dolomite-rich rocks above. The upper limit for talc is, however, given by reaction (14) in these rocks.

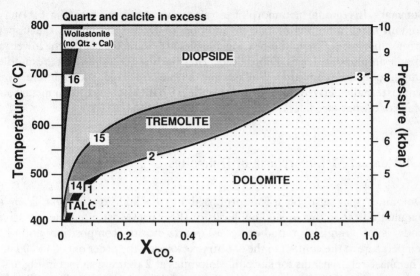

Fig. 6.3. PTX phase relationships in calcsilicate marbles containing excess quartz and calcite. The *vertical axis* represents an orogenic geotherm characteristic for kyanite-type terrains

The **tremolite**-producing reaction (2) will consume all dolomite and the resulting assemblage, after completion of reaction (2), is Cal + Qtz + Tr. Here, tremolite will be removed from the rocks at higher grades by reaction (15), which occurs at significantly lower temperatures than reaction (5), which removes tremolite in dolomite-excess rocks. The tremolite field for these rocks is smaller than for dolomite-rich marbles.

Diopside is usually produced by reaction (15). For the same reason as above, diopside may appear in calcsilicate marbles at lower temperature than in dolomite marbles (~650 °C). The highest grade assemblage in calcsilicate marbles is Cal + Qtz + Di (diopside field of Fig. 6.3). In contrast to dolomite-rich rocks, there is very little overlap between the tremolite and the diopside field in calcsilicate marbles.

A little complication may occur if reaction (15) consumes all quartz, leaving a rock with the assemblage Cal + Tr + Di (composition field 1c in Fig. 6.1). Tremolite is then removed from the rocks at higher grade by reaction (5). This reaction produces new metamorphic dolomite, the rock again becomes saturated with dolomite and the phase relations shown in Fig. 6.2 apply. This conversion from quartz-saturated to dolomite-saturated conditions as metamorphism progresses affects only rocks of the restricted composition range (1c in Fig. 6.1).

Wollastonite does not form in regional metamorphic rocks under closed system conditions. Even under granulite facies conditions the assemblage Cal + Qtz remains stable. Figure 6.3 shows that wollastonite may only form from reaction (16) by interaction of the marble with an H_2O-rich fluid. Such externally derived fluids may invade the marbles along fractures or shear zones.

Summary. In regional metamorphic terrains, quartz-rich marbles (Fig. 6.1) may contain small amounts of talc in addition to Dol + Cal + Qtz in the upper greenschist facies. Tremolite-bearing marbles are characteristic for the lower to middle amphibolite facies. Diopside-bearing marbles are stable from the middle amphibolite facies onwards. Wollastonite marbles are diagnostic for interaction with externally derived fluids or for specific pT path after maximum metamorphic conditions (see later).

6.4 Contact Metamorphism of Dolomites

Dolomites may also be metamorphosed by the addition of heat from a magmatic heat source at shallow levels of the crust. Most often the intrusive body is of granitic, granodioritic or quartz-dioritic composition and the temperature at the contact to the country rocks is on the order of 600 to 700 °C. The phase relationships for siliceous dolomites at 2 kbar are shown in Fig. 6.4. The figure is valid for rocks with dolomite + calcite present throughout (like the companion Fig. 6.2). It will be used to discuss progressive metamorphism of dolomites in a typical contact aureole of a granitic pluton at a depth of about 7 km (2 kbar). The size (thickness) of the distinct mineralogical zones in the aureole depends on the heat given off by the pluton which, in turn, depends on the size of the intrusion, the composition of the magma and its temperature. Typically, contact effects of shallow level intrusions can be recognized in the country rocks to distances ranging from a few hundred metres to 1 or 2 km.

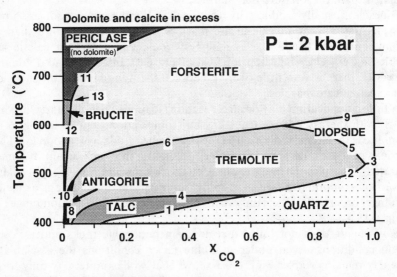

Fig. 6.4. TX phase relationships in dolomite marbles containing excess dolomite and calcite at a constant pressure of 2 kbar (contact metamorphic aureoles)

Reactions (1), (2) and (3) (Table 6.1) define the upper thermal limit for the sedimentary Dol + Cal + Qtz assemblage (as in metamorphism along orogenic geotherms). However, the invariant point Tlc + Tr + Qtz (+ Dol + Cal) is at $X_{CO_2} = 0.6$ and reaction (1) has a low dT/dX slope in Fig. 6.4. These features potentially result in the development of a recognizable zone with **talc**-bearing dolomites in the outer (cooler) parts of the aureole. Such a Tlc zone is diagnostic for temperatures below 450 °C.

A zone with **tremolite**-bearing dolomites may develop from reactions (2), (4) or the reaction at the invariant point. The low temperature limit for tremolite coincides with the upper limit for talc (~450 °C). All tremolite will subsequently be removed from dolomite marbles by reactions (5) and (6). The maximum temperature for tremolite in dolomites is given by the position of the invariant point Tr + Di + Fo (+ Dol + Cal) at about 600 °C. Reaction (6), which is metastable in Fig. 6.2, produces forsterite in tremolite-bearing marbles in low pressure aureoles.

Forsterite appears in marbles in modally recognizable amounts at about 570 to 600 °C, depending on the quartz content of the original dolomite. The assemblage Fo + Dol + Cal is characteristic for dolomites close to the contact to the intrusion. Reaction (6), if run to completion, results in the assemblage Dol + Cal + Fo in dolomite-rich rocks of composition (1a) in Fig. 6.1. The rock remains dolomite-saturated. However, reaction (6) may also consume all dolomite present in the rock and the resulting assemblage is Cal + Fo + Tr. This applies to rocks of the composition field (1b) in Fig. 6.1. The assemblage cannot be represented in Fig. 6.4 because it lacks dolomite. In this assemblage tremolite will be removed from the marble by reaction (7). The equilibrium curve for this reaction emerges from invariant point Tr + Fo + Di (+ Dol + Cal) in Fig. 6.4 and follows the tremolite-forsterite boundary inside the forsterite field at slightly higher temperatures (maximum $\Delta T \sim 15$ °C). This means that the maximum temperature for tremolite in rocks of composition (1b) roughly coincides with that of composition (1a). However, the high temperature assemblage for these compositions is Cal + Di + Fo instead of Dol + Cal + Fo. The stability field for the Cal + Di + Fo assemblage is about coincident with the field labelled *forsterite* in Fig. 6.4.

Diopside, if present in dolomite-excess rocks at all, is restricted to a narrow T interval near 600 °C under fluid-present conditions. It will form by reaction (5) or in the invariant point Qtz + Tr + Di (+ Dol + Cal). Diopside may also occur in situations where H_2O had no access to the marbles and the Dol + Cal + Qtz assemblage is replaced by Di + Dol + Cal according to reaction (3) near 520 °C. Diopside will be removed from the marbles by reaction (9) in the temperature interval 600 to 620 °C. Rocks of the composition range (1a) will, after completion of the reaction, contain Dol + Cal + Fo, rocks of the composition range (1b) the assemblage Cal + Di + Fo (see also above).

The assemblage Dol + Cal + Fo represents the high-T assemblage in most contact aureoles. In many contact aureoles periclase- or brucite-bearing marbles occur close to the contact to the intrusives or in marble inclusions in the intrusives themselves (xenoliths, roof pendants). **Periclase** forms usually from decomposition of dolomite according to reaction (11). **Brucite** may form by direct decomposition of dolomites by reaction (12) or, more commonly, by retrograde hydration of previously formed periclase by the reversed reaction (13). It is apparent from Fig. 6.4, that both periclase and brucite can only form by interaction of forsterite-marble with an externally derived H_2O-rich fluid phase. Under closed system conditions, dolomite remains stable in "normal" contact aureoles (around granitoid intrusions).

Periclase and brucite are most often found in pure dolomites (SiO_2-absent rocks) because such rocks do not have the capacity to create and control their own (relatively CO_2-rich) fluid phase. Pure dolomites are, therefore, most exposed to periclase formation by H_2O flushing from magmatic fluids released by the intrusives close to the contact.

A stability field for **antigorite** is present in Fig. 6.4 which is also not accessible for progressively metamorphosed siliceous dolomites under closed system conditions. However, antigorite may form in dolomites at peak conditions by interaction of the marble with an external fluid. Reaction (8) replaces Tr + Dol + Cal by Atg + Dol + Cal. Antigorite may be destroyed in more H_2O-rich fluids by reaction (10). Antigorite also forms by the reversed reaction (10) during retrograde replacement of the Fo + Dol + Cal assemblage.

Summary. In low pressure contact aureoles siliceous dolomites may develop the following mineralogical zonation (from low to high T, listing only the silicates): (1) dolomite-rich compositions (1a): Qtz, Tlc, Tr, (Di), Fo. (2) dolomite-poor compositions (1b): Qtz, Tlc, Tr, Tr + Fo, Fo + Di or Qtz, Tlc, Tr, Di, Fo + Di. Periclase may form in pure dolomite close to the igneous heat source. Periclase, brucite and antigorite also often form by interaction of the marble with an externally controlled H_2O-rich fluid.

6.5 Contact Metamorphism of Limestones

Rocks of the composition field (2) of Fig. 6.1 undergo progressive transformations which can be described by the phase relationships shown in Fig. 6.5 when heated in contact aureoles of granitoid intrusions. The figure is valid for rocks containing quartz and calcite at all temperatures, except where indicated. The three low-T reactions (1), (2) and (3) limit the sedimentary Qtz + Dol + Cal assemblage as in the previous cases. **Talc** is subsequently removed by reaction (14).

Tremolite has a distinctly lower maximum temperature of occurrence compared with dolomite-rich rocks. Tremolite is replaced by diopside by reaction (15), which has its temperature maximum at 540°C. Tremolite occurs in a relatively narrow T interval of about 80°C. For the sake of completeness, if reaction (15) consumes all quartz, the resulting assemblage will be Cal + Tr + Di. For this to happen, rock compositions in the narrow composition range (1c) of Fig. 6.1 are required. For a discussion of the faith

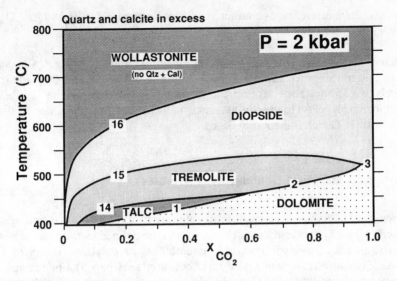

Fig. 6.5. TX phase relationships in calcsilicate marbles containing excess quartz and calcite at a constant pressure of 2 kbar (contact metamorphic aureoles)

of this assemblage as metamorphism progresses see foregoing section [reactions (7), (5), and (9)].

It follows from above that **diopside** first appears in calcsilicate marbles further away from the intrusive contact than in dolomite marbles (if it occurs at all in the latter). The assemblage Cal + Di + Qtz is typical for calcsilicate marbles of the CMS-HC system (diopside field of Fig. 6.5) near the contact to the intrusives.

Wollastonite may form in contact metamorphic calcsilicate marbles even under closed-system conditions. In siliceous limestones with the sedimentary assemblage Qtz + Cal and an H_2O-rich pore fluid, for example, wollastonite will form in small but petrographically detectable amounts at temperatures around 600 °C. In rocks which contained initially dolomite and consequently underwent a series of CO_2-producing reactions, the fluid may have the composition of the T maximum of reaction (15) when entering the diopside field under closed-system conditions. Here a temperature of about 700 °C is required for the onset of the wollastonite-producing reaction (16). Such high temperatures are rarely reached even at the immediate contact to the intrusives. Thus, wollastonite forms by interaction of the Cal + Qtz + Di assemblage with an externally controlled H_2O-rich fluid phase.

If the wollastonite reaction (16) runs to completion the composition field (2) of Fig. 6.1 is subdivided into two subfields. Rocks of the field (2a) contain Cal + Wo + Di, reaction (16) has used up all quartz. Rocks of the field (2b) contain Qtz + Wo + Di, reaction (16) has consumed all calcite. These rocks are devoid of carbonates and, consequently, cannot be called marbles any longer. At a typical temperature of about 620 °C at the immediate contact to granitoid intrusions, three assemblages characterize calcsilicate marbles: $Cal + Qtz + Di$, $Cal + Di + Wo$ and $Qtz + Di + Wo$.

Summary. Calcsilicate marbles of the composition field (2) (Fig. 6.5) may contain talc in the outer aureole, tremolite in a middle zone and diopside in relatively wide inner zone around the intrusive body. Wollastonite is often present in marbles close to the contact and usually forms from interaction of the marble with H_2O-rich (magmatic) fluids.

6.6 Isograds and Zone Boundaries in Marbles

As discussed in Chapter 3, isograds and boundaries for diagnostic assemblages in mixed volatile systems such as carbonate rocks are defined by the positions of polybaric traces of isobaric invariant assemblages or polybaric traces of T maxima of isobaric univariant equilibria (if confused see Chap. 3). For example, the invariant point involving $Qtz + Tr + Di + Dol + Cal$, which marks the first occurrence of diopside in marbles in Figs. 6.2, 6.4 (triple point of Tr, Di, Qtz fields), 6.3 and 6.5 (triple point of Tr, Di, Dol fields), changes its TX position as a function of pressure. The assemblage $Qtz + Tr + Di + Dol + Cal$ can, therefore, be represented on a pT diagram as a curve. This curve will mark the first occurrence of diopside in pT space and corresponds to a mappable diopside-in isograd in the field.

In Fig. 6.6 some important isograds and boundaries for common assemblages in marbles are summarized. The figure also shows three typical geotherms: the Ky-type geotherm used in Figs. 6.2 and 6.3, a Sil-type geotherm, a scenario for contact metamorphism and a decompression path which will be used below.

Let us return to the **diopside-in isograd** above. The isograd shown in Fig. 6.6 represents coexistence of $Dol + Cal + Tr + Qtz + Di$. The assemblage coexists with increasingly H_2O-rich fluids ($X_{CO_2} = 0.96$ at 2 kbar, $X_{CO_2} = 0.76$ at 8 kbar) as pressure increases along the curve. This is a general relationship and holds for all polybaric traces of isobaric invariant assemblages. It can be seen from Fig. 6.6, that a diopside-in isograd mapped in a contact aureole at 2 kbar marks the 500 °C isotherm, in a Sil-geotherm terrain it coincides with the 600 °C isotherm and in the Ky-type terrain the diopside-in isograd is coincident with the 670 °C isotherm (as previously shown).

The assemblage $Dol + Cal + Qtz + Tlc + Tr$ defines the **talc-out isograd** and **tremolite-in isograd**. Its temperature is about 500 °C and rather independent of the precise position of the geotherms in orogenic metamorphic terrains.

The upper limit for $Tr + Cal + Qtz$ (**Tr + Cal + Qtz-out isograd**) coincides with the diopside-in isograd at pressures above about 5 kbar. Below this pressure (marked with a filled square on Fig. 6.6) the isograd occurs at slightly higher T than the diopside-in isograd.

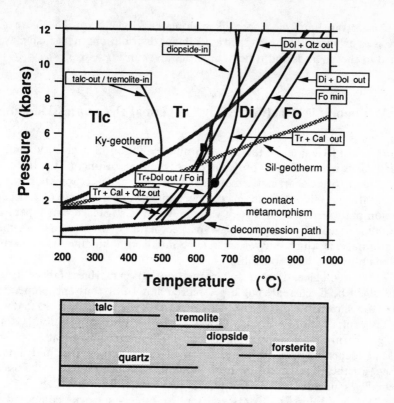

Fig. 6.6. Isograds and mineral zone boundaries in siliceous dolomites and limestones projected into the PT plane. Four geological scenarios for metamorphism are shown as orogenic gradients in kyanite terrains and sillimanite terrains, an isobaric gradient at 2 kbar for contact metamorphic terrains and a typical isothermal decompression path for returning rocks of the middle amphibolite facies on the Ky geotherm to the surface

The **Dol + Qtz-out isograd** is defined by the T maximum of reaction (3) at a given pressure (pure CO_2 fluid). It marks the maximum conditions for the sedimentary Dol + Cal + Qtz assemblage.

The **Tr + Cal-out isograd** represents the T maximum of reaction (5) at a given pressure (> 3.5 kbar). Along the curve X_{CO_2} is constant at 0.5. Below this pressure (filled circle in Fig. 6.6) the upper limit of Tr + Cal is given by the polybaric trace of the isobaric invariant assemblage Dol + Cal + Tr + Di + Fo. Along this curve X_{CO_2} increases with decreasing pressure.

The same curve also marks approximately the **Fo + Cal-in isograd**. At higher pressures, the **Fo-min** boundary is taken at the approximate minimum temperature where petrographically detectable amounts of forsterite have been produced by reaction (9) ($X_{CO_2} = 0.6$).

The **Di + Dol-out isograd** is defined by the pT position of equilibrium (9) in pure CO_2 fluids.

The distribution of silicates in marbles which contain dolomite and calcite in excess (dolomite marbles) is shown on the small figure at the bottom of Fig. 6.6.

The temperature limits have been taken along the Sil-geotherm. The figure may serve as an example of how mapped assemblage distributions in the field may be related to the isograds and zone boundaries given in Fig. 6.6.

6.7 Metamorphic Reactions Along Isothermal Decompression Paths

As outlined in Chapter 3, the metamorphic evolution of a given volume of crustal rocks involved in an orogenic process follows distinct pTt paths. Such pTt paths are often characterized by a nearly isothermal decompression (uplift) section after equilibration at maximum temperature. In Fig. 6.6, a decompression path is shown for a rock volume which has reached its temperature maximum at 650°C along the Ky-type geotherm. Below we shall examine transformations which may occur in such middle amphibolite facies marbles during uplift and decompression.

For this geological situation, phase relationships in marbles of the CMS-HC system will be discussed using a pX representation at constant temperature (650°C) as shown in Fig. 6.7. The figure is valid for all carbonate rocks that were initially composed of Dol + Cal + Qtz. Marbles metamorphosed along the Ky-type geotherm may contain a number of different assemblages at 650°C and 7.2 kbar, depending on the composition of the rock and the fluid initially present in the pore space (or introduced during metamorphism). Possible assemblages include: Dol + Cal + Qtz, Dol + Cal + Tr, Cal + Qtz + Tr, and Cal + Tr + Di (Fig. 6.7). Consider, for example, a rock composed of Dol + Cal + Qtz + Tr [reaction (2)]. The position of this rock is labelled with M

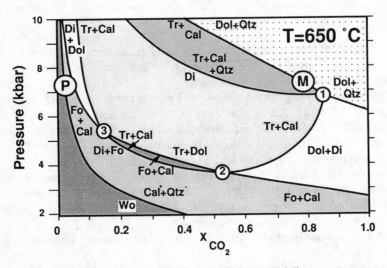

Fig. 6.7. PX phase relationships for siliceous dolomites and limestones at a constant temperature of 650°C

in Fig. 6.7. This rock will travel along a complex path through Fig. 6.7 upon decompression to a final pressure of say 2 kbar before the onset of cooling. The precise path depends on the modal abundance of Dol, Cal and Qtz in the initial rock.

Diopside will appear, however, irrespective of rock composition slightly below 7 kbar. Diopside production will take place in invariant point (1) or by reaction (15). Tremolite will be removed by reaction (5). Ultimately, forsterite may appear in the marbles either in invariant point (2) or from reaction (9) at pressures below 4 kbar.

In Cal + Qtz marbles (e.g. at point labelled P in Fig. 6.7) wollastonite may be produced by reaction (16) in detectable amounts at pressures below about 4 kbar. The effect may further be enhanced by introducing H_2O-rich fluids during decompression. Wollastonite observed in granulite facies marbles can usually be related to rapid uplift along a path of nearly isothermal decompression together with interaction of the marbles with late H_2O-rich fluids.

Summary. Rapid uplift along isothermal decompression paths may drastically modify marble assemblages that originally equilibrated on a given orogenic geotherm. Assemblages produced by reactions during isothermal decompression may be difficult to distinguish from assemblages produced on the orogenic geotherm (maximum pT assemblages). The safest way to detect the presence of decompression assemblages in marbles is by an integral evaluation of the metamorphic evolution of the entire terrain (e.g. comparison with thermobarometry data from metapelites).

6.8 Marbles Beyond the CMS-HC System

In the following section the consequences of additional components in marbles will be discussed.

6.8.1 Fluorine

As mentioned earlier, marbles are often extremely poor in iron. Consequently, hydrous Mg silicates such as Tlc and Tr are usually very close to their end-member compositions. However, metamorphic fluids often have a small partial pressure of hydrofluoric acid (HF). HF is strongly partitioned into solid hydrate phases. In addition, X_{Fe} and X_F in hydrate silicates (e.g. talc and tremolite) are negatively correlated, which means that at a given p_{HF} the replacement of hydroxyl groups by fluorine rapidly decreases with increasing X_{Fe}. The consequence of the $F(OH)_{-1}$ exchange in talc and tremolite is a general increase of their respective stability fields on the phase diagrams presented above. The net effect on the talc-tremolite boundary is an increase of the talc field.

In forsterite marbles a finite HF pressure may lead to the formation of minerals of the humite group. The two types of humite group minerals occasionally found in forsterite grade marbles are clinohumite (low p_{HF}) and chondrodite (high p_{HF}). The humites are not stable in pure CO_2-H_2O fluids. The stability fields for the two minerals roughly coincides with the forsterite fields on the pTX diagrams above. However, their occurrence is restricted to the H_2O-rich side of the phase diagrams. The precise location of the phase boundaries involving Chu or Chn depends on the prevailing HF pressure during metamorphism. Typical assemblages in dolomite marbles are: Dol + Cal + Fo + Chu (clinohumite marbles) or Dol + Cal + Chu + Chn (chondrodite marbles, humite marbles). Chn and Fo do not occur together. Chu and Chn may also be stabilized by the presence of titanium, but both minerals contain always some fluorine. It is strongly advised to analyze minerals for fluorine when studying marble samples by electron microprobe.

6.8.2 Aluminium

Dolomites often contain various amounts of Mg-chlorite (clinochlore). The mineral is stable at all metamorphic grades except in the granulite facies and in low pressure contact aureoles where it may decompose according to the two spinel-producing reactions:

$$2\,Dol + Chl \Rightarrow 2\,Cal + 3\,Fo + Spl + 2\,CO_2 + 4\,H_2O \quad \text{or} \tag{17}$$

$$3\,Chl + 2\,Cal \Rightarrow 2\,Di + 5\,Fo + 3\,Spl + 2\,CO_2 + 12\,H_2O. \tag{18}$$

At 2 kbar, reaction (17) has its T maximum at 610°C which marks the chlorite-out isograd in rocks containing Dol + Cal + Fo + Chl. Reaction (18) has its maximum temperature at 640°C. The temperature corresponds to the chlorite-out isograd in rocks containing Cal + Di + Fo + Chl.

The maximum temperature for the two reactions increases with increasing pressure, which means that chlorite remains stable in rocks metamorphosed along orogenic geotherms up to granulite facies conditions.

In contact aureoles the mineral clintonite (Cli), a trioctahedral brittle mica, occasionally occurs in dolomite marbles of the forsterite zone. Typical assemblages are (for example): Cal + Fo + Di + Spl + Cli and Cal + Fo + Chu + Spl + Cli.

The presence of aluminium in calcsilicate marbles may lead to the appearance of a number of additional minerals such as margarite, zoisite, grossular, scapolite and anorthite. The phase relationships for such rocks will be discussed in Chapter 8 on metamorphism of marls.

6.8.3 Potassium

Potassium is usually stored in sedimentary dolomites in the form of autigenic K-feldspar. The mineral is the first mineral of the sedimentary assemblage removed from marbles undergoing progressive metamorphism. The relevant reaction consumes K-feldspar and replaces it by phlogopite (biotite).

$$3\,Dol + Kfs + H_2O \Rightarrow Phl + 3\,Cal + 3\,CO_2. \tag{19}$$

For natural systems, equilibrium for this reaction runs approximately parallel to and a few degrees below reactions (1) and (2) on the phase diagrams above. After completion of reaction (19), dolomites contain the assemblage $Dol + Cal + Qtz + Phl$. The assemblage undergoes subsequently reactions (1) or (2) as discussed above. Reaction (19) increases X_{CO_2} in the fluid and may run to completion at X_{CO_2} greater than that of the invariant point $Tlc + Tr + Qtz + Dol + Cal$. Thus, rocks undergoing reaction (19) are often guided around the talc field and talc will not form in phlogopite-bearing marbles. Such rocks contain the following assemblages (with increasing grade): $Dol + Cal + Qtz + Kfs$, $Dol + Cal + Qtz + Phl$, $Dol + Cal + Tr + Phl$ (phlogopite marbles). Phlogopite remains stable in rocks containing excess Dol and Cal to very high grades. For example, the assemblage $Dol + Cal + Fo + Chu$ (fluorine) $+ Spl$ (aluminium) $+ Phl$ (potassium) often occurs in contact metamorphic marbles near the contact to the intrusive body. Note that also phlogopite is significantly affected by the $F(OH)_{-1}$ exchange.

The phase relationships involving muscovite, phlogopite and K-feldspar in calcsilicate rocks will be discussed in Chapter 8 on metamorphism of marls.

6.8.4 Sodium

Sodium may be present as NaCl in the metamorphic fluid (albite in low grade dolomites is rare). Its presence complicates the phase relationships in Al-bearing dolomite marbles because it continuously modifies the composition of tremolite with increasing temperature. The compositional changes of calcic amphibole are closely described by the tremolite-pargasite exchange reaction (# Mg 2Si = Na 3Al; # denotes a vacancy). The exchange reaction produces pargasitic amphiboles from tremolite at high temperature. Consequently, the upper limit of amphibole in marbles is displaced to higher temperatures compared to the pure CMS-HC system.

6.9 Thermobarometry in Marbles

There are only two useful fluid-independent equilibria available for geological pT estimates.

6.9.1 Calcite-Aragonite Phase Transition

$CaCO_3$ undergoes a phase transition from trigonal calcite to orthorhombic aragonite at high pressures. At 400°C the transition pressure is close to 10 kbar, at 800°C aragonite is stable at pressures greater than 15 kbar. However, aragonite is easily converted to calcite during decompression of carbonate-bearing high-pressure rocks. Occasionally, the mineral is preserved in rocks that were metamorphosed along a subduction-type geotherm and rapidly returned to the surface without being substantially heated. The presence of aragonite in metamorphic rocks indicates minimum pressures in excess of 10 kbar.

6.9.2 Calcite-Dolomite Miscibility Gap

In rocks containing excess dolomite, the solubility of $MgCO_3$ in calcite is a function of the temperature and to a lesser degree also of the pressure. The temperature dependence of the $MgCa_{-1}$ exchange in calcite (and because of the asymmetry of the miscibility gap with some restrictions also the $CaMg_{-1}$ exchange in dolomite) represents a widely used geological thermometer. A large number of calibrations exist in the geological literature (e.g. Bickle and Powell 1977; Powell et al. 1984).

References and Further Reading

Anovitz LM, Essene EJ (1987) Phase equilibria in the system $CaCO_3$– $MgCO_3$– $FeCO_3$. J Petrol 28:389–414

Bickle MJ, Powell R (1977) Calcite-dolomite geothermometry for iron-bearing carbonates. The Glockner area of the Tauern window, Austria. Contrib Mineral Petrol 59:281–292

Bowen NL (1940) Progressive metamorphism of siliceous limestones and dolomite. J Geol 48:225–274

Bucher-Nurminen K (1982) On the mechanism of contact aureole formation in dolomitic country rock by the Adamello intrusion (N Italy). Am Mineral 67:1101–1117

Castelli D (1991) Eclogitic metamorphism in carbonate rocks: the example of impure marbles from the Sesia-Lanzo zone, Italian Western Alps. J Metamorph Geol 9:61–78

Ferry JM (1976) P, T, f_{CO_2} and f_{H_2O} during metamorphism of calcareous sediments in the Waterville-Vassalboro area, south-central Maine. Contrib Mineral Petrol 57:119–143

Flowers GC, Helgeson HC (1983) Equilibrium and mass transfer during progressive metamorphism of siliceous dolomites. Am J Sci 283:230–286

Greenwood HJ (1962) Metamorphic reactions involving two volatile components. Carnegie Inst Wash Year Book Annu Rep Direc Geophys Lab 61:82–85

Misch PM (1964) Stable association wollastonite-anorthite and other calc-silicate assemblages in amphibolite-facies crystalline schists of Nanga Parbat, northwest Himalayas. Beitr Mineral Petrogr (later Contrib Mineral Petrol 10:315–356

Powell R, Condliffe DM, Condliffe E (1984) Calcite-dolomite geothermometry in the system $CaCO_3$-$MgCO_3$-$FeCO_3$: an experimental study. J Metamorph Geol 2:33–41

Rice JM (1977a) Contact metamorphism of impure dolomitic limestone in the Boulder Aureole, Montana. Contrib Mineral Petrol 59:237–259

Rice JM (1977b) Progressive metamorphism of impure dolomitic limestone in the Marysville aureole, Montana. Am J Sci 277:1–24

Skippen GB (1971) Experimental data for reactions in siliceous marbles. J Geol 70:451–481

Skippen GB (1974) An experimental model for low pressure metamorphism of siliceous dolomitic marbles. Am J Sci 274:487–509

Skippen GB, Trommsdorff V (1975) Invariant phase relations among minerals on T-X_{fluid} sections. Am J Sci 275:561–572

Skippen G, Trommsdorff V (1986) The influence of NaCl and KCl on phase relations in metamorphosed carbonate rocks. Am J Sci 286:81–104

Slaughter J, Kerrick DM, Wall VJ (1975) Experimental and thermodynamic study of equilibria in the system CaO-MgO-SiO_2-H_2O-CO_2. Am J Sci 275:143–162

Tilley CE (1951) A note on the progressive metamorphism of siliceous limestones and dolomites. Geol Mag 88:175–178

Tracy RJ, Hewitt DA, Schiffries CM (1983) Petrologic and stable-isotopic studies of fluid-rock interactions south-central Connecticut. I. The role of infiltrations in producing reaction assemblages in impure marbles. Am J Sci 283A:589–616

Trommsdorff V (1966) Progressive Metamorphose kieseliger Karbonatgesteine in den Zentralalpen zwischen Bernina und Simplon. Schweiz Mineral Petrogr Mitt 46:431–460

Trommsdorff V (1972) Change in T-X during metamorphism of siliceous dolomitic rocks of the central Alps. Schweiz Mineral Petrogr Mitt 52:567–571

Valley JW, Peterson EU, Essene EJ, Bowman JR (1982) Fluorphlogopite and fluortremolite in Adirondack marbles and calculated C-O-H-F fluid compositions. Am Mineral 67:545–557

7 Metamorphism of Pelitic Rocks (Metapelites)

7.1 Metapelitic Rocks

Metapelites are probably the most distinguished family of metamorphic rocks. Typical metapelites include well-known metamorphic rocks such as, for example: chlorite-kyanite-schists, staurolite-garnet micaschists, chloritoid-garnet micaschists, kyanite-staurolite schists, biotite-garnet-cordierite gneisses, sillimanite-biotite gneisses and orthopyroxene-garnet granulites. Many distinct metamorphic minerals are found in metapelitic rocks (e.g. staurolite, chloritoid, kyanite, andalusite, sillimanite and cordierite).

In many metamorphic terrains, characteristic minerals in metapelites show a regular spatial distribution that can be readily related to the intensity of metamorphism. The distribution pattern of metamorphic minerals in metapelites reflects the general metamorphic style and structure of almost all metamorphic terrains in any orogenic belt. Mineral zones and phase relationships in metapelites often permit a fine-scale analysis of the intensity and nature of metamorphism in a given area. A number of well-established and calibrated geological thermometers and barometers can be applied to mineral assemblages found in metapelitic rocks.

7.2 Pelitic Sediments

7.2.1 General

Metapelites are metamorphic rocks derived from clay-rich sediments including unconsolidated sediments such as mud and clay, and consolidated sediments, e.g. shales, claystones and mudstones. After incipient metamorphism, all these sedimentary rocks are collectively termed argillite. Weak metamorphism transforms these rocks into slates. In general, pelitic sediments are characterized by very small grain size (often $<2\,\mu m$) and by a mineralogy which is dominated by clay minerals.

The term pelitic rock is used in metamorphic geology in a general practice to designate fine-grained clay-rich sediments. It includes sediments that fall under the general terms mudstone and shale used by sedimentary geologists. Clay-rich pelites often develop a sequence of characteristic minerals and mineral assemblages during prograde metamorphism. Mudstones with a high proportion of silt (siltstones) and less abundant clay transform to metamorphic equivalents with less interesting and diagnostic mineral assemblages. They are often referred to as semi-pelites in the metamorphic literature (e.g. semi-pelitic gneisses). Shales represent > 80% of all sedimentary rocks (Table 2.2).

7.2.2 The Chemical Composition

Compositions of two types of pelites are listed in Table 2.3. The characteristic compositional features of pelitic rocks are best represented by the analysis of typical pelagic clay. Aluminium is very high, total iron is up to 10 wt%, there is a fair amount of magnesium, however, CaO is extremely low. The water content of pelites is high (5 mol H_2O per kg rock in the case of the pelagic clay in Table 2.3). From the viewpoint of a metamorphic geologist this is a positive aspect because it can be expected that H_2O released during metamorphism helps to maintain the rocks in chemical equilibrium. Prograde metamorphism of pelitic sediments starts with rocks at a maximum hydrated state. This is often not the case in other rock compositions (e.g. ultramafic and mafic rocks). Shales deposited on platforms often deviate from pure clay compositions as a result of higher silt fractions (higher SiO_2, more quartz and feldspar) or the presence of carbonate minerals (higher CaO and CO_2). All transitions between pure clay compositions and arkoses or marls exist. This Chapter deals with carbonate-free, CaO-pure claystones and shales (such as pelagic clay in Table 2.3).

7.2.3 Mineralogy

The mineralogy of clays and shales is dominated, as expected, by clay minerals. Clay minerals are aluminous sheet silicates of variable compositions (important representatives are; montmorillonite, smectite, kaolinite). Sericite and paragonite of detrital or autigenic origin are the next important group of minerals in shales. Abundant chlorite carries much of the Mg and Fe found in shales. All sheet silicates together often represent more than 50 vol% of the rock. Quartz occurs in various modal proportions (typically 10–30 vol%). As a consequence, most of the metamorphic equivalents of shales contain free quartz despite the fact that many prograde mineral reactions in pelitic rocks consume quartz. Most shales also contain detrital and/or autigenic feldspars and sporadically also a substantial amount of zeolites. An additional complication arises from high modal proportions of Fe-oxides-hydroxides (hematite-limonite-goethite) or sulfides in shales. This has the consequence that Redox-reactions and sulfur-involving reactions can be important in metapelitic rocks. Titanite or rutile are minor minerals in shales. Carbonates (absent in pure clays) and organic carbonaceous matter (oil shale) may complicate the picture.

7.3 Pre-Metamorphic Changes in Pelitic Sediments

During compaction and diagenesis, primary clays and shales undergo significant chemical and mechanical changes. The very large (e.g. > 50%) porosity of clays is continuously reduced during burial and compaction. Typical shales may still contain several % pore space filled with formation water when metamorphism starts (at approximately 200°C and say about 6 km depth = 1.6 kbar). The original clay minerals, such as smectite, are replaced by illite (a precursor mineral of the white K-micas) and chlorite. The lattice ordering of the sheet silicates, particularly of illite, progressively increases with increasing temperature and pressure. Illite "crystallinity" can in fact be used as a measure of the degree of diagenetic and very low-grade metamorphic recrystallization. Carbonaceous matter undergoes a series of reactions which ultimately destroy the organic compounds. The organic carbon compounds are replaced finally by graphite or are completely transferred to the vapour phase as CO_2 or CH_4 under oxidizing and reducing conditions respectively. The optical reflectivity of the carbonaceous matter is also used as a sensitive indicator of the thermal regime during diagenesis and incipient metamorphism. Compaction and recrystallization during burial and diagenesis also produce a distinct fissility and cleavage (slaty cleavage) resulting from a pressure-induced parallel orientation of the sheet silicate minerals.

Consequently, shales at the beginning of metamorphism have been transformed to slates and phyllites. The mineralogy most typically includes: illite (muscovite), chlorite, quartz and feldspar (K-feldspar and albite), sulfide and organic material or hematite. The chemical composition of the sediments remains more or less preserved during pre-metamorphic processes with the exception of H_2O loss.

Below, we shall explore what happens to such slates and phyllites if they experience an increase in temperature and pressure during a tectono-metamorphic event such as a typical orogenic cycle.

7.4 Orogenic Intermediate-Pressure Metamorphism of Pelitic Rocks (± Ky Geotherm)

Metapelites formed by this type of metamorphism include the classic metamorphic rocks mentioned above (Sect. 7.1). The distinct zonal pattern of mineral assemblages in metapelitic rocks that is produced by intermediate-pressure orogenic metamorphism has been found and extensively studied in a great number of metamorphic terrains from many orogenic belts (fold belts, mountain belts, mobile belts on Precambrian shields) ranging in age from the Precambrian to the Tertiary. The type of metamorphism is also referred to as Barrovian metamorphism after its first description from the Scottish Highlands by G. Barrow. It corresponds roughly to a prograde metamorphic path along a kyanite-type geotherm of the phase diagrams shown in this book.

7.4.1 Chemical Composition and Chemographies

Phase relationships in metapelites can be discussed by means of selected simple chemical subsystems. By reference to the composition of pelagic clays (Table 2.3) the components occur in the following sequence of abundance: SiO_2, Al_2O_3, $FeO-Fe_2O_3$, H_2O, MgO, K_2O, Na_2O, TiO_2, CaO. It is convenient to separate this complex ten-component system into a series of subsystems. Metapelites are often approximated by a simple six-component system including the six most abundant components above and representing total iron as FeO. This system can be graphically represented in AFM diagrams as outlined in Chapter 2. Na_2O is mainly stored in feldspar (albite) and paragonite. Reactions involving Na-bearing minerals will be discussed separately below. CaO is typically low in pure clays. The component is stored in metapelitic rocks in garnet (grossular component) and in plagioclase (anorthite component). Complications arising from CaO and other minor components in metapelites will be treated in Section 7.9. Metamorphism of calcareous clay (marls) such as the platform shale listed in Table 2.3 is outlined in Chapter 8.

7.4.2 Mineral Assemblages at the Beginning of Metamorphism

The AFM chemography of shales at the onset of metamorphism is represented in Fig. 7.1. The requirements given by the AFM projection are met by many shales. As outlined above most shales contain modally abundant quartz and illite (white mica, muscovite, sericite) and excess H_2O. Because pelitic sediments are at a maximum hydrated state (at least when metamorphosed for the first time!), they will experience a series of dehydration reactions when heat is added to the rocks during prograde metamorphism. Thus, during much of the prograde metamorphic evolution of metapelites the rocks may remain water-saturated.

The composition of typical shales falls into the dark shaded area of Fig. 7.1. The recalculated composition of the pelagic clay (Table 2.3) has the following AFM coordinates: A = 0.41; F = 0.62. It is represented by point P in Fig. 7.1. The composition P can be regarded as reference "normal" pelite composition. Such a shale contains kaolinite/pyrophyllite + chlorite in addition to quartz, illite and water at the beginning of metamorphism. The TS component in chlorite is buffered by the assemblage, whereas the FM component is constrained by the bulk composition. This follows from the fact that the point P is in the univariant field kaolinite + chlorite. Shale compositions falling into the divariant chlorite field contain chlorite as the only AFM phase. The composition of the chlorite is given by the bulk composition and can vary within the black field representing the range of compositional variation observed in low temperature chlorites (chlorite solid solution). Shales with A values below about 0.17 (clinochlore composition) contain the univariant assemblage chlorite + feldspar. Such rocks will not develop the characteristic and diagnostic mineral assemblages that one usually associates with metapelites unless the

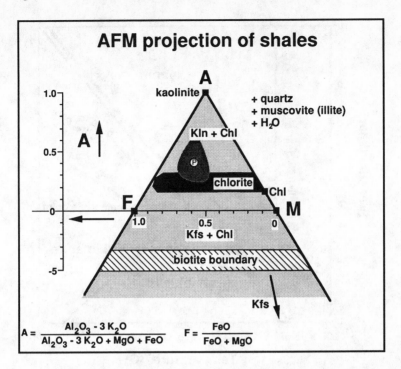

Fig. 7.1. AFM diagram for low-grade shales. Pelagic clays and mudstones fall into the shaded area and the average pelagic clay (Table 2.3) projects to point P. AFM coordinates of minerals and rocks can be calculated from the equations given at the bottom of the figure (molar basis)

bulk composition is much more magnesian than for the compositions of the shaded area. If the A coordinate is below –0.5, diagnostic assemblages will not develop in such rocks irrespective of their F coordinate. This compositional restriction may be designated biotite boundary. Biotite-gneisses and schists, and garnet-biotite gneisses are typical metamorphic products of low-Al shales/siltstones. They are often referred to as semi-pelitic gneisses and shales.

7.4.3 Phase Relationships in the ASH System

The two most abundant components (SiO_2 and Al_2O_3), together with H_2O, permit the representation of phase relationships among kaolinite, pyrophyllite, the three aluminosilicate polymorphs (kyanite, sillimanite, andalusite) and quartz (and its polymorphs). The ASH system is a subsystem of the KFMASH system portrayed by AFM diagrams. It describes all phase relations and reactions in the A apex of an AFM diagram. Reaction equilibria in the ASH system are shown in Fig. 7.2 and listed in Table 7.1. Only reactions affecting bulk compositions with excess quartz and H_2O (as for AFM diagrams) are shown.

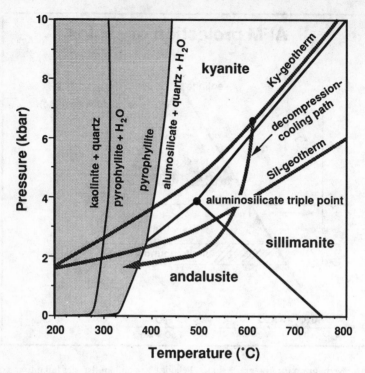

Fig. 7.2. Equilibria in the ASH system calculated from the RB88 database

Table 7.1. Reactions in the ASH system

Sillimanite, kyanite, andalusite	Al_2SiO_5
Quartz	SiO_2
Pyrophyllite	$Al_2Si_4O_{10}(OH)_2$
Kaolinite	$Al_2Si_2O_5(OH)_4$

(1)   And = Ky
(2)   And = Sil
(3)   Sil = Ky
(4)   Kln + 2 Qtz = Prl + H_2O
(5)   Prl = Ky + 3 Qtz + H_2O
(6)   Prl = And + 3 Qtz + H_2O

Prograde metamorphism, for example along the Ky-geotherm, will replace kaolinite by pyrophyllite at about 300°C. The stable mineral on the A apex of an AFM diagram becomes pyrophyllite. At about 400°C in pure H_2O fluid, pyrophyllite reaches its upper thermal stability and decomposes to aluminosilicate + quartz (kyanite along the Ky-geotherm, andalusite along low pressure geotherms). At higher temperatures, the stable divariant assemblage in the ASH system is quartz + H_2O + aluminosilicate. The nature of the stable aluminosilicate polymorph depends on pressure and temperature (along the Ky-geotherm: $Qtz + Ky + H_2O$). The stable mineral on the A apex of an AFM diagram is kyanite, sillimanite or andalusite, depending on the pT conditions selected for the AFM diagram. Andalusite is not stable at pressures greater than 4 kbar, sillimanite is not stable below about 500 °C. Aluminosilicates are not stable below about 350–400 °C in the presence of water and quartz [reactions (5) and (6), Table 7.1]. Along a sillimanite-type geotherm, pyrophyllite decomposes to andalusite + quartz at about 380 °C. Andalusite is replaced by sillimanite at about 530 °C.

Note, however, that in rocks containing abundant organic material (a situation that is not uncommon in pelites) the fluid phase may contain much CH_4 at low temperature. Under this usual condition, the equilibrium of any dehydration reaction is displaced to lower temperatures compared with the pure H_2O fluid case shown in Fig. 7.2. Therefore, pyrophyllite usually forms at much lower temperature than shown in Fig. 7.2. Pyrophyllite is common at temperatures below 200 °C, whereas kaolinite is extremely rare in metapelites (> 200 °C).

The polymorphic transitions [reactions (1), (2) and (3)] require a complete reconstruction of the aluminosilicate structure. Because these "solid-solid" reactions involve only small changes in free energy, metastable persistence of one polymorph in the field of another more stable polymorph is very common. For example, prograde metamorphism along the Ky-geotherm to conditions near 600 °C and 6.5 kbar produces a rock with the assemblage kyanite + quartz (+ H_2O). From these pT conditions the rock follows a typical decompression and cooling path as shown in Fig. 7.2. Excess water conditions cannot be taken for granted along this path (retrograde metamorphism). However, reactions (1), (2) and (3) are water-absent reactions. At about 6 kbar, coarse-grained kyanite that has formed along the prograde path becomes less stable than sillimanite. Under equilibrium conditions all kyanite should now be replaced by sillimanite. This is often not the case in real rocks. Fine sillimanite needles may grow on kyanite or precipitate from aluminosilicate-saturated fluids in fractures and nodules. Most of the originally formed kyanite remains in the rock. At about 550 °C and 3 kbar along the decompression and cooling path, both kyanite and sillimanite become less stable than andalusite. However, in most rocks, no spontaneous formation of andalusite from kyanite or sillimanite occurs in the matrix of the rock. Andalusite, like sillimanite, may form in nodules, veins and fissures by direct precipitation from a fluid phase. Thus, all three aluminosilicates (kyanite, sillimanite and andalusite) can often be observed at a given locality or even in hand-specimen or thin-section scale. Obviously, this does not necessarily mean that the rocks have been metamorphosed at or near the aluminosilicate triple point. The three aluminosilicates often found at one locality are normally not isochronous and formed each at distinct sections along a metamorphic pTt-path.

The observation of chronic metastable persistence of one polymorph mineral in the stability field of another polymorph mineral suggests that water, although not a species in the stoichiometric reaction equation, needs to be present in the rock as a solvent and catalyst. If water is present along the cooling and decompression path, metastable kyanite (for example) can be dissolved and stable sillimanite may precipitate. This is further supported by the frequent observation of sillimanite or andalusite in veins, fractures and nodules in otherwise massive kyanite-bearing rocks, that is in structures that are often clearly late in the metamorphic history.

7.4.4 Metamorphism in the FASH System

Addition of FeO to the ASH system introduced above permits discussion of phase relationships among a number of further minerals: Fe-chlorite (daphnite, amesite), chloritoid, staurolite, almandine garnet, Fe-cordierite, hercynite-spinel. Additional minerals in the FASH system include Fe-orthoambiboles (Oam) and Fe-orthopyroxene (Opx). Phase relationships among Oam- and Opx-involving assemblages will be presented in Sections 7.5. and 7.6.

Some important equilibria in the FASH system are shown in Fig. 7.3. The phase relationships on this figure can be directly applied to Fe-rich pelites. A comparison of Fig. 7.3 with Fig. 7.2 shows that equilibria in the ASH system are at slightly different pT locations. This is because the HP90 database has been used to calculate Fig. 7.3, whereas RB88 was used for Fig. 7.2 (see Sect. 3.8.1.). Note especially the different coordinates of the aluminosilicate triple point (500 °C, 4 kbar: Fig. 7.2; 580 °C, 5 kbar: Fig. 7.3). The triple point version of Fig. 7.3, will be used throughout this chapter.

The inset in the upper left corner of Fig. 7.3 shows on top the chemography of the FASH system projected from quartz and H_2O onto the FeO-Al_2O_3 line. The stable phases in the ASH system, pyrophyllite and aluminosilicate project onto the right hand side of the chemography. Fe-chlorite has the lowest $X_{Al_2O_3}$ of all minerals of interest here. This means that the decomposition of chlorite at higher temperature produces an iron oxide in addition to a product silicate (the * on Chl indicates that this position in the binary chemography will be occupied by an iron oxide at higher temperature). However, stable Fe-oxides in meta-pelitic rocks are either hematite or/and magnetite that both contain trivalent iron. Consequently, terminal chlorite decomposition depends on the oxygen activity in the rocks. Two important chlorite-involving REDOX reactions are shown in Fig. 7.3 and they will be explained below. The stoichiometric equations of all reaction equilibria shown in Fig. 7.3 are listed in Tables 7.2 and 7.1.

Let us now consider prograde metamorphism along the Ky-geotherm of a pelitic rock containing the assemblage chlorite + pyrophyllite in addition to excess quartz and H_2O.

The chlorite-pyrophyllite pair is replaced at about 220 °C by chloritoid by reaction (7). Fe-rich metapelites now contain the assemblages chlorite-chloritoid or chloritoid-pyrophyllite depending on $X_{Al_2O_3}$ of the bulk rock. In "normal" metapelites chloritoid will usually appear at higher temperatures (see below)

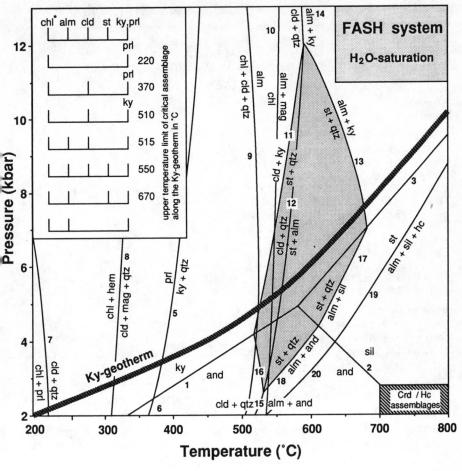

Fig. 7.3. Equilibria in the FASH-system. *Reaction numbers* refer to Tables 7.1 and 7.2. *Shaded field* Coexistence of staurolite + quartz. Cordierite- and hercynite-involving reactions are shown on later figures

At about 300 °C, chlorite reacts with hematite, the later is often present in low-grade slates and phyllites, and produces magnetite and more chloritoid [reaction (8)]. The oxygen activity is controlled by the hematite-magnetite pair through the reaction: $6\ Fe_2O_3 = 4\ Fe_3O_4 + O_2$. This is known as the hematite-magnetite oxygen buffer (Sect. 3.6.2.4). It defines relatively high oxygen activities at a given temperature. Since chlorite is modally more abundant than Fe-oxides in low-grade phyllites, all hematite will normally be consumed by reaction (8). The assemblage chlorite-chloritoid-magnetite is now stable above about 300 °C.

At 370 °C, pyrophyllite is replaced by kyanite and chloritoid-kyanite becomes the diagnostic assemblage in Al-rich compositions [reaction (5)].

Table 7.2. Reactions in the FASH system (in addition to reactions in Table 7.1)

Chlorite (daphnite)	$Fe_5Al_2Si_3O_{10}(OH)_8$
Chloritoid	$FeAl_2SiO_6(OH)_2$
Staurolite	$Fe_4Al_{18}Si_{7.5}O_{44}(OH)_4$
Almandine	$Fe_3Al_2Si_3O_{12}$
Hercynite	$FeAl_2O_4$
Magnetite	$FeFe_2O_4$
Hematite	Fe_2O_3

Reactions with excess quartz + H_2O

(7)	Chl + 4 Prl = 5 Cld + 2 Qtz + 3 H_2O
(8)	Chl + 4 Hem = Cld + 4 Mag + 2 Qtz + 3 H_2O
(9)	Chl + Cld + 2 Qtz = 2 Alm + 5 H_2O
(10)	3 Chl = 3 Alm + 2 Mag + 12 H_2O (+ QFM)
(11)	8 Cld + 10 Ky = 2 St + 3 Qtz + 4 H_2O
(12)	23 Cld + 7 Qtz = 2 St + 5 Alm + 19 H_2O
(13)	75 St + 312 Qtz = 100 Alm + 575 Ky + 150 H_2O
(14)	3 Cld + 2 Qtz = Alm + 2 Ky + 3 H_2O
(15)	3 Cld + 2 Qtz = Alm + 2 And + 3 H_2O
(16)	8 Cld + 10 And = 2 St + 3 Qtz + 4 H_2O
(17)	75 St + 312 Qtz = 100 Alm + 575 Sil + 150 H_2O
(18)	75 St + 312 Qtz = 100 Alm + 575 And + 150 H_2O

Decomposition of staurolite in qtz–free rocks

(19)	2 St = Alm + 12 Sil + 5 Hc + 4 H_2O
(20)	2 St = Alm + 12 And + 5 Hc + 4 H_2O

At about 510 °C, almandine garnet is formed by chlorite-chloritoid decomposition [reaction (9)]. The two new diagnostic assemblages are chlorite-almandine and almandine-chloritoid. At roughly the same temperature, chloritoid and kyanite react according to reaction (11) and form staurolite. The first occurrence of staurolite and almandine-garnet marks the transition from the greenschist facies to the amphibolite facies in pelitic rocks. The temperature at the beginning of the amphibolite facies is slightly above 500 °C. Chloritoid-staurolite and staurolite-kyanite are the diagnostic assemblages in the lower amphibolite facies.

Chlorite decomposes in quartz-saturated rocks at about 540 °C by reaction (10) under production of almandine-garnet and magnetite. The oxygen activity along reaction (10) in Fig. 7.3 is controlled by the QFM buffer. It is implicitly assumed that oxygen activities in metapelites at the beginning of the amphibolite facies are close to those defined by the QFM assemblage. However, this appears to be reasonable. The left hand side of the chemographies on the inset of Fig. 7.3 is now represented by magnetite. The stable assemblages at this stage depending on the $X_{Al_2O_3}$ of the bulk rock are: Mag + Alm, Alm + Cld, Cld + St and St + Ky.

At about 550 °C, chloritoid reaches its upper thermal stability in quartz-saturated rocks and decomposes to almandine-garnet and staurolite [reaction (12)]. This means that the stability field of chloritoid, which was the most characteristic mineral in the greenschist facies, extends into the lower amphibolite facies. The disappearance of chloritoid-bearing assemblages marks the upper boundary of the lower amphibolite facies. The greater part of the amphibolite facies is characterized by the assemblages St + Ky, St + Alm and Alm + Mag.

Finally, at 670°C staurolite decomposes in quartz-bearing rocks and the characteristic high-grade assemblage is almandine-garnet + kyanite [reaction (13)]. The pT field with stable staurolite + quartz is shaded in Fig. 7.3. It follows from Fig. 7.3 that metamorphism along prograde paths at very low and very high pressures will bypass the staurolite + quartz field. Chloritoid is replaced directly by almandine-garnet + andalusite at pressures below 2500 bars [reaction (15)] and by almandine-garnet + kyanite at pressures greater than 12 kbar [reaction (14)]. Staurolite in quartz-free rocks survives to much higher temperatures than in quartz-bearing rocks. Reactions (19) and (20) describe staurolite decomposition in quartz-free rocks. The product assemblage is almandine-garnet + aluminosilicate + hercynite-spinel. In high-temperature contact metamorphism a number of Fe-cordierite- and hercynite-involving assemblages may appear in Fe-rich metapelites. The pT area of cordierite and hercynite reactions that are stable in the pure FASH system is shown as a *box* in Fig. 7.3. Cordierite-spinel assemblages will be discussed separately in Section 7.5.

The sequence of diagnostic assemblages and their pT limits during metamorphism along other prograde pT paths can be taken from Fig. 7.3 directly. For example, the temperature interval with stable staurolite + quartz is much smaller along a low-pressure metamorphic pT path (e.g. Sil-geotherm of Fig. 7.2) than along the Ky-geotherm of Fig. 7.3. In fact, extensive zones with staurolite-bearing metapelites are most characteristic for typical Barrovian-style (Ky-geotherm) regional metamorphic terrains, in agreement with Fig. 7.3. The sequence of characteristic assemblages along the Ky-geotherm is shown, together with their upper temperature limit, in the inset of Fig. 7.3. In Al-rich compositions, the prograde sequence of two-mineral assemblages is: Chl + Prl, Cld + Prl, Cld + Ky, St + Ky and Alm + Ky. These diagnostic assemblages can also be used to define characteristic metamorphic zones in a Barrovian-style regional metamorphic terrain.

7.4.5 Mica-Involving Reactions

Micas are a very important group of minerals in metapelitic rocks. Typical metapelitic rocks are, for this reason, micaschists and micagneisses. The principal micas in metapelites are muscovite, paragonite and biotite. The minerals muscovite and paragonite are commonly designated white micas (dioctahedral micas). Their distinction is difficult with microscopic techniques (use X-ray patterns). All micas show at least compositional variations along the $FeMg_{-1}$ (FM) and $MgSiAl_{-1}Al_{-1}$ (TS) exchange vectors. Sedimentary and diagenetic illite rapidly recrystallizes to K-white mica with various amounts of

Table 7.3. Additional important reactions in the KNFASH system (in addition to reactions in Table 7.1 and 7.2)

Annite (biotite)	$KFe_3AlSi_3O_{10}(OH)_2$
Muscovite	$KAl_3Si_3O_{10}(OH)_2$
Paragonite	$NaAl_3Si_3O_{10}(OH)_2$
K-feldspar	$KAlSi_3O_8$
Albite	$NaAlSi_3O_8$
Jadeite	$NaAlSi_2O_6$

All reactions with excess quartz + H_2O

(21)	$3\ Chl + 8\ Kfs = 5\ Ann + 3\ Ms + 9\ Qtz + 4\ H_2O$
(22)	$1\ Ms + 3\ Chl + 3\ Qtz = 4\ Alm + 1\ Ann + 12\ H_2O$
(23)	$1\ Ms + 1\ Ann + 3\ Qtz = 1\ Alm + 2\ Kfs + 2\ H_2O$
(24)	$Pa + Qtz = Ab + Als + H_2O$
(25)	$Ms + Qtz = Kfs + Als + H_2O$
(26)	$Jd + Qtz = Ab$

Discontinuous reactions in the KFMASH–system

(27)	$Ctd = St + Grt + Chl$
(28)	$Grt + Chl = St + Bi$
(29)	$St + Chl = Bt + Als$
(30)	$St = Grt + Bt + Als$
(31)	$St + Bt = Grt + Als$
(32)	$St + Chl = Als + Grt$
(33)	$Grt + Chl = Bt + Als$

TS and FM component. Low-temperature (high-pressure) K-white mica contains a significant amount of TS component. Such micas are termed phengites. Increasing temperature removes much of the TS component and high-grade white micas are close to the muscovite end-member composition.

Reaction (21) in Table 7.3 replaces the sedimentary assemblage K-feldspar + chlorite by biotite + muscovite. Equilibrium conditions for this reaction in the pure KFASH system are shown in Fig. 7.4. The product biotite is Fe-rich biotite (often green biotite under the microscope). The first prograde biotite appears at about 400 °C.

Reaction (22) limits the presence of chlorite in rocks containing excess muscovite. The assemblage is replaced by garnet + biotite at a temperature of about 520 °C. Biotite is removed from rocks containing much muscovite by reaction (23) at about 590 °C in the pure KFASH system (Fig. 7.4). Following a metamorphic path along the Ky-geotherm in Fig. 7.4, the sedimentary (diagenetic) Na-white mica, paragonite, breaks down in the presence of quartz at about 620 °C [reaction (24)]. The product assemblage is albite + kyanite. It follows that paragonite is present in large portions of the stability field of staurolite + quartz. Staurolite – kyanite – paragonite schists are one of the classical metamorphic rocks of the mid-amphibolite facies zone of the Central Alps. Muscovite remains stable along the Ky-geotherm within the pT window of Fig. 7.4.

Reaction (25) represents the upper limit of muscovite in the presence of quartz. The product assemblage is K-feldspar + sillimanite (or andalusite,

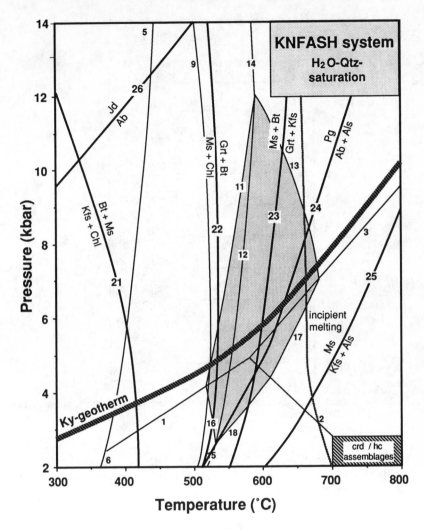

Fig. 7.4. Mica reactions. *Reaction numbers* refer to Tables 7.1, 7.2 and 7.3

respectively). However, because reaction (24) produces albite (plagioclase) and reaction (25) produces K-feldspar, the assemblage at high temperatures is Kfs – Ab – Qtz. The assemblage will, in the presence of excess water, begin to melt at rather low temperatures (approximate position of the solidus is given in Fig. 7.4). Complications arising from partial melting in metapelites will be discussed separately in Section 7.6.

Reaction (26) limits the stability field of albite in the presence of quartz. Albite decomposes at high pressures to jadeite (sodium pyroxene). High-pressure metamorphism of metapelites will be discussed also in Section 7.8. The sequence of stable two-phase assemblages in the pure KFASH system along the

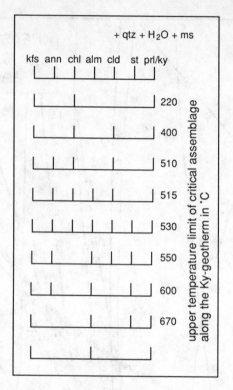

Fig. 7.5. Sequence of assemblages in the KFASH system projected from muscovite, quartz and H_2O onto the $FeO-Al_2O_3$ binary

Ky-geotherm is given in Fig. 7.5 together with the approximate upper temperature limits for the respective assemblages in the presence of excess quartz, muscovite and H_2O.

7.4.6 Metamorphism in the KFMASH System (AFM System)

The six-component system $K_2O-FeO-MgO-Al_2O_3-SiO_2-H_2O$ is often used to discuss metamorphism in metapelites. AFM diagrams are popular graphic representations of phase relationships in this system (see Chap. 2 and above). In contrast to the discussion above, the effects of the FM component in Fe-Mg-bearing minerals on mineral equilibria is now considered. The sequence of mineral assemblages in metapelitic rocks during prograde metamorphism along a medium pressure geotherm (Barrovian-type, Ky geotherm) is shown in Fig. 7.6 in a series of AFM projections. Prograde metamorphism will now be discussed, step by step, by reference to Fig. 7.6. As outlined above, all mineralogical changes occurring in the A apex of the AFM figures are described in the ASH system. The FASH system, also discussed above, refers to the FA

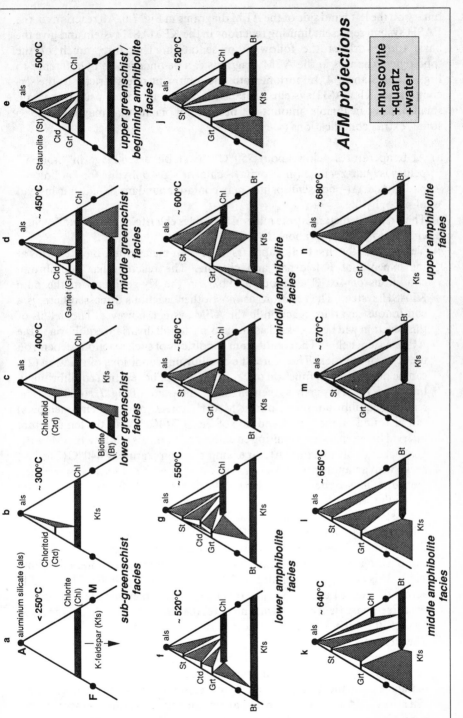

Fig. 7.6. Sequence of assemblages in the KFMASH system (AFM system) shown on AFM projections

binary on the left hand side of the AFM diagrams in Fig. 7.6. All reactions in the FASH system represent limiting reactions in the KFMASH system and give the basic framework of the following presentation. Therefore, much of the topological changes in the AFM diagrams can be understood by reference to Figs. 7.2, 7.3 and 7.4. Important note: the treatment refers strictly to the six-component KFMASH system. Metapelitic rocks, however, usually contain small but not negligible amounts of other components that may give rise to some "extra" complications (see Sect. 7.9).

a) At temperatures below about 250°C, the stable assemblage in "normal" pelites is quartz – illite (muscovite) – chlorite – pyrophyllite (or kaolinite) – paragonite. Al-poor semi-pelites may contain abundant K-feldspar instead of pyrophyllite.

b) The first truly metamorphic mineral formed is **chloritoid**. In "normal" pelites chloritoid appears at about 300°C.

c) Near 400°C, the first **biotite** appears in Al-poor metapelites. Biotite forms at the expense of K-feldspar and chlorite. The reaction has equilibrium conditions of 420°C at about 3.5 kbar on the Ky-geotherm in the pure KFASH system. The reaction, as most other reactions discussed here, is a continuous divariant reaction in the AFM system, however. The details of this reaction will be discussed as an example for all divariant equilibria in the AFM system below. The equilibrium conditions of such reactions depend on the X_{FeO} of the rock. Therefore, the equilibrium conditions of reaction (21) can be discussed as a function of X_{FeO} (Fig. 7.7). The reaction equilibrium in the pure KFASH system is given on the left hand side of Fig. 7.7 (420°C). The reaction equilibrium in the pure KMASH system (pure Mg end-members) can be found on the right hand side of Fig. 7.7 (460°C). In the temperature interval between the two limiting equilibria all minerals (Chl + Kfs + Ms + Bt) may occur stably together. At, for example, a temperature of 440°C (Fig. 7.7) rocks with an appropriate A-coordinate may consist of biotite + muscovite, biotite + muscovite + K-feldspar + chlorite, or K-feldspar + chlorite depending on the X_{FeO} of the rock. The composition of biotite and chlorite in the assemblage biotite + muscovite + K-feldspar + chlorite is given by the filled circles. It can be seen from Fig. 7.7 that the three-phase field biotite + K-feldspar + chlorite moves across the entire AFM diagram from Fe-rich compositions to Mg-rich composition within a temperature interval of only 40°C. This can also be seen in Fig. 7.6b–e, where at Fig. b Kfs-Chl is stable across the entire range of X_{FeO}, at (c) and (d) some intermediate positions of the three-phase field are shown, and at (e) the entire range of X_{FeO} is covered by Chl-Bt (in the presence of excess muscovite + quartz).

d) The **first garnet** appears in metapelites at temperatures of around 450°C. This temperature is in conflict with the information given for the pure FASH system. However, garnets in natural rocks preferentially incorporate manganese in the form of spessartine component at low temperatures. Therefore, Mn-Fe-garnet appears at significantly lower temperatures than pure almandine. In addition, natural garnets in metapelites also always

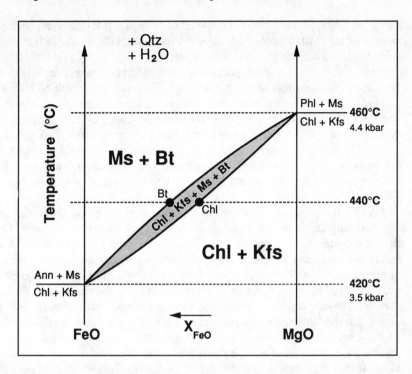

Fig. 7.7. pTX$_{FeO}$ diagram representing the Kfs-Ms-Bt-Chl assemblage [*reaction (21);* Table 7.3]. The pT gradient corresponds to the Ky geotherm

contain small amounts of Ca that further lowers the temperature of arrival of garnet in prograde metamorphism. This has been taken into account in Fig. 7.6.

e) At 500°C, the K-feldspar-chlorite assemblage disappears also in Mg-rich compositions. The **first staurolite** grows. Its appearance marks the transition to the **amphibolite facies**. New diagnostic assemblages at this stage are: Ky – St – Chl and St – Cld – Chl.

f) Fe-rich chlorite begins to be replaced by **garnet + biotite** between 500 and 520°C. The new assemblage garnet + biotite remains stable to very high grades, and the Fe-Mg partitioning between the two minerals is an often used a geothermometer. Note, however, the succession of critical mineral assemblages shown in Figs. 7.6f–n probably has a better temperature resolution than the Grt-Bt thermometer. Because garnet is stabilized by small amounts of Mn and Ca (that are preferentially fractionated into the first prograde garnet), also the Grt + Bt assemblage may occur in natural rocks at significantly lower temperatures. The first Grt + Bt pair evidently forms at temperatures as low as 470°C.

g) Fe-chloritoid breaks down to **garnet + staurolite** and the diagnostic three phase assemblage garnet + staurolite + chloritoid is present in a narrow temperature interval and in special bulk compositions (about 550°C).

h) **Chloritoid disappeared** from the rocks. The terminal reaction that removes chloritoid is the first discontinuous AFM reaction in the sequence [reaction (27), Table 7.3]. The equilibrium temperature of reaction (27) is only a few degrees C above equilibrium (12) in the pure FASH system (Fig. 7.3).

i) Fe-rich biotite decomposes to garnet + K-feldspar. From about 600°C, K-feldspar + garnet + biotite constitutes a stable assemblage in semi-pelitic gneisses. Three assemblages cover the full range of Fe-Mg variations in Al-poor rocks: K-feldspar + garnet (in Fe-rich compositions), K-feldspar + garnet + biotite (in intermediate compositions), and K-feldspar + biotite (in Mg-rich compositions). Consequently, the continuous reaction (23) has a much steeper slope in a TX_{FeO} diagram than reaction (21) shown in Fig. 7.7.

j) The AFM discontinuous reaction (28) has removed the garnet + chlorite tie-line and replaced it with a two-phase field between staurolite + biotite. The **first staurolite + biotite** appears at temperatures slightly above 600°C and marks the beginning of **middle amphibolite facies**. The discontinuous nature of reaction (28) makes it well suited for isograd mapping. In metamorphic terrains the chlorite + garnet zone is separated from the staurolite + biotite zone by a sharp isograd. The temperature at the boundary is about 600°C and it is rather insensitive on pressure (along Sil- or Ky geotherms).

k) The discontinuous AFM reaction (29) replaced the staurolite + chlorite tie-line by the new assemblage **kyanite (sillimanite) + biotite**.

l) Mg-chlorite decomposes to biotite + kyanite (sillimanite). All chlorite disappears from quartz- + muscovite-saturated rocks at about 650°C. Note, however, that some additional complications turn up in extremely Mg-rich composition because talc and Mg-rich cordierite are additional stable minerals near the AM binary at various conditions. The phase relationships in extremely Mg-rich pelites will be discussed in Sections 7.5. and 7.7; they are irrelevant for "normal" pelite compositions (Fig. 7.1) in regional metamorphism.

m) Fe-rich staurolite breaks down and forms garnet + kyanite (sillimanite). The new diagnostic assemblage is kyanite + staurolite + garnet, which is restricted to a small range of bulk compositions, however.

n) Staurolite has disappeared from quartz-saturated rocks. The new diagnostic assemblage is **garnet + biotite + kyanite (sillimanite)**. The appearance of the assemblage by the discontinuous AFM reaction (30) also marks the beginning of the **upper amphibolite facies**. Note that this boundary also coincides with the production of the **first melt** in rocks of suitable composition at H_2O-saturated conditions. Note that metamorphism along a higher pressure path shows an inverted sequence of reactions limiting the assemblages Chl + Grt and Chl + St respectively. Reaction (32) (Table 7.3) breaks down the St + Chl tie line at slightly lower temperature than reaction (33), which limits the occurrence of Grt + Chl. The pressure effect on the phase relationships in metapelitic rock can been used to delineate bathograds and bathozone as outlined in Chapter 4.

7.5 Low-Pressure Metamorphism of Pelites

In low-pressure metamorphic terrains (<5 kbar at $700°C$), the sequence of stable mineral assemblages in metapelitic rocks is quite different from those found in orogenic Barrovian type terrains. The heat source in low-pressure terrains is normally magmatic intrusions that transfer heat from deeper to shallower levels of the crust. Diapiric plutons may rise up to 3 km below the surface before the magma looses dissolved H_2O and cools below the solidus. Thermal contact aureoles develop around such intrusive bodies. The pressure is often in the range of 1 to 4 kbar (3–12 km depth). The maximum temperature to which sediments can be heated depends on a number of factors, including the composition of magma and the size of the intrusion. However, maximum contact temperatures at shallow-level "wet" granitic and granodioritic plutons rarely exceed 650°C. On the other hand, at somewhat deeper levels in the upper crust large-scale accumulation of magma may result in temperatures up to 750° C or so. Intrusion of large amounts of hot and dry magmas from the mantle may result in extensive high-temperature contact aureoles with temperatures in the metamorphic envelope reaching 900 to 1000°C. High-temperature contact aureoles are typically found around large mafic intrusions (gabbros, troctolites), intrusions of charnockite and mangerite batholiths, and anorthosite complexes that occur in Precambrian continental crust.

The consequences of isobaric heating of pelitic sediments of "normal" composition in a low-pressure contact aureole (for example, at 2 kbar) can be derived from the mineral equilibria shown in Fig. 7.8.

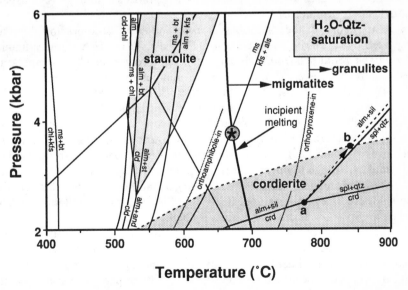

Fig. 7.8. Low-pressure, high-temperature reactions and phase relationships in metapelites and metamorphism in the KFASH-system (*reactions* listed in Tables 7.1–7.4)

7.5.1 KFASH System

Let us first consider equilibria in the pure KFASH system in order to derive a first overview. At low temperatures, the sequence of assemblages is the same as in orogenic metamorphism. Important here is reaction (21). It divides the cooler parts of a contact aureole into an outer zone with chlorite + K-feldspar and a higher-grade zone with biotite + muscovite. The temperature of the zone boundary is at about 400 °C. Around very shallow level intrusions, such as for example in the Oslo rift zone (P ~ 500 bar), this is, in fact, the only mineralogical change taking place in metapelitic sediments. The maximum temperature that can be reached at contacts to H_2O-rich granitic magmas that stop a few km under the surface is below 500 °C (magma temperature < 800 °C, country rock temperature $T° = 200$ °C, $\Delta T < 600$ °C; rule of thumb: maximum T in sediments $= T° + 1/2\ \Delta T$). The final assemblage is biotite + chlorite + muscovite + quartz.

The reactions at higher temperature replace the following assemblages present from lower grade (Fig. 7.5): Kfs + Bt, Bt + Chl, Chl + Cld, Cld + And (all assemblages with excess muscovite and quartz). Note that andalusite is the characteristic aluminosilicate polymorph in low pressure terrains. Its maximum pressure stability limit varies with composition (Fe^{3+}) and structural details (e.g. dislocation density) but is around 3.5 to 5 kbar.

Slightly above 500 °C, three reactions remove chlorite and chloritoid from the assemblages. All reactions produce garnet. No staurolite is formed in contact metamorphism below about 3 kbar. At about 550 °C, biotite disappears from rocks with excess muscovite. At 600 °C, muscovite breaks down in the presence of quartz and the diagnostic high-grade assemblage K-feldspar + andalusite is produced.

At about 650 °C, the assemblage garnet + andalusite becomes unstable and is replaced by cordierite by reaction (35) (Table 7.4). The first stable sillimanite appears at about 650–700 °C. In most contact aureoles such high temperatures are not realized even at the immediate contact to the intrusives. Andalusite + cordierite + K-feldspar + biotite represent the highest grade assemblage close to the contact of granitoid shallow-level intrusions. Metastable sillimanite occasionally can be found close to the contact to shallow level intrusives as a result of structural disorder and chemical impurities that stabilizes sillimanite relative to andalusite. Cordierite + sillimanite + K-feldspar + quartz is the stable assemblage at high temperatures in the pure KFAS system.

At slightly higher pressures (> 2.5 kbar), a stability field for the assemblage spinel (hercynite) + quartz appears at very high temperatures (above 770 °C). The pair Spl + Qtz is diagnostic for very high temperatures.

The sequence of successive mineral assemblages in low-pressure metamorphism of iron-rich metapelites can be summarized as follows. At low temperatures, the sequence is identical to the one found in orogenic metamorphism. Between about 500 and 550 °C, cordierite appears in various assemblages involving chlorite, muscovite and biotite. Staurolite is absent and also garnet may not be present. Andalusite is the characteristic aluminosilicate.

Table 7.4. Cordierite–spinel–orthopyroxene–orthoamphibole–reactions (KFASH system)

Cordierite	$Fe_2Al_3[AlSi_5]O_{18}n\ H_2O$
Spinel (hercynite)	$FeAl_2O_4$
Orthopyroxene (ferrosilite)	$FeSiO_3$
Orthoamphibole (ferro–anthophyllite)	$Fe_7Si_8O_{22}(OH)_2$
Sapphirine (ferro–sapphirine)	$Fe_2Al_4SiO_{10}$
Osumilite (ferro–osumilite)	$KFe_2Al_3[Al_2Si_{10}]O_{30}$

All reactions with excess quartz + H_2O

(35)   2 Alm + 4 Als + 5 Qtz + n H_2O = 3 Crd
(36)   2 Spl + 5 Qtz = Crd
(37)   Alm + 2 Als = 3 Spl + 5 Qtz
(38)   Oam = 7 Opx + Qtz + H_2O
(39)   14 Chl + 57 Qtz = 8 Oam + 7 Crd + 48 H_2O
(40)   Ann + 3 Qtz = 3 Opx + Kfs + H_2O

Some important cordierite reactions involving sheet silicates
(in the order of increasing grade)

Continuous reactions

(41)   2 Chl + 8 And/Sil + 11 Qtz = 5 Crd + 8 H_2O
(42)   Chl + Ms + 2 Qtz = Crd + Bt + 4 H_2O
(43)   3 Crd + 2 Ms = 8 Sil + 2 Bt + 7 Qtz
(44)   3 Crd + 2 Bt = 4 Grt + 2 Ms + 3 Qtz
(45)   3 Crd + 4 Bt + 3 Qtz = 6 Grt + 4 Kfs + 4 H_2O
(46)   2 Bt + 6 Ms + 15 Qtz = 3 Crd + 8 Kfs + 8 H_2O
(47)   6 Sil + 2 Bt + 9 Qtz = 3 Crd + 2 Kfs + 2 H_2O
(48)   2 Grt + 4 Ms + 9 Qtz = 3 Crd + 4 Kfs + 5 H_2O

Discontinuous reactions

(49)   Chl + Sil + Ms + Qtz = Crd + Bt + H_2O
(50)   Bt + Sil + Qtz = Crd + Grt + Kfs + H_2O
(51)   Bt + Grt + Qtz = Crd + Opx + Kfs + H_2O
(52)   Bt + Grt + Qtz = Opx + Sil + Kfs + H_2O
(53)   Bt + Sil + Qtz = Crd + Opx + Kfs + H_2O
(54)   Opx + Sil + Qtz = Crd + Grt
(55)   Crd + Grt + Sil = Spl + Qtz
(56)   Crd + Sil + Kfs + Qtz = Osm + Opx

Above about 600°C, metapelites contain the diagnostic assemblage cordierite + K-feldspar + biotite ± andalusite (at higher T sillimanite).

7.5.2 KFMASH System

The iron-magnesium exchange in the Fe-Mg minerals has profound effects in high-temperature metamorphism of metapelitic rocks. The cordierite-involving reactions [reactions (35), (36) and (37)] are fluid-absent reactions and as such strongly dependent on the compositions of the participating minerals (geological thermometers and barometers). Cordierite may contain molecular fluid species such as H_2O and CO_2 in the channels of the structure. The number of moles of water in cordierite (n in Table 7.4) varies between 0 and 2. The amount depends on the total pressure, the H_2O pressure, the composition of cordierite and the temperature. The presence of water in the cordierite stabilizes the mineral relative to the respective reactant assemblage. The H_2O is relatively loosely bound and can be compared with molecular water stored in many zeolite minerals. However, the reactions involving cordierite can also be formulated for H_2O-free environments (n = 0). In addition, the iron-magnesium content of the bulk rock is strongly fractionated between cordierite and garnet and spinel. Cordierite in metapelites is much more magnesian than coexisting garnet or spinel.

As a consequence, the stability field for cordierite is greatly expanded relative to the pure FAS system. The shaded area in the lower right pT space of Fig. 7.8 shows the approximate expansion for the cordierite field for Fe-rich bulk compositions ($X_{Fe} \sim 0.8$). Cordierite occurs now at about 530°C (instead of 680°C in pure FAS system) in a 2 kbar contact aureole. In addition, the Grt + Als + Qtz = Crd reaction (35) intersects a number of mica- and chlorite-involving reactions and this brings about some important cordierite-sheet silicate reactions (and orthoamphibole reactions). These reactions are not shown in Fig. 7.8 because their equilibrium coordinates vary strongly with X_{Fe} of the rock. Some of these reactions are listed in Table 7.4, however. Reactions (41) and (42) are continuous cordierite-producing dehydration reactions that are terminated by the discontinuous reaction (49) which intersects the 2 kbar isobar at about 530°C. Above this temperature cordierite + biotite may be present in metapelites. Reactions (43) and (44) are water-absent reactions. The equilibria are characterized by very low dp/dT slopes. Equilibrium conditions are independent of temperature and the assemblages have potential as geobarometers. The assemblages cordierite + muscovite and cordierite + biotite occur on the low-pressure side of the equilibria.

At higher temperatures, a number of continuous dehydration reactions (45–48) produce K-feldspar (that may enter a melt phase together with quartz and H_2O). The reactions are linked by the discontinuous reaction (50) which ultimately produces the characteristic assemblages cordierite + garnet + K-feldspar + biotite and cordierite + garnet + K-feldspar + aluminosilicate (andalusite or sillimanite). The assemblages are diagnostic for high-grade cordierite gneisses. The assemblages are widespread in low- to medium-pressure terrains and typical pT conditions deduced for such gneisses are in the range of 700 ± 50° C at pressures from 2 to 5 kbar. Cordierite – garnet – K-feldspar – biotite – gneisses are transitional between upper-amphibolite and granulite facies conditions. The assemblage often indicates equilibration under conditions of reduced water-pressure. Very often cordierite-gneisses do not contain K-feldspar, however. Biotite is then the only K-bearing mineral in the rock and biotite assemblages *cannot* be represented in AFM diagrams (Kfs projection), biotite becomes an "extra" phase and typical assemblages involve four (or even more) "AFM" minerals. The representative assemblage in such rocks is Crd + Grt + Sil + Bt. It forms at pT conditions similar to the Kfs-bearing Crd-gneisses. They are most often found in terrains that were metamorphosed at 3–5 kbar and 650–750°C and low water pressure.

Some additional important boundaries are shown in Fig. 7.8. At about 600–650°C, orthoamphibole starts to form by low-pressure metamorphism of pelites. Orthoamphibole-bearing assemblages are produced from chlorite- and biotite-consuming reactions. As example, reaction (39) is formulated in Table 7.4. Reaction (39) metastable in the pure FASH system, but it is important in natural systems. It generates cordierite-**anthophyllite** rocks that are widespread in Precambrian shield areas. Note that the term **orthoamphibole** is used here for all Fe-Mg-amphiboles and includes also **cummingtonite** and **gedrite**. More than one type of Fe-Mg-amphibole may occur in high-grade rocks. Gedrites often occur in rocks all the way up to granulite-facies conditions. The precise conditions of the first occurrence of Fe-Mg-amphiboles in metapelites depends on the rock- and fluid composition. The orthoamphibole-in curve in Fig. 7.8 represents an approximate boundary above which metapelites (and related rocks) may contain various Fe-Mg-amphiboles in a number of associations formed by several reactions. Note that there is little overlap of the orthoamphibole and muscovite fields, respectively. In addition, biotite appears to be distinctly more stable than the alternative assemblage K-feldspar + orthoamphibole in metapelites. Kfs + Oam have not been reported from metapelitic rocks to our knowledge, even if the mineral pair is predicted to be stable by the thermodynamic datasets (Chap. 3.8.1). A typical stable natural orthoamphibole assemblage at low- to intermediate-pressures (3–5 kbar) is Bt + Oam + Crd ± Grt.

7.5.3 Cordierite-Garnet-Spinel Equilibria

The three FAS reactions (35), (36) and (37) define a stable invariant point (point a in Fig. 7.8). The assemblage spinel + quartz is a diagnostic high-temperature assemblage and the temperature of about 770°C given by the invariant point a represents the lowest possible temperature for this assemblage. In Fe-rich metapelites the corresponding equilibrium conditions of the three reactions are approximately given by the dashed curves in Fig. 7.8. It can be seen that the invariant point has moved to position b and the spinel + quartz assemblage requires at least 840°C. The dashed arrow in Fig. 7.8 represents the equilibrium conditions of the discontinuous reaction (51) as a function of decreasing X_{Fe}.

The equilibrium constant of reaction (35) can be written (assuming pure Als, Qtz and H_2O) as: $\ln K_{pT} = 3 \ln a_{Crd} - 2 \ln a_{Alm}$. Coexisting garnet and cordierite in medium-pressure high-grade gneisses may, for example, contain 80 mol% almandine and 40 mol% "dry" Fe-cordierite, respectively. The corresponding activities using a simple ideal site mixing solution model are: $a_{Alm} = 0.51$ and $a_{Crd} = 0.16$ and, hence, the calculated $\ln K_{pT}$ for this Grt-Crd pair is -4.159. On the schematic pT diagram (Fig. 7.9) the pT equilibria (35), (36) and (37) are contoured for ln K-values. The garnet-cordierite pair from above is constrained to the ln K = -4.16 contour of equilibrium (35). It can be seen that the field

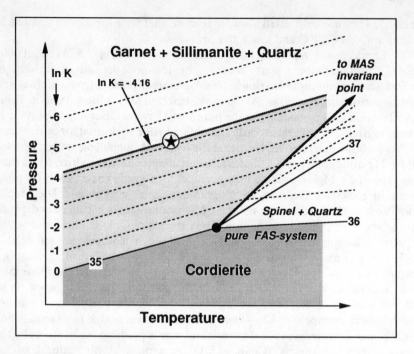

Fig. 7.9. Equilibria of continuous reactions involving Grt - Crd - Sil - Qtz. *Reaction stoichiometries* listed in Table 7.4

for stable cordierite (light shading) is *enlarged* relative to the pure FAS system (dark shading).

It also follows from the reaction equation (Table 7.4) that the presence of H_2O, which can be accommodated in the cordierite structural channels, will lead to a further stabilization of cordierite.

In addition, because there are two ferro-magnesian minerals involved in the transfer reaction (35) an Fe-Mg-exchange reaction can be written between Grt and Crd: $FeMg_{-1}$ (Crd) = $FeMg_{-1}$ (Grt). The exchange reaction must be at equilibrium simultaneously with the transfer reaction (35). The equilibrium conditions of the Fe-Mg exchange are virtually dependent on the temperature alone. Simultaneous solution of the two equilibria (using data by Perchuk and Lavrent'eva 1983, and Aranovich and Podlesskii 1983) gives P = 6 kbar and T = 740 °C for the garnet-cordierite pair of the composition above coexisting with sillimanite and quartz. The garnet-cordierite pairs in gneisses with excess sillimanite and quartz define then a unique equilibration point in pT space (star in Fig. 7.9). The phase relationships among garnet-cordierite-spinel may also be represented, for example, by an isothermal pressure-composition diagram such as Fig. 7.10 (T ~ 800 °C). At a temperature of about 800 °C the equilibrium pressure of the discontinuous reaction (51) is about 2.7 kbar. All three minerals tend to become more Mg-rich as a result of the continuous reactions (35), (36) and (37). There is a large pressure range over which the assemblage Grt-Crd-Sil-

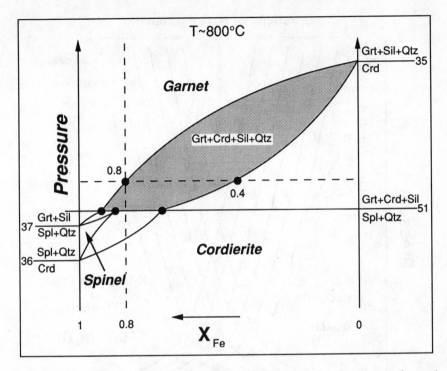

Fig. 7.10. pX_{FeO}-diagram showing equilibria among Grt - Crd - Sil - Qtz. *Reaction numbers* and *reaction stoichiometries* listed in Table 7.4

Qtz may occur (shaded field in Fig. 7.10). Pelitic rocks often have an X_{Fe} of about 0.8 (normal pelites). This composition is shown as a vertical dashed line in Fig. 7.10. Such a rock containing cordierite at low pressure will produce spinel by reaction (36) as pressure increases. The spinel will be removed from the rock by reaction (51) at a specific pressure. Above this pressure cordierite decomposes to garnet by the reaction (35) until all cordierite is used up. In more Mg-rich rock compositions, the spinel fields are by-passed and cordierite is directly replaced by garnet. With decreasing X_{Fe} cordierite decomposes at progressively higher pressures. Note: the pure MAS version of reaction (35) is metastable.

7.6 Very High-Temperature Metamorphism of Pelites – Metapelitic Granulites

7.6.1 Partial Melting and Migmatites

At about 650–700 °C, partial melting begins to be important in rocks containing feldspar and quartz under water-saturated conditions (Fig. 7.8). The curve of incipient melting marks the onset of partial melting in "granitic" systems and

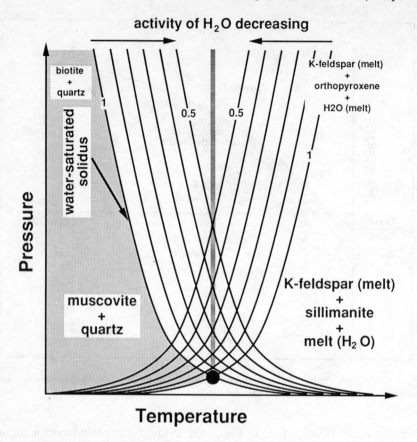

Fig. 7.11. Effects of water-pressure variations on the melt-present and melt-absent muscovite + quartz breakdown reaction (25) [and the corresponding biotite + quartz reaction (40)]. *Reaction stoichiometries* listed in Table 7.3 and 7.4

the beginning of migmatite formation. This means that above this curve Kfs component produced by dehydration reactions (Table 7.4) might dissolve in a melt phase that may leave the site of production in some cases. The melt segregates and removes Kfs and Ab produced by mica-involving dehydration reactions. The rocks left behind may be devoid of alkali feldspar and consequently also alkalis and "granite" component. Particularly important is the decomposition of muscovite in the presence of quartz. The melt-absent reaction (25) that produces andalusite or sillimanite and Kfs at low pressures intersects the minimum melt curve at about 4 kbar. At higher pressures, muscovite decomposes to sillimanite and a Kfs-bearing melt. The situation is schematically shown in Fig. 7.11. The melt-absent decomposition of muscovite (+ Qtz) is an ordinary dehydration reaction and has a characteristic curve shape on a pT diagram with a positive dp/dT slope at the pressures of interest here. At $a_{H_2O} = 1$, Ms + Qtz has its maximum stability. The equilibrium of reaction (25) can be contoured for decreasing a_{H_2O}, similar to, for example, increasing dilution of the aqueous fluid with CO_2 (Fig. 7.11). With decreasing a_{H_2O}, K-feldspar + sillimanite forms at progressively lower temperature. The formation of sillimanite by reaction (25) is occasionally referred to as the second sillimanite isograd in the geological literature (in contrast to sillimanite from the phase transitions Ky= Sil and And= Sil respectively, that may or may not form because of metastable survival of And or Ky!). The solidus for muscovite-granite at H_2O-saturated conditions has a negative slope in a pT diagram at the pressures of interest. At the minimum melt curve shown on the Figs. 7.8 and 7.11 quartz + feldspar begin to melt. Muscovite of the solidus assemblage dissolves in the melt and contributes Kfs component and H_2O to the melt, leaving sillimanite as a solid residue. Compared with the melt-absent reaction, decreasing a_{H_2O} has an opposite effect on the Ms + Qtz breakdown curve and the minimum temperature of muscovite melting increases with decreasing a_{H_2O}. The consequences for the muscovite-out isograd in Qtz-bearing rocks are as follows: at pressures below the intersection of the water-saturated melt-present and melt-absent Ms-breakdown curves (black dot in Fig. 7.11), muscovite disappears by reaction (25) which forms Kfs + Als + H_2O. At pressures above that intersection, muscovite disappears by partial melting. The intersection marks the highest possible temperature for stable Ms + Qtz. The dark shaded area in Fig. 7.11 marks the muscovite field in Qtz-bearing rocks. Decreasing a_{H_2O} displaces the intersection point of the two Ms-out reactions towards higher pressures whilst the net temperature effect is minimal. Muscovite cannot be found in Qtz-bearing rocks outside the light shaded area irrespective of fluid composition. The quantitative position of the intersection point of the two reactions is shown in Fig. 7.8 (~4 kbar, 680 °C).

The melt produced by prograde metamorphism of mica schists is close to eutectic composition of the granite system and H_2O-saturated. The further fate of the produced melt phase depends primarily on the amount of melt produced in the partial melting process but also on accompanying deformation and other factors. Small amounts of melt may not migrate over long distances and will be found later as **leucocratic quartzo-feldspathic bands, pods, lenses and patches**. If much anatectic "granitic" melt is produced by local in-situ partial melting of gneisses and schists, the melt may collect to larger masses and may cross cut the primary gneissic foliation or bedding, thus forming veins and irregular masses of **discordant granite** and the melt may eventually leave the site of production and collect in a larger magma chamber of crustal granitic melt that may rise as a pluton to shallow levels in the crust by buoyancy forces. The light Qtz-Fsp material in gneisses representing the anatectic melt phase often shows randomly oriented minerals and fabrics typical of magmatic rocks (leucosome). The restite material from which the "granitic" melt has been extracted has a high modal proportion of mafic minerals (mainly biotite and garnet) and has a dark gneissic appearance (melanosome, restite). The two components of gneisses that underwent partial melting and anatexis, the leucosome and the restite, may be present in countless proportions and arranged in an infinite number of different light-dark patterns. These anatectic gneisses are commonly termed migmatites. Migmatite is one of the prime rock materials found in continental crust.

7.6.2 Granulites

Orthoamphibole that may have formed previously decomposes to orthopyroxene and quartz [reaction (38)] at about 750–800°C. The appearance of orthopyroxene in quartz-bearing rocks marks the transition from upper amphibolite facies to granulite facies conditions. However, the diagnostic granulite facies assemblage orthopyroxene + quartz may originate from a number of different reactions. The most important of them is reaction (40), which eventually removes the last remaining sheet silicate, biotite, from metapelites. In general, biotite breaks down in the presence of quartz at about 800°C to orthopyroxene + K-feldspar. Biotite replacement by orthopyroxene takes place over a fairly wide T interval, depending on the composition of biotite and fluid. Biotite decomposition produces Kfs that together with the H_2O released by reaction (40) commonly enters a melt phase. The melt-absent version of reaction (40) is metastable at pressures above a few hundred bars and in the presence of H_2O-rich fluids relative to corresponding biotite melting reaction. The relationships are about analogous to those of the muscovite breakdown shown in Fig. 7.11, which can also be used to portray biotite breakdown. Kfs may dissolve in a melt phase (H_2O-rich conditions) or it may remain in the rock (dry conditions).

Fig. 7.12a shows a representative AFM chemography of phase relationships in high-grade metapelites. The chemography shall be typical for conditions of about 5 kbar and 750°C and a H_2O-activity smaller than 1. The assemblages contain excess K-feldspar and quartz. The Fig. shows that under these conditions "normal" Fe-rich metapelites contain garnet + sillimanite or garnet + sillimanite + cordierite. More Mg-rich rocks contain orthopyroxene + cordierite + garnet, whereas biotite is restricted to fairly Mg-rich compositions. Biotite may occur in assemblages with cordierite + orthopyroxene. Note that the stable biotite + orthopyroxene assemblage shown in Fig. 7.12a together with excess K-feldspar and quartz requires equilibrium of reaction (40). The Bt + Opx pair can in fact be used to estimate H_2O activity provided that an independent pT estimate can be made for the assemblage from geothermobarometry. Fig. 7.12a depicts a stage in metamorphism where Fe-rich biotite has been replaced by orthopyroxene [reaction (40)] whereas Mg-rich biotite still persists. It is also immediately evident from Fig. 7.12a that a great number of other possible assemblages can be expected involving the minerals garnet, sillimanite, cordierite, orthopyroxene and biotite in rocks containing excess K-feldspar + quartz. Using topology given in Fig. 7.12a, the five minerals are connected by five discontinuous reactions at a fixed water pressure. The phase relations among the five most common minerals in high-grade metapelites are shown in Fig. 7.12b, together with a possible pT path of metamorphism that produced the assemblages shown in Fig. 7.12a. At "low-temperature" (= higher

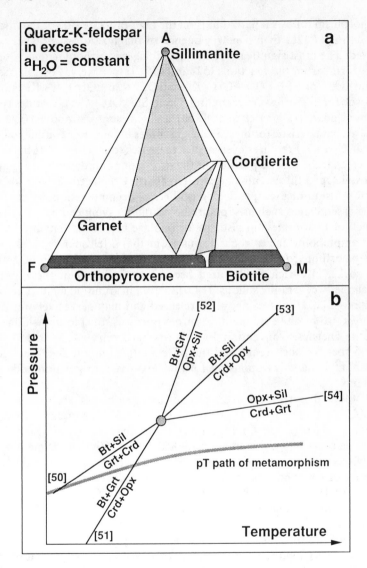

Fig. 7.12. a AFM projection (from K-feldspar) of typical granulite facies metapelites at a characteristic pT coordinate and $a_{H_2O} < 1$. **b** Qualitative pT diagram, showing five discontinuous high-grade (granulite-facies) reactions (listed on Table 7.4)

amphibolite facies here) the typical assemblage is biotite + sillimanite. Garnet occurs as an additional mineral in "normal" metapelites. The high-grade K-feldspar + sillimanite assemblage is also present because Kfs is an excess phase above the Ms-out isograd. As discussed above, low- to intermediate-pressure metamorphism replaces biotite + sillimanite with garnet + cordierite [reaction (50)]. At higher grade the garnet + biotite tie-line is replaced by the new orthopyroxene + cordierite assemblage shown in Fig. 7.12b by reaction (51). If metamorphism follows a path that is on the high-pressure side of the invariant point of Fig. 7.12b, then a new sequence of diagnostic assemblages can be observed. The most significant of the assemblages is orthopyroxene + sillimanite. It forms from the reactions (52) and (54), respectively. Possible assemblages include Opx + Sil + Qtz + Grt + Kfs and Opx + Sil + Qtz + Bt + Kfs; both are diagnostic for high-pressure granulites. Opx + Sil + Qtz + Grt granulites typically form at pressures greater than 8 kbar and temperatures above 800 °C. The rocks are characteristic for metapelitic granulites in the lower continental crust. The pair Opx + Ky requires even higher pressures to form (p > 10 kbar). Note, however, that the precise position of the invariant point depends on a_{H_2O}. The diagnostic Opx + Sil assemblage is often (partly) replaced at later stages of metamorphism, for example during isothermal decompression, by cordierite-bearing assemblages such as sapphirine + cordierite symplectites.

Much of the biotite and also the assemblage biotite + sillimanite found in higher amphibolite facies rocks forms from the K-feldspar + garnet and K-feldspar + orthopyroxene assemblages by retrograde rehydration of granulite facies rocks. H_2O-rich fluids often become available when the rocks cool through the "wet" granite solidus. H_2O dissolved in granitoid melts that formed during migmatization (see above) is released and may pervasively retrograde granulites to biotite + sillimanite + garnet gneisses or biotite + sillimanite + cordierite gneisses. Conventional geothermobarometry will yield 650–700 °C and 3–5 kbar for such rocks and the pT conditions mimic the "wet" granite solidus that actually is responsible for the pervasive retrogression under these conditions.

At still higher temperatures, diagnostic assemblages are spinel + quartz (> 850 °C) that may form by reaction (55), sapphirine + quartz (> 900 °C), osumilite + quartz (> 950 °C). Pigeonite (ternary pyroxene) may also appear in high-grade rocks at temperatures above 850–900 °C. High-grade metamorphic rocks with assemblages that are diagnostic for crustal temperatures above 850–900 °C are rare and have been reported only from Precambrian terrains where massive heat transfer from the mantle to the middle continental crust took place by means of dry magmatic intrusions (e.g. contact aureoles around anorthosite complexes).

Metamorphism of metapelitic rocks is strongly affected by partial melting reactions at very high temperatures. Increased formation of partial melts has severe consequences for the bulk composition of the rocks. The rocks become depleted in "granite" component, that is in feldspar and quartz. Fe is also strongly fractionated into the melt phase and consequently the residuum is enriched in Mg and Al. Repetition of dehydration reactions along a prograde pTt path also inevitably leaves the rocks devoid of an aqueous fluid phase. Accordingly, extreme high-grade "metapelites" are often quartz-free Al-Mg-rich restites left behind by partial melting processes and they consist of various assemblages among distinct minerals including orthopyroxene, cordierite, sillimanite (kyanite), sapphirine, spinel, garnet, corundum and others. Water-deficient conditions are often reflected in disequilibrium micro-structures including symplectites, replacement structures and coronites.

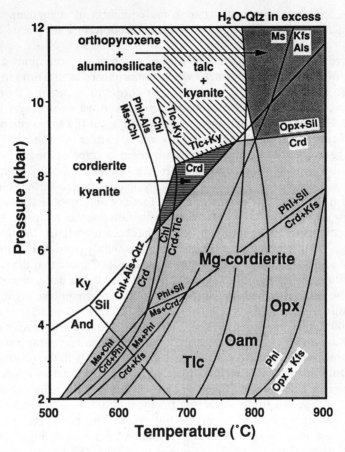

Fig. 7.13. Tentative pT grid for Mg-rich metapelites

7.7 Metamorphism of Very Mg-Rich "Pelites"

Extremely Mg-rich primary sedimentary shale compositions are fairly unusual (see Sect. 7.2). However, Mg-rich micaschists and other rocks with remarkably high Mg/Fe ratios can be found occasionally intercalated in other schists and gneisses in many metamorphic terrains and orogenic belts. The unusual bulk rock composition may have a number of feasible origins including evaporitic sediments, fluid/rock interaction (metasomatism) during metamorphism or restites after partial melting. Nevertheless, the assemblages of these rocks resemble those of "normal" metapelites and the rock compositions may show continuous transitions to more Fe-rich metapelites. A number of special features are, however, special to Mg-rich schists and some selected equilibria in the pure KMASH system are shown in Fig. 7.13. Note that the phase diagram is valid for rocks that contain quartz and for conditions $p_{tot} = p_{H_2O}$. The latter condition has the consequence that some equilibria at high temperatures become metastable relative to melt producing reactions (see also Sects. 7.5 and 7.6).

Mg-cordierite can be found over a wide range of pT conditions and the mineral can be found in rocks that were metamorphosed at pressures in excess of 8 kbar. This is in sharp contrast to Fe-cordierite occurring in Fe-rich metapelites. The pT range for Mg-cordierite in various assemblages is light shaded in Fig. 7.13. Note that there is some overlap of the Mg-cordierite and kyanite fields, respectively (horizontally ruled pattern in Fig. 7.13). The assemblage cordierite + kyanite has been observed in amphibolite facies schists from Barrovian metamorphic terrains. It is restricted to a small area in pT space. Other observed natural assemblages are, in fact, diagnostic for even more restricted pT ranges. For example, in the Tertiary amphibolite facies Ticino mountains of the Central Swiss Alps, the assemblage cordierite + kyanite + quartz + chlorite + phlogopite (Mg-biotite) has been reported by Wenk (1968). From Fig. 7.13 it is evident that the assemblage is tightly restricted to pressures between 7 and 8 kbar and a temperature close to 670 °C [note that the geologic interpretation of such numbers (pairs of numbers) is a theme not covered by this book]. The diagnostic high-pressure granulite facies assemblage orthopyroxene + aluminosilicate (dark shaded area in Fig. 7.13) is restricted to temperatures > 800 °C and pressures > 8 kbar. The diagnostic granulite facies assemblage orthopyroxene (enstatite) + quartz requires minimum temperatures of 800 °C. However, one must be aware that this is true only in the presence of a pure H_2O-fluid. Also note that in extremely Mg-rich rocks talc may appear in a number of new assemblages. Talc is not present in Fe-rich metapelites and the mineral does not substitute much iron for magnesium. It therefore promptly disappears with increasing X_{Fe} of the bulk rock. The most interesting and diagnostic assemblage is talc + kyanite which is restricted to high pressures above 8 or 9 kbar at temperatures in the range of 700 ± 50 °C. The so-called white schist (Schreyer 1977) assemblage Tlc + Ky appears in high-pressure low-temperature metamorphism associated with lithosphere subduction (see below) but also formed in many Precambrian granulite facies terrain by isobaric cooling of lower crustal granulites. In the lower right corner of Fig. 7.13 the position of the metastable phlogopite + quartz breakdown curve is shown. The reaction takes place at about 100 °C above the corresponding annite + quartz breakdown. Both biotite-consuming Opx-forming reactions are metastable relative to the corresponding melt-producing reactions. As discussed above, the maximum temperature for biotite + quartz is about 850 °C.

7.8 High-Pressure-Low Temperature Metamorphism

High-pressure low-temperature metamorphism (HPLT) is most commonly associated with subduction of oceanic lithosphere along destructive plate margins. The few pelitic sediments that are found associated with HPLT

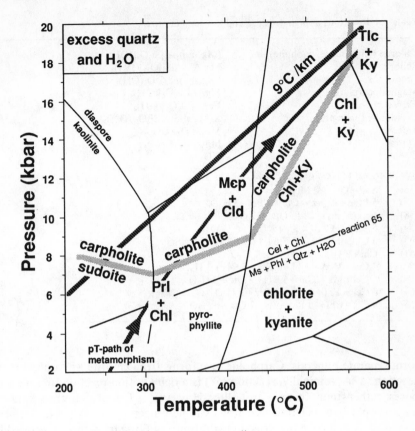

Fig. 7.14. Tentative pT grid for high-pressure metapelites

ophiolite complexes and metapelites in continental HPLT complexes show very distinct metamorphic minerals and characteristic assemblages that are unique for HPLT metamorphism.

Some relationships in HPLT rocks are summarized in Fig. 7.14 and a selection of relevant continuous reactions are listed in Table 7.5. The calculated (RB88) phase relationships in the ASH system is shown in Fig. 7.14 as a reference frame. The constant $9\,°C\,km^{-1}$ geotherm shall represent the approximate boundary of the geologically accessible pT space. A characteristic pT path of subduction zone metamorphism is also given in Fig. 7.14. This type of metamorphism is initially characterized by extremely steep dp/dT slopes that continuously decrease as HPLT metamorphism progresses (see also Sect. 3.5).

The mineral sudoite, a dioctahedral chlorite, appears to be typical in very low grade though high-pressure rocks. With increasing pressure the distinctive mineral carpholite may replace sudoite. The H_2O-conserving reaction (56) describes this transition in terms of the end-member phase components given in Table 7.5. There is, as in the case of talc, the unusual situation that Fe-carpholite and Mg-carpholite apparently do not mix over the entire FM range, but rather

Table 7.5. Reactions at high pressure and low temperature

Chlorite (clinochlore, daphnite)	$(Mg, Fe)_5Al_2Si_3O_{10}(OH)_8$
Sudoite	$Mg_2Al_3[AlSi_3O_{10}](OH)_8$
Chloritoid	$(Mg, Fe)Al_2SiO_6(OH)_2$
Magnesiocarpholite	$MgAl_2[Si_2O_6](OH)_4$
Ferrocarpholite	$FeAl_2[Si_2O_6](OH)_4$
Celadonite (phengite)	$K(Mg, Fe)AlSi_4O_{10}(OH)_2$
Talc	$Mg_3Si_4O_{10}(OH)_2$
Pyrope	$Mg_3Al_2Si_3O_{12}$
Coesite	SiO_2

Reactions with excess quartz $+ H_2O$

(56)   Sud + Qtz = 2 Mcp
(57)   5 Mcp + 9 Qtz = Chl + 4 Prl + 2 H_2O
(58)   Fcp (Mcp) = Cld + Qtz + H_2O
(59)   5 Mcp = Chl + 4 Ky + 3 Qtz + 6 H_2O
(60)   3 Chl + 14 Qtz = 5 Tlc + 3 Ky + 7 H_2O
(61)   Chl + 3 Qtz = 2 Cld + Tlc + H_2O
(62)   4 Cel + Mcp = Chl + 4 Kfs + 3 Qtz + 2 H_2O
(63)   5 Cel + Ms = Chl + 6 Kfs + 2 Qtz + 2 H_2O
(64)   3 Cel = Phl + 2 Kfs + 3 Qtz + 2 H_2O
(65)   4 Cel + Chl = 3 Phl + Ms + 7 Qtz + 4 H_2O

form separate minerals. Carpholite also forms from chlorite + pyrophyllite by reaction (57). Nominally, reaction (57) is a dehydration reaction that decomposes rather than forms carpholite. However, pT paths in that kind of metamorphism are characterized by extremely steep dp/dT slopes and it may therefore be that pT paths cross dehydration equilibria from the dehydrated to the hydrated side upon pressure increase even if temperature also slightly increases. The effect is also shown in Fig. 3.13, where the path "dehydration by decompression" is followed in the other direction in HPLT metamorphism ("hydration by compression"). For this reason carpholite may even form by hydration of the chlorite + kyanite assemblage by reaction (59). Note, however, that the carpholite formation by these mechanisms requires that free H_2O fluid is available and carpholite-forming reactions will cease when water is used up.

Ferrocarpholite decomposes to chloritoid [reaction (58)] whereas magnesiocarpholite is replaced by chlorite + kyanite [reaction (59)] if the rocks are not carried to greater depth but rather follow a "normal" clockwise pT path through the greenschist facies. If HPLT metamorphism continues, carpholite decomposition produces magnesiochloritoid [Mg-version of reaction (58)]. Chloritoid may, in fact, become extremely Mg-rich in HPLT rocks, and chloritoid with $X_{Mg} \sim 0.5$ is not uncommon. It is a general observation that the typical Fe-rich AFM-phases chloritoid, staurolite and garnet become increasingly magnesian at very high pressures. The carpholite boundaries are given tentatively in Fig. 7.14. Talc is also a characteristic mineral in HPLT metapelites and it often occurs together with potassium white mica (the two minerals are difficult to distinguish under the microscope). Chloritoid + talc forms from the continuous chlorite-consuming reaction (61). The two minerals finally combine to form pyrope component in garnet at very high pressures. The diagnostic white schist assemblage Tlc + Ky results from the chlorite-breakdown reaction (60) or analogous chloritoid- and carpholite-involving reactions.

The micas in high pressure metapelites also show some characteristic diagnostic features. Particularly notable is the absence or scarcity of biotite in HPLT metapelites. Biotite persists only in extremely Mg-rich high-pressure rocks in the form of phlogopite. The main reason for this is that potassium white mica experiences significant tschermak substitution with increasing pressure.

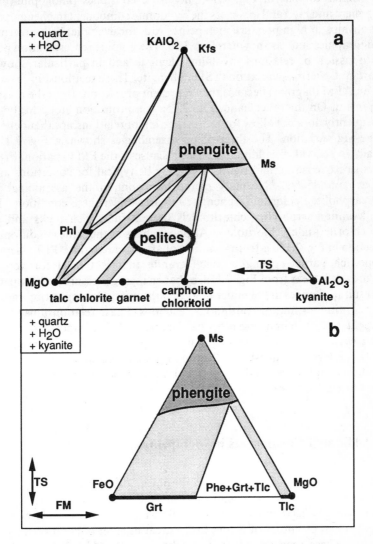

Fig. 7.15a, b. Isothermal isobaric composition phase diagrams of high-pressure metamorphism of metapelites

Thus, Mg-Fe-free muscovite changes its composition systematically towards celadonite (Table 7.5) with increasing pressure. Celadonite-rich potassic white micas are commonly termed phengite. Phengite is the characteristic mica in HPLT rocks. Celadonite component is produced by a number of continuous reactions such as (62) through (65) (Table 7.5). The equilibria of all four continuous reactions have low dp/dT slopes in the pT range of interest. Celadonite component is always on the high-pressure side of the equilibrium curve. As an example, the equilibrium of reaction (65) is shown for a given value of the equilibrium constant in Fig. 7.14. HPLT metamorphism that follows a pT path similar to that in Fig. 7.14, therefore, consumes phlogopite-(biotite) component and H_2O and increases the celadonite component in phengite. Phase relationships in high-pressure metapelites can, for example, be represented in chemographies such as shown in Fig. 7.15. The graphic representation permits the discussion of relations involving phengite and in particular shows the important tschermak variation (TS) in phengite. These relationships cannot be seen on AFM diagrams because they represent projections from the muscovite component. On the other hand, Fig. 7.15a is a projection along $MgFe_{-1}$ and consequently does not allow for representation of relationships that depend on the Fe-Mg variation. The three-phase assemblages shown in Fig. 7.15, will contain, in general, an additional phase because of the FM variation. Fig. 7.15 shows isobaric isothermal figures that may be typical for conditions around 13 kbar and 450°C. "Normal" metapelites contain the assemblage chlorite + carpholite + chloritoid + phengite + quartz under these conditions. Phengite + kyanite + carpholite + chloritoid is present in Al-rich rocks and phengite + chlorite + talc + chloritoid in Al-poor schists at the same conditions. The projection in Fig. 7.15b is from kyanite + quartz. It shows the HPLT assemblage pyrope-rich garnet + phengite + talc + kyanite that is typical for very high pressures. It is clear from Fig. 7.15a, b that changing pressure and temperature will result in changes in the mineralogy and the mineral composition caused by many continuous and discontinuous reactions. Phase relationships in HPLT metapelites offer an immense potential for analysis of metamorphic conditions and successive sequences of mineral assemblages that must be worked out for each HPLT terrain in question. Ultra-high-pressure metamorphism of metapelites and crustal schists and gneisses ultimately produces coesite from quartz and diamond from graphite.

7.9 Additional Components in Metapelites

Until now we have discussed metamorphism of pelites, step by step, by increasing the considered chemical complexity. The six-component AFM system typically accounts for more than 95 wt% of pure shales. However, the compositional inventory of pelites lists a great number of additional chemical elements. These may be stored somewhere in "AFM" minerals that form complex solid solutions or they may give rise to separate minerals. Such "extra" phases in turn may give valuable information about the metamorphic evolution of a terrain.

With respect to a "normal" pelite, it is obvious that the widespread and abundant chemical elements calcium, sodium, manganese, titanium and ferric iron will also be present to some degree in the even most mature pelitic sediments. These elements are found in any chemical analysis of a pelite and, hence, they must be found in some of the minerals that make up the rock. In the most common "AFM minerals" one finds many components that are not part of the AFM system.

Garnet often incorporates Ca, Mn, Fe^{3+} and OH in significant amounts. The spessartine component in prograde low-grade garnet is particularly important. Plagioclase is often present in metapelites and the mineral carries a great deal of the bulk rock's Na and Ca content. If no plagioclase is present in low- to medium-grade metapelites, the calcium content of the rock may be found entirely in garnet as grossular component or in a separate Ca-mineral, especially zoisite, clinozoisite or epidote. Biotite usually contains Ti, Mn, halogens and Fe^{3+}. Biotite can take up more than 5 wt% TiO_2 and it may be the only significant Ti-bearing mineral in the rock. If the bulk rock TiO_2-content exceeds Ti-saturation of biotite, some extra Ti-phase will be present (usually titanite or ilmenite). Staurolite in metapelitic rocks tends to grasp Mn and particularly Zn. Staurolite may contain several wt% zinc. Cordierite often accommodates Li and Be that replace Mg, Fe and Al on regular crystallographic cation sites and Na, CO_2 and H_2O in the structural channels. These "extra" elements in cordierite are difficult or impossible to analyze with standard microprobe techniques. Muscovite, as discussed above, is often phengitic. Spinel may contain a large variety of "non-AFM elements" including Fe^{3+}, Zn, Mn, Cr and V. Spinels often contain high concentrations of zinc (several wt%).

There are two major effects of extra components in "AFM minerals". The equilibrium conditions of mineral reactions can be displaced in pT space relative to the pure AFM system and, in the most extreme, the topology of phase relationships can be altered and inverted. In other words, mineral assemblages that are not stable in the pure AFM system can become conceivable as a result of strong fractionation of one "extra" element (e.g. zinc) into one "AFM mineral". Strong partitioning of manganese into garnet, for example causes the first garnet to appear in prograde metamorphism at a temperature that is about 50–100°C lower than in the pure AFM system. The other major effect of the presence of extra components is that it causes more minerals in an assemblage to appear than are usually expected in the pure AFM case. For example, in a typical general AFM situation, the minerals chlorite + biotite + staurolite can be found in a rock with muscovite + quartz in excess. In such a rock it will not be unusual to find also garnet as an additional mineral and as a true member of the assemblage owing to the garnet's manganese content. The coexistence of the four minerals does not mean, in this case, that the rock equilibrated at the conditions of the AFM discontinuous reaction (28) (Table 7.3). Garnet is in reality not a co-planar phase in terms of an AFM diagram but rather defines a phase volume in the AFM + Mn space.

Particularly difficult are high-grade assemblages involving spinel because the mineral often deviates significantly from pure Mg-Al-spinel – hercynite solutions and because it is usually involved in fluid-absent reactions which are especially sensitive to small deviations from the pure AFM system. Therefore, great care must be taken in the interpretation of spinel-bearing assemblages.

In conclusion, the observed common deviation of metapelites from the pure AFM composition may have significant consequences on phase relationships. The interpretation of phase relationships in metapelites, therefore, requires common petrologic sense and a great deal of professional experience. These difficulties, however, certainly do not invalidate the use of an AFM model for the prediction of phase relationships in metapelites.

References and Further Reading

Aranovich LY, Podlesskii KK (1983) The Cordierite-garnet-sillimanite-quartz equilibrium: experiments and applications. In: Saxena SK (ed) Kinetcs and equilibrium in mineral reactions. Springer, Berlin Heidelberg New York, pp 173–198

Aranovich LY, Podlesskii KK (1989) Geothermobarometry of high-grade metapelites: simultaneously operating reactions. In: Daly St, Cliff RA, Yardley BWD (eds) Evolution of metamorphic belts. Geological Society Special Publication. Blackwell, Oxford, pp 45–62

Ashworth JR (1985) Migmatites. Blackie, Glasgow, 301 pp

Barrow G, (1893) On an intrusion of muscovite biotite gneiss in the S.E. highlands of Scotland and its accompanying metamorphism. Q J Geol Soc Lond 49:330–358

Barrow G (1912) On the geology of lower Deeside and the southern highland border. Proc Geol Assoc 23:268–284

Bhattacharya A, Sen SK (1986) Granulite metamorphism, fluid buffering and dehydration melting in the Madras charnockites and metapelites. J Petrol 27:1119–1141

Black PM, Maurizot P, Ghent ED, Stout MZ (1993) Mg-Fe carpholites from aluminous schists in the Diaphot region and implications for preservation of high-pressure/low-temperature schists, northern new Caledonia. J Metamorph Geol 11:455–460

Blenkinsop TG (1988) Definition of low-grade metamorphic zones using illite crystallinity. J Metamorph Geol 6:623–636

Bohlen SR (1987) Pressure-temperature-time paths and a tectonic model for the evolution of granulites. J Geol 95:617–632

Bohlen SR (1991) On the formation of granulites. J Metamorph Geol 9:223–230

Bohlen SR, Wall VJ, Boettcher AL (1981) Geobarometry in granulites. In: Newton RC, Navrotsky A, Wood BJ (eds) Thermodynamics of minerals and melts. Advances in physical geochemistry, vol 1.Springer, Berlin Heidelberg New York, pp 141–172

Bucher-Nurminen K, Droop GTR (1983) The metamorphic evolution of garnet-cordierite-sillimanite gneisses of the Gruf-Complex, eastern Pennine Alps. Contrib Mineral Petrol 84:215–227

Carmichael DM (1968) On the mechanism of prograde metamorphic reactions in quartz-bearing pelitic rocks. Contrib Mineral Petrol 20:244–267

Carmichael DM (1970) Intersecting isograds in the Whetstone Lake area, Ontario. J Petrol 11:147–181

Chamberlain CP, Lyons JB (1983) Pressure, temperature, and metamorphic zonation studies of pelitic schists in the Merrimack synclinorium, south-central New Hampshire. Am Mineral 68:530–540

Chatterjee ND (1972) The upper stability limit of the assemblage paragonite + quartz and its natural occurrences. Contrib Mineral Petrol 34:288–303

Chinner GA (1967) Chloritoid and the isochemical character of Barrow's zone. J Petrol 8:268–282

Chopin C (1981) Talc-phengite, a widespread assemblage in high-grade pelitic blueschists of the western Alps. J Petrol 22:628–650

Chopin C (1983) Magnesiochloritoid, a key-mineral for the petrogenesis of high-grade pelitic blueschists. Bull Minéral 106:715–717

Chopin C (1984) Coesite and pure pyrope in high-grade blueschists of the western Alps: a first record and some consequences. Contrib Mineral Petrol 86:107–118

Chopin C, Schreyer W (1983) Magnesiocarpholite and magnesiochloritoid: two index minerals of pelitic blueschists and their preliminary phase relations in the model system MgO-Al₂O₃-SiO₂-H₂O. Am J Sci 283A:72–96

Chopin C, Henry C, Michard A (1991) Geology and petrology of the coesite-bearing terrain, Dora Maira massif, Western Alps. Eur J Mineral 3:263–291

Clarke GL, Powell R, Guirand M (1989) Low-pressure granulite facies metapelitic assemblages and corona textures from MacRobertson Land, east Antarctica: the importance of Fe₂O₃ and TiO₂ in accounting for spinel-bearing assemblages. J Metamorph Geol 7:323–336

Connolly JAD, Cesare B (1993) C-O-H-S fluid composition and oxygen fugacity in graphitic metapelites. J Metamorph Geol 11:379–388

Currie KL, Gittins J (1988) Contrasting sapphirine parageneses from Wilson Lake, Labrador and their tectonic implications. J Metamorph Geol 6:603–622

Curtis CD (1985) Clay mineral precipitation and transformation during burial diagenesis. Philos Trans R Soc Lond A315:91–105

Dempster TJ (1985) Garnet zoning and metamorphism of the Barrovian type area, Scotland. Contrib Mineral Petrol 89:30–38

Dougan TW (1983) Textural relations in melanosomes of selected specimens of migmatitic pelitic schists: implications for leucosome generating processes. Contrib Mineral Petrol 83:82–98

Droop GTR (1985) Alpine metamorphism in the south-east Tauern Window, Austria. 1. P-T variations in space and time. J Metamorph Geol 3:371–402

Droop GTR, Bucher-Nurminen K (1984) Reaction textures and metamorphic evolution of sapphirine-bearing granulites from the Gruf complex, Italian Central Alps. J Petrol 25:766–803

Dunoyer de Segonzac G (1970) The transformation of clay minerals during diagenesis and low grade metamorphism: a review. Sedimentology 15:281–346

Earley D III, Stout JH (1991) Cordierite-cummingtonite facies rocks from the Gold Brick District, Colorado. J Petrol 32; 6:1169–1201

Ellis DJ (1987) Origin and evolution of granulites in normal and thickened crusts. Geol 15:167–170

Ellis DJ, Sheraton JW, England RN, Dallwitz WB (1980) Osumilite-sapphirine-quartz-granulites from Enderby Land, Antarctica – mineral assemblages and reactions. Contrib Mineral Petrol 72:123–143

Elvevold S, Andersen T (1993) Fluid evolution during metamorphism at increasing pressure: carbonic- and nitrogen-bearing fluid inclusions in granulites from Øksfjord, north Norwegian Caledonides. Contrib Mineral Petrol 114:236–246

Enami M (1983) Petrology of pelitic schists in the oligoclase-biotite zone of the Sanbagawa metamorphic terrain, Japan: phase equilibria in the highest grade zone of a high-pressure intermediate type of metamorphic belt. J Metamorph Geol 1:141–161

Engel AEJ, Engel CE (1958) Progressive metamorphism and granitization of the major paragneiss, northwest Adirondack mountains, New York, part 1. Geol Soc Am Bull 69:1369–1414

Evans BW, Guidotti CV (1966) The sillimanite-potash feldspar isograd in western Maine, USA. Contrib Mineral Petrol 12:25–62

Evans NH, Speer JA (1984) Low-pressure metamorphism and anatexis of Carolina Slate Belt Phyllites in the contact aureole of the Lilesville pluton, North Carolina, USA. Contrib Mineral Petrol 87:297–309

Foster CT Jr (1991) The role of biotite as a catalyst in reaction mechanismus that form sillimanite. Can Mineral 29; 4:943–964

Fransolet A-M, Schreyer W (1984) Sudoite, di/trioctahedral chlorite: a stable low-temperature phase in the system MgO- Al_2O_3-SiO_2- H_2O. Contrib Mineral Petrol 86:409–417

Frey M (1987) The reaction-isograd kaolinite + quartz = pyrophyllite + H_2O, Helvetic Alps, Switzerland. Schweiz Mineral Petrogr Mitt 67:1–11

Frey M, Teichmüller M, Teichmüller R, Mullis J, Künzi B, Breitschmid A, Gruner U, Schwizer B (1980) Very low-grade metamorphism in external parts of the central Alps: illite crystallinity, coal rank and fluid inclusion data. Eclogae Geol Helv 73; 1:173–203

Frost BR, Frost CD (1987) CO_2, melts, and granulite metamorphism. Nature 327:503–506

Frost RB, Frost CD, Touret JLR (1989) Magmas as a source of heat and fluids in granulite metamorphism. In: Bridgewater D (ed) Fluid movements – element transport, and the composition of the deep crust. Kluwer Dordrecht, pp 1–18

Giaramita MJ, Day HW (1991) The four-phase AFM assemblage staurolite- aluminum-silicate-biotite-garnet: extra components and implications for staurolite-out isograds. J Petrol 32:1203–1230

Giaramita MJ, Day HW (1992) Buffering in the assemblage staurolite- aluminium silicate-biotite-garnet-chlorite. J Metamorph Geol 9; 4:363–378

Grant JA (1981) Orthoamphibole and orthopyroxene relations in high-grade metamorphism of pelitic rocks. Am J Sci 281:1127–1143

Grant JA (1985) Phase equilibria in low-pressure partial melting of pelitic rocks. Am J Sci 285:409–435

Grew ES (1980) Sapphirine and quartz association from Archean rocks in Enderby Land, Antactica. Am Mineral 65:821–836

Grew ES (1988) Kornerupine at the Sar-e-Sang, Afghanistan, whiteschist locality: implications for tourmaline-kornerupine distribution in metamorphic rocks. Am Mineral 73; 3–4:345–357

Guiraud M, Holland T, Powell R (1990) Calculated mineral equilibria in the greenschist-blueschist-eclogite facies in the system Na_2O-FeO-MgO-Al_2O_3-SiO_2-H_2O: methods, results and geological applications. Contrib Mineral Petrol 104:85–98

Harley SL (1983) Regional geobarometry-geothermometry and metamorphic evolution of Endreby Land, Antarctica. Antarctic Earth sciences. Cambridge University Press, Cambridge 25–30

Harley SL (1984) Comparison of the garnet-orthopyroxene geobarometer with recent experimental studies, and applications to natural assemblages. J Petrol 25:697–712

Harley SL (1986) A sapphirine-cordierite-garnet-sillimanite granulite from Enderby Land Antarctica: implications for FMAS petrogenetic grids in the granulite facies. Contrib Mineral Petrol 94:452–460

Harley SL (1989) The origins of granulites: a metamorphic perspective. Geol Mag 126:215–247

Harley SL, Fitzsimons IC (1991) P-T evolution of metapelitic granulites in a polymetamorphic terrane: the Rauer Group, East Antarctica. J Metamorph Geol 9:231–244

Harte B, Hudson NFC (1979) Pelite facies series and the temperatures and pressures of Dalradian metamorphism in E. Scotland. Geological Society, London, pp 323–337

Hesse R, Dalton E (1992) Diagenetic and low-grade metamorphic terranes of Gaspé Peninsula related to the geological structure of the Taconian and Acadian orogenic belts, Quebec Appalachians. J Metamorph Geol 9; 6:775–790

Holdaway MJ, Mukhopadhyay B (1993) Geothermobarometry in pelitic schists: a rapidly evolving field. Am Mineral 78:681–693

Hollister LS (1966) Garnet zoning: an interpretation based on the Rayleigh fractionation model. Science 154:1647–1651

Kerrick DM (1988) Al_2SiO_5-bearing segregations in the Lepontine Alps, Switzerland: aluminum mobility in metapelites. Geology 16:636–640

Kisch HJ (1980) Incipient metamorphism of Cambro-Silurian clastic rocks from the Jämtland Supergroup, central Scandinavian Caledonides, western Sweden: illite crystallinity and 'vitrinite' reflectance. J Geol Soc Lond 137:271–288

Kisch HJ (1992) Development of slaty cleavage and degree of very low-grade metamorphism: a review. J Metamorph Geol 9; 6:735–750

Kohn MJ, Spear FS (1993) Phase equilibria of margarite-bearing schists and chloritoid + hornblende rocks from western New Hampshire, USA. J Petrol 34:631–651

Lang HM (1991) Quantitative interpretation of within-outcrop variation in metamorphic assemblage in staurolite-kyanite-grade metapelites, Baltimore, Maryland. Can Mineral 29; 4:655–672

Lonker SW (1980) Conditions of metamorphism in high-grade pelites from the Frontenac Axis, Ontario, Canada. Can J Sci 17:1666–1684

Lonker SW (1981) The P-T-X relations of the cordierite-garnet-sillimanite-quartz equilibrium. Am J Sci 281:1056–1090

Loomis TP (1983) Compositional zoning of crystals: a record of growth and reaction history. Advances in physical geochemistry. In: Saxena SK (ed) Kinetics and equilibrium in mineral reactions. Advances in physical geochemistry; vol 3. Springer, Berlin Heidelberg New York, pp 1–60.

Loomis TP (1986) Metamorphism of metapelites: calculations of equilibrium assemblages and numerical simulations of the crystallization of garnet. J Metamorph Geol 4:201–229

Mather JD (1970) The biotite isograd and the lower greenschist facies in the Dalradian rocks of Scotland. J Petrol 11:253–275

Miyashiro A, Shido F (1985) Tschermak substitution in low- and middle-grade pelitic schists. J Petrol 26:449–487

Mohr DW, Newton RC (1983) Kyanite-staurolite metamorphism in sulfidic schists of the Anakeesta formation, Great Smoky Mountains, North Carolina. Am J Sci 283:97–134

Munz IA (1990) Whiteschists and orthoamphibole-cordierite rocks and the P-T-t path of the Modum Complex, S. Norway. Lithos 24:181–200

Newton RC, Smith JV, Windley BF (1980) Carbonic metamorphism, granulites and crustal growth. Nature 288:45–50

Nichols GT, Berry RF, Green DH (1992) Internally consistent gahnitic spinel-cordierite-garnet equilibria in the FMASHZn system: geothermobarometry and applications. Contrib Mineral Petrol 111:362–377

Pattison DRM (1987) Variations in Mg/(Mg + Fe), F, and (Fe, Mg) Si = 2Al in pelitic minerals in the Ballachulish thermal aureole, Scotland. Am Mineral 72; 2–3:255–272

Powell R, Holland T (1990) Calculated mineral equilibria in the pelite system, KFMASH (K$_2$O-FeO-MgO-Al$_2$O$_3$-SiO$_2$-H$_2$O). Am Mineral 75:367–380

Reinhardt J (1987) Cordierite-anthopyllite rocks from north-west Queensland, Australia: metamorphosed magnesian pelites. J Metamorph Geol 5; 4:451–472

Robinson D, Warr LN, Bevins RE (1990) The illite 'crystallinity' technique: a critical appraisal of its precision. J Metamorph Geol 8:333–344

Schreyer W (1973) Whiteschist: a high pressure rock and its geological significance. J Geol 81:735–739

Schreyer W (1977) Whiteschists: their compositions and pressure–temperature regimes based on experimental, field and petrographic evidence. Tectonophysics 3:127–144

Schumacher JC, Hollocher KT, Robinson P, Tracy RJ (1990) Progressive metamorphism and melting in central Massachusetts and southwestern New Hampshire, USA. In: Ashworth JR, Brown M (eds) High-temperature metamorphism and crustal anatexis. The Mineralogical Society Series 2. Unwin Hynman, London, pp 198–234

Selverstone J, Spear FS (1985) Metamorphic P-T paths from pelitic schists and greenstones from the south-west Tauern window, eastern Alps. J Metamorph Geol 3:439–465

Selverstone J, Spear FS, Franz G, Morteani G (1984) High-pressure metamorphism in the SW Tauern window, Austria: P-T paths from hornblende-kyanite-staurolite schists. J Petrol 25:501–531

Shaw DM (1956) Geochemistry of pelitic rocks. Part III. Major elements and general geochemistry: Geological Society of Am Bull 67:919–934

Spear FS, Cheney JT (1989) A petrogenetic grid for pelitic schists in the system SiO_2-Al_2O_3-FeO- MgO-K_2O-H_2O. Contrib Mineral Petrol 101:149–164

Spear FS, Kohn MJ, Florence FP, Menard T (1990) A model for garnet and plagioclase growth in pelitic schists: implications for thermobarometry and P-T path determinations. J Metamorph Geol 8:683–696

Symmes GH, Ferry JM (1992) The effect of whole-rock MnO content on the stability of garnet in pelitic schists during metamorphism. J Metamorph Geol 10; 2:221–238

Theye T, Seidel E, Vidal O (1992) Carpholite, sudoite, and chloritoid in low-grade high-pressure metapelites from Crete and the Peloponnese, Greece. Eur J Mineral 4:487–507

Thompson AB (1982) Dehydration melting of pelitic rocks and the generation of H_2O-undersaturated granitic liquids. Am J Sci 282:1567–1595

Touret J, Dietvorst P (1983) Fluid inclusions in high-grade anatectic metamorphites. J Geol Soc Lond 140:635–649

Tracy RJ, Robinson P (1983) Acadian migmatite types in pelitic rocks of Central Massachusetts. In: Atherton MP, Gribble CD (eds) Migmatites, melting and metamorphism. Nantwich, Shiva, pp 163–173

Tracy RJ, Robinson P (1988) Silicate-sulfide-oxide-fluid reaction in granulite-grade pelitic rocks, central Massachusetts. Am J Sci 288-A:45–74

Vidal O, Goffe B, Theye T (1992) Experimental study of the stability of sudoite and magnesiocarpholite and calculation of a new petrogenetic grid for the system FeO-MgO-Al_2O_3-SiO_2-H_2O. J Metamorph Geol 10; 5:603–614

Waters DJ (1991) Hercynite-quartz granulites: phase relations and implications for crustal processes. Eur J Mineral 3:367–386

Weaver BL, Tarney J (1983) Elemental depletion in Archaean granulite facies rocks. In: Atherton MP, Gribble CD (eds) Migmatites, melting and metamorphism. Nantwich, Shiva, pp 250–263.

Wenk E (1968) Cordierit im Val Verzasca. Schweiz Mineral Petrogr Mitt 48:455–457

Yardley BWD (1977) The nature and significance of the mechanism of sillimanite growth in the Connemara Schists, Ireland. Contrib Mineral Petrol 65:53–58

Yardley BWD, Leake BE, Farrow CM (1980) The metamorphism of Fe-rich pelites from Connemara, Ireland. J Petrol 21:365–399

8 Metamorphism of Marls

8.1 General

Marls are carbonate-bearing pelitic sediments covering a large range of composition between "impure" carbonate rocks and "true" pelites. In the Anglo-American literature, this group of sedimentary rocks is better known as argillaceous carbonate rocks, calcareous sediments, calcic pelitic rocks, or calc-pelites.

Unmetamorphosed marls are composed of mixtures of clay minerals (smectite, illite, kaolinite, chlorite etc.), carbonates (mainly calcite and/or dolomite), quartz and feldspars in varying proportions. In metamorphosed marls (metamarls), most minerals occurring in metapelites and metacarbonates may be present, plus additional Ca-Al-bearing silicates like Ca-amphiboles, epidote-group minerals, lawsonite, margarite, scapolite and vesuvianite. Such a large group of possible mineral constituents requires a very complex chemical system: K_2O-Na_2O-CaO-FeO-MgO-Al_2O_3-SiO_2-H_2O-CO_2. The approach to an understanding of such a complex system involves the study of simpler subsystems. In this chapter, three such subsystems will be discussed, one dealing with Al-poor and two with Al-rich metamarls. In all three examples, quartz and calcite will be considered to be present in excess, which is true for most low- and medium-temperature metamarls (although quartz and/or calcite may be used up at high temperature). Therefore, SiO_2 will be treated as an excess component. In addition, a H_2O-CO_2-bearing fluid phase is assumed to have been present during metamorphism. Chemographic relationships of any metamarl system can then be projected from SiO_2, H_2O and CO_2.

8.2 Orogenic Metamorphism of Al-Poor Marls

Al-poor marls are common in many orogenic belts. In the northern Appalachians, as an example, the prograde metamorphism of such rocks has been studied in great detail (e.g. Ferry 1976, 1983a, b, 1992; Hewitt 1973; Zen 1981). For the Vassalboro Formation in Maine, Ferry mapped five mineral zones, each characterized by an index mineral and separated by reaction-isograds. With increasing metamorphic grade, the following index minerals were encountered: ankerite, biotite, Ca-amphibole, zoisite and diopside. A summary of mineral assemblages and related reactions is given by Yardley (1989, pp. 143–145). Below we shall refer to these reactions and present a simplified model for the progressive metamorphism of Al-poor marls similar to those described by Ferry.

8.2.1 Phase Relationships in the KCMAS-HC System

For the system $K_2O–CaO–MgO–Al_2O_3-SiO_2-H_2O-CO_2$, phase relationships are considered among anorthite, calcite, clinochlore, diopside, dolomite, K-feld-

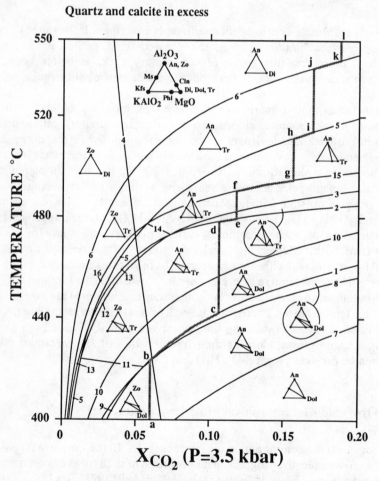

Fig. 8.1. TX phase relationships for the KCMAS-HC system with excess quartz and calcite at $0 < X_{CO_2} < 0.2$ and a constant pressure of 3.5 kbar. Pure end-member mineral compositions are used. The *inset* shows the chemography projected from quartz, calcite, H_2O and CO_2 onto the $KAlO_2-MgO-Al_2O_3$ plane. Compatibility diagrams are shown for some selected divariant fields only. The prograde path *a-l* is emphasized and discussed in Section 8.2.2

Table 8.1. Reactions in the KCMAS–HC system

Anorthite	$CaAl_2Si_2O_8$
Calcite	$CaCO_3$
Clinochlore	$Mg_5Al_2Si_3O_{10}(OH)_8$
Diopside	$CaMgSi_2O_6$
Dolomite	$CaMg(CO_3)_2$
K–feldspar	$KAlSi_3O_8$
Muscovite	$KAl_3Si_3O_{10}(OH)_2$
Phlogopite	$KMg_3AlSi_3O_{10}(OH)_2$
Quartz	SiO_2
Tremolite	$Ca_2Mg_5Si_8O_{22}(OH)_2$
Zoisite	$Ca_2Al_3Si_3O_{12}(OH)$

All reactions with excess quartz, calcite and fluid

$(1) \quad Ms + 8\ Dol + 3\ Qtz + 4\ H_2O = Phl + Cln + 8\ Cal + 8\ CO_2$

$(2) \quad 5\ Ms + 3\ Cln + 7\ Qtz + 8\ Cal = 5\ Phl + 8\ An + 12\ H_2O + 8\ CO_2$

$(3) \quad Cln + 7\ Qtz + 3\ Cal = Tr + An + 3\ H_2O + 3\ CO_2$

$(4) \quad 2\ Zo + CO_2 = 3\ An + Cal + H_2O$

$(5) \quad 5\ Phl + 24\ Qtz + 6\ Cal = 5\ Kfs + 3\ Tr + 2\ H_2O + 6\ CO_2$

$(6) \quad Tr + 2\ Qtz + 3\ Cal = 5\ Di + H_2O + 3\ CO_2$

$(7) \quad Cln + 6\ Cal + 4\ CO_2 = 5\ Dol + An + Qtz + 4\ H_2O$

$(8) \quad Phl + 3\ Cal + 3\ CO_2 = Kfs + 3\ Dol + H_2O$

$(9) \quad Ms + 5\ Dol + 3\ Qtz + 3\ H_2O = Kfs + Cln + 5\ Cal + 5\ CO_2$

$(10) \quad 5\ Dol + 8\ Qtz + H_2O = Tr + 3\ Cal + 7\ CO_2$

$(11) \quad 8\ Kfs + 3\ Cln = 3\ Ms + 5\ Phl + 9\ Qtz + 4\ H_2O$

$(12) \quad 5\ Phl + 5\ Cln + 49\ Qtz + 16\ Cal = 5\ Ms + 8\ Tr + 12\ H_2O + 16\ CO_2$

$(13) \quad 3\ Cln + 21\ Qtz + 10\ Cal = 3\ Tr + 2\ Zo + 8\ H_2O + 10\ CO_2$

$(14) \quad 5\ Ms + 3\ Tr + CO_2 = 5\ Phl + 5\ An + 14\ Qtz + Cal + 3\ H_2O$

$(15) \quad Ms + 2\ Qtz + Cal = Kfs + An + H_2O + CO_2$

$(16) \quad 3\ Ms + 6\ Qtz + 4\ Cal = 3\ Kfs + 2\ Zo + 2\ H_2O + 4\ CO_2$

spar, muscovite, phlogopite, quartz, tremolite and zoisite. The chemography of these phases is shown as inset in Fig. 8.1 projected from $SiO_2 + CaO + H_2O + CO_2$. Reaction equilibria in the KCMAS-HC system with excess quartz, calcite and H_2O-CO_2 are listed in Table 8.1 and depicted in Fig. 8.1 in an isobaric TX-section. The pressure of 3.5 kbar has been chosen in accordance with Ferry (1983a, b), derived from metapelites near the sillimanite isograd. The range of $X_{CO_2} < 0.2$ has been chosen according to Ferry (1983b, Table 2).

The petrogenetic grid in Fig. 8.1 will now be considered in some detail. First, we shall characterize some of the reactions followed by an analysis of the stability fields of some selected phases. Mixed-volatile reactions with steep (at $X_{CO_2} < 0.05$) to gentle (at $X_{CO_2} > 0.1$) positive slopes are predominant. Reactions (6) and (10) were already encountered in Chapter 6 discussing the metamorphism of dolomites and limestones. Reactions (11) and (14) with a gentle negative slope are dehydration reactions, and reaction (11) was already mentioned in Chapter 7 dealing with the metamorphism of pelites. Reaction (4), with a steep negative slope and crossing the whole diagram of Fig. 8.1, is a special case of a mixed-volatile reaction of the type $a + CO_2 = b + H_2O$ (cf. Fig. 3.17). This reaction is dividing the diagram into zoisite-bearing assemblages to the left and anorthite-bearing assemblages to the right. Because diopside, dolomite and tremolite coincide in the chemographic projection with excess quartz, calcite and fluid (see inset of Fig. 8.1), the TX stability fields of these three phases do not overlap. Diopside is stable above reaction (6), tremolite is stable between reactions (6) and (10), and dolomite is stable below reaction (10). K-feldspar is stable over the whole diagram because of its single corner position in the chemography. For clinochlore, reaction (7) defines its lower and reactions (3) and (13) its upper stability limits. Phlogopite is stable between reactions (8) and (5). For muscovite, finally, reactions (15) and (16) limit its TX stability field towards high temperatures and very low X_{CO_2}.

The sequence of index minerals observed in the Vassalboro Formation by Ferry, i.e. ankerite, biotite, Ca-amphibole, zoisite, diopside is approximated in the KCMAS-HC system by dolomite, phlogopite, tremolite, zoisite, diopside. In Fig. 8.1, in terms of $T-X_{CO_2}$ values, this means an increase in temperature and, after leaving the tremolite zone, and in order to reach the zoisite zone, a decrease in X_{CO_2} caused by an infiltration of H_2O. This conclusion is in accordance with the findings of Ferry (1976, 1983a, b). Furthermore, the relative positions of reactions (1) – (6) in TX space of Fig. 8.1 closely match the sequence of reaction-isograds mapped by Ferry (1976). The main difference between the isograd reactions as formulated by Ferry (1976) and reactions (1) – (6) of Table 8.1 concerns plagioclase. Albite and intermediate plagioclase are reactants in reactions (2) and (3), as formulated by Ferry, but the albite component in plagioclase has not been considered in the KCMAS-HC system dealt with here.

Summary. The progressive metamorphism of Al-poor marls described by Ferry (1976, 1983a, b) can be modelled in the system KCMAS-HC, at least in a qualitative way. A quantitative treatment would require two additional components, FeO and Na_2O, or the use of activity terms. Such a procedure is, however, beyond the scope of this book.

8.2.2 Prograde Metamorphism in the KCMAS-HC System at Low X_{CO_2}

In this section, the prograde metamorphism of a rock consisting of Ms + Dol with excess Qtz + Cal and an initial fluid composition of $X_{CO_2} = 0.06$ will be considered. The prograde path in TX_{CO_2} space is shown in Fig. 8.1 by a curve running from a to l and the changing mineral modal composition is displayed in Fig. 8.2. It is important to note that this calculation, using the computer program THERIAK (De Capitani and Brown 1987), is valid only for a specific bulk composition. The prograde path can be divided into several steps as follows.

a-b) (Fig. 8.1) The starting mineral assemblage Dol + Ms + Qtz + Cal is heated up at a constant X_{CO_2} of 0.06.

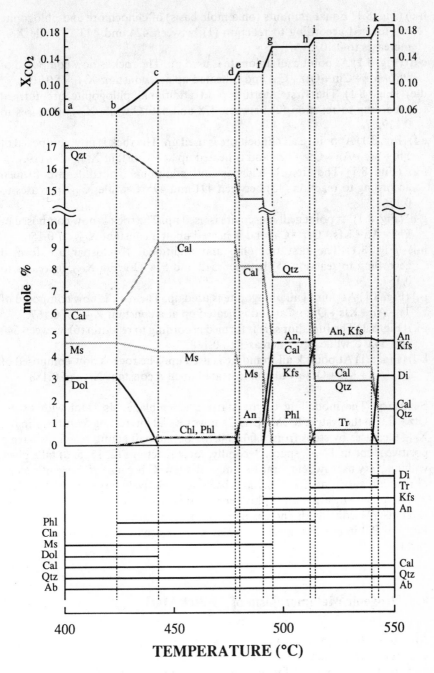

Fig. 8.2. Modal composition (mol%), fluid composition (X_{CO_2}) and mineral distribution along the prograde path *a-l* shown in Fig. 8.1

b-c) (Fig. 8.1) Equal amounts (on a mole basis) of clinochlore and phlogopite are formed according to reaction (1) between 424 and 443°C while X_{CO_2} increases to 0.107.

c-d) (Fig. 8.1) At point c all dolomite is used up. The rock is now composed of Phl + Ms + Cln + Qtz + Cal and is heated up at a constant X_{CO_2} of 0.107.

d-e) (Fig. 8.1) The first anorthite and additional phlogopite are formed according to reaction (2) between 478 and 480°C while X_{CO_2} increases to 0.119.

e-f) (Fig. 8.1) At point e all clinochlore is used up. The rock is now composed of Phl + Ms + An + Qtz + Cal and is heated up at a constant X_{CO_2} of 0.119.

f-g) (Fig. 8.1) The first K-feldspar and additional anorthite are formed according to reaction (15) between 491 and 495°C while X_{CO_2} increases to 0.156.

g-h) (Fig. 8.1) At point g all muscovite is used up. The rock is now composed of Phl + An + Kfs + Qtz + Cal and is heated up at a constant X_{CO_2} of 0.156.

h-i) (Fig. 8.1) The first tremolite and additional K-feldspar are formed according to reaction (5) between 512 and 514°C while X_{CO_2} increases to 0.167.

i-j) (Fig. 8.1) At point i all phlogopite is used up. The rock is now composed of Tr + An + Kfs + Qtz + Cal and is heated up at a constant X_{CO_2} of 0.167.

j-k) (Fig. 8.1) The first diopside is formed according to reaction (6) between 540 and 543°C while X_{CO_2} increases to 0.188.

k-l) (Fig. 8.1) At point k all tremolite is used up. The rock is now composed of Di + An + Kfs + Qtz + Cal and is heated up at a constant X_{CO_2} of 0.188.

Summary. The modelling of the starting assemblage Ms + Dol with excess Qtz + Cal in the system KCMAS-HC shows several interesting features. Firstly, X_{CO_2} increases by steps from 0.06 to 0.188 because all acting reactions have a positive slope in TX_{CO_2} space. Secondly, most reactions (2, 15, 5, 6) take place within a very narrow temperature range of a few °C because of the gentle slope of reaction curves at $X_{CO_2} > 0.1$ and because a relatively large amount of fluid was assumed. In a metamorphic terrain, this would lead to sharp reaction-isograds. Thirdly, with increasing metamorphic grade sheet silicates (Cln, Ms, and Phl in our example) are replaced by feldspars and chain silicates (Tr and Di).

8.3 Orogenic Metamorphism of Al-Rich Marls

Al-rich marls are present in many orogenic belts. In the Alps, as an example, such rocks are widespread, both in platform sediments of the Helvetic realm and in deep-sea metasediments of the Penninic realm. Their mineralogic transformation has been studied in some detail (e.g. Frey 1978; Bucher et al. 1983; Frank 1983).

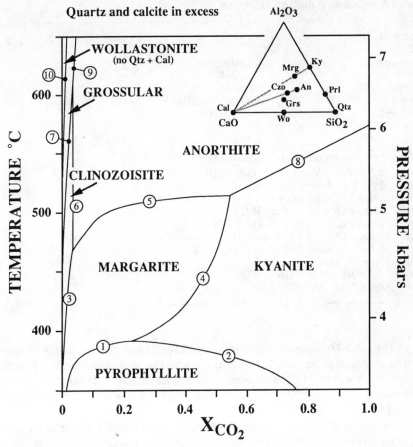

Fig. 8.3. PTX phase relationships for the CAS-HC system with excess quartz and calcite. The *vertical axis* represents an orogenic geotherm characteristic for kyanite-type terrains. The *inset* shows the chemography projected from H_2O and CO_2 onto the CaO-Al_2O_3-SiO_2 plane. Two collinearities corresponding to reactions (4) and (6) are indicated

8.3.1 Phase Relationships in the CAS-HC System

The system CaO-Al_2O_3-SiO_2-H_2O-CO_2 demonstrates phase relationships among anorthite, calcite, clinozoisite, grossular, kyanite, margarite, pyrophyllite, quartz and wollastonite. The chemography of these phases is shown as inset in Fig. 8.3 as projected from $H_2O + CO_2$. Note the presence of several collinearities among three phases in this system, e.g. Ky-Prl-Qtz, Ky-Mrg-Cal, An-Czo-Cal. Reaction equilibria in the CAS-HC system with excess quartz, calcite and H_2O-CO_2 are listed in Table 8.2 and depicted in Fig. 8.3 in a polybaric TX-section along the Ky-geotherm. According to the phase rule (cf. Chap. 3.8.1), each divariant assemblage consists of one single phase (in addition to Qtz + Cal + Fluid).

Table 8.2. Reactions in the CAS–HC system

Anorthite	$CaAl_2Si_2O_8$
Calcite	$CaCO_3$
Clinozoisite	$Ca_2Al_3Si_3O_{12}(OH)$
Grossular	$Ca_3Al_2Si_3O_{12}$
Kyanite	Al_2SiO_5
Margarite	$CaAl_4Si_2O_{10}(OH)_2$
Pyrophyllite	$Al_2Si_4O_{10}(OH)_2$
Quartz	SiO_2
Wollastonite	$CaSiO_3$

All reactions with excess quartz, calcite and fluid

- (1) $2\,Prl + Cal = Mrg + 6\,Qtz + H_2O + CO_2$
- (2) $Prl = Ky + 3\,Qtz + H_2O$
- (3) $3Mrg + 5\,Cal + 6\,Qtz = 4\,Czo + H_2O + 5\,CO_2$
- (4) $Mrg + CO_2 = 2\,Ky + Cal + H_2O$
- (5) $Mrg + 2\,Qtz + Cal = 2\,An + H_2O + CO_2$
- (6) $2\,Czo + CO_2 = 3\,An + Cal + H_2O$
- (7) $2\,Czo + 3\,Qtz + 5\,Cal = 3\,Grs + H_2O + 5\,CO_2$
- (8) $Ky + Qtz + Cal = An + CO_2$
- (9) $An + Qtz + 2\,Cal = Grs + 2\,CO_2$
- (10) $Qtz + Cal = Wo + CO_2$

In the CAS-HC system, the assemblage Prl-Qtz-Cal represents the characteristic assemblage in Al-rich marls at the onset of metamorphism. The fluid composition will be H_2O-rich because sheet silicates have not yet reacted with carbonates (liberating CO_2), and the interstitial water is therefore not yet diluted by CO_2.

Margarite will form during prograde metamorphism according to reaction (1) at temperatures below 390 °C, and this temperature will be lower than 350 °C for very H_2O-rich fluid compositions ($X_{CO_2} < 0.02$). However, as a consequence of the steep equilibrium position of reaction (1) at $X_{CO_2} < 0.05$ (Fig. 8.3), a noticeable amount of margarite will be produced between 375 and 390 °C only. Therefore, the first occurrence of margarite should represent a mappable reaction-isograd at low greenschist facies conditions. Assuming a low pyrophyllite modal content and closed system behaviour, reaction (1) will consume all pyrophyllite before reaching the invariant point involving Prl + Mrg + Ky (+ Qtz + Cal). The metamarl will now enter the divariant margarite field, the assemblage Mrg + Qtz + Cal being characteristic for the greenschist facies. It should be pointed out that, at such low metamorphic grade, the Ca-Al mica margarite will be very fine-grained and may be easily missed in thin sections, and X-ray diffraction work is needed (see e.g. Frey 1978, p.110).

Next, the rock will meet reaction (5) between 470 and 510 °C, depending on fluid composition, producing **anorthite component in plagioclase**. As discussed in Section 8.3.2, however, in margarite-bearing rocks plagioclase will be formed already at considerably lower temperatures, if Na is considered as an additional component.

Clinozoisite is formed by reaction (3), but the steep course of this reaction in PTX space and the corresponding small change in X_{CO_2} allows only for very minor modal clinozoisite. At the invariant point involving Czo + Mrg + An (+ Qtz + Cal), anorthite component in plagioclase is produced at the expense of all other solid phases present. Depending on whether clinozoisite or margarite is used up first, reaction boundary (6) or (5) would be followed respectively. Because of the small amount of clinozoisite formed by reaction (3), it is conceivable to assume that this phase will be used up first. Following reaction (5), more anorthite component in plagioclase is produced. Another and more effective way of clinozoisite formation is by interaction of a metamarl with an externally derived H_2O-rich fluid driving reactions (3) or (6).

Grossular and wollastonite do not form in orogenic metamorphic rocks under closed–system conditions. According to Fig. 8.3, these minerals may be generated by reactions (7) or (9) in the case of grossular and by reaction (10) in the case of wollastonite by interaction with a H_2O-rich fluid. Such metamarls were described, e.g. by Trommsdorff (1968) and by Gordon and Greenwood (1971).

Summary. In orogenic metamorphic terrains, relatively Al-rich but quartz- and calcite-bearing metamarls (Fig. 8.3) will show two index minerals with prograde metamorphism: pyrophyllite under subgreenschist and margarite under greenschist facies conditions. Grossular- and wollastonite-bearing metamarls are diagnostic for interaction with an externally derived H_2O-rich fluid.

8.3.2 Phase Relationships in the KNCAS-HC System

Addition of K_2O and Na_2O to the CAS-HC system discussed above allows discussion of the phase relations among four extra phases: muscovite, paragonite, K-feldspar and albite. The chemography for the system KNCAS-HC is shown as inset in Fig. 8.4. Grs and Wo have been omitted for simplicity, and all remaining phases considered in the CAS-HC system are coincident in the Al_2O_3 corner (projection from excess Qtz + Cal + H_2O + CO_2). The six additional equilibria are listed in Table 8.3 and depicted in Fig. 8.4. Phase relationships in this figure indicate the following maximum thermal stability for pure end–member white mica in the presence of excess quartz and calcite: 515°C for margarite, 540°C for paragonite, and 590°C for muscovite. These temperature values are valid only for the kyanite geotherm. Note that equilibria (5), (12), and (15) are all mixed volatile reactions with identical coefficients for all the phases involved, except for feldspars: 2 mol An are produced from Mrg + 2 Qtz + Cal, 1 mol each of Ab and An from Pg + 2 Qtz + Cal, and 1 mol each of Kfs and An from Ms + 2 Qtz + Cal.

The petrogenetic grid of Fig. 8.4 is not yet applicable to natural rocks because of solid solution of white micas and feldspars. For simplicity, only the effect of plagioclase solid solution will be quantified below. Anorthite involving reactions from the grid of Fig. 8.4 may be contoured for different An content

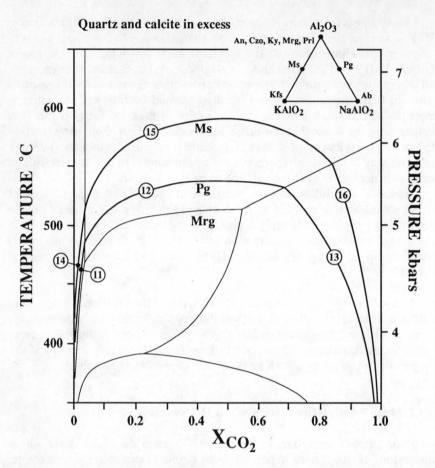

Fig. 8.4. PTX phase relationships for the KNCAS-HC system with excess quartz and calcite. Pure end–member mineral compositions are used. The *vertical axis* represents an orogenic geotherm characteristic for kyanite-type terrains. Note the maximum thermal stabilities for margarite, paragonite and muscovite. The inset shows the chemography projected from quartz, calcite, H_2O and CO_2 onto the $KAlO_2$-$NaAlO_2$-Al_2O_3 plane

Table 8.3. Reactions in the KNCAS–HC system (in addition to reactions of Table 8.2

Muscovite	$KAl_3Si_3O_{10}(OH)_2$
Paragonite	$NaAl_3Si_3O_{10}(OH)_2$
K–feldspar	$KAlSi_3O_8$
Albite	$NaAlSi_3O_8$

All reactions with excess quartz, calcite and fluid

(11) $3\,Pg + 6\,Qtz + 4\,Cal = 3\,Ab + 2\,Czo + 2\,H_2O + 4\,CO_2$
(12) $Pg + 2\,Qtz + Cal = Ab + An + H_2O + CO_2$
(13) $Pg + Qtz = Ab + Ky + H_2O$
(14) $3\,Ms + 6\,Qtz + 4\,Cal = 3\,Kfs + 2\,Czo + 2\,H_2O + 4\,CO_2$
(15) $Ms + 2\,Qtz + Cal = Kfs + An + H_2O + CO_2$
(16) $Ms + Qz = Kfs + Ky + H_2O$

Quartz and calcite in excess

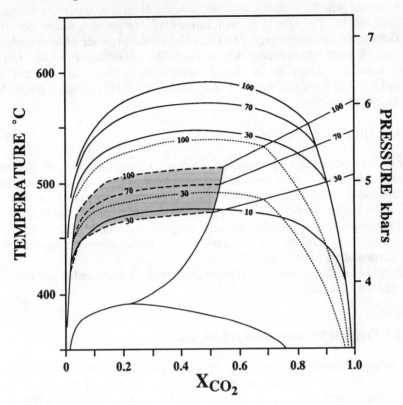

Fig. 8.5. PTX phase relationships for the KNCAS-HC system with excess quartz and calcite. The *vertical axis* represents an orogenic geotherm characteristic for kyanite-type terrains. Polybaric divariant equilibria for a particular composition of plagioclase, An_x, are shown as *full lines* (muscovite-involving reactions), *stippled lines* (paragonite-involving reactions) or *dashed lines* (margarite-involving reactions). *Figures* refer to mol% An-content in the plagioclase. Most clinozoisite-involving equilibria from Fig. 8.4 have been omitted for clarity. The stability field for the assemblage Mrg + Pl (+ Qtz + Cal) is emphasized in *grey colour*

in the plagioclase as depicted in Fig. 8.5. The stability field of Pl + Mrg + Qtz + Cal is emphasized in grey colour. Its low-temperature boundary is fixed by reaction (5) contoured for a plagioclase composition of An_{30}, based on field observations by Frey and Orville (1974). The effect of margarite-paragonite solid solution would shift the grey–coloured field to higher temperatures, but would also expand it (because of increasing solubility of paragonite component in margarite with increasing temperature). Compared with Fig. 8.4, the stability field of Mrg + Qtz + Cal in Fig. 8.5 is reduced considerably, with an upper temperature limit of about 470°C.

In contrast to margarite-bearing rocks, albite does occur in paragonite-bearing metamarls of the greenschist facies (e.g. Ferry 1992). According to

Fig. 8.5, the equilibrium curve according to reaction (12) for Pl(An30) is stable up to about 490°C, that is about 40°C lower than the maximum thermal stability of Pl + Pg + Qtz + Cal. An interesting feature is noticed for the breakdown of the assemblage Pg + Qtz + Cal in the presence of intermediate plagioclase composition (about An_{50} – An_{70}). The assemblage Pg + Qtz + Cal is now breaking down at a lower temperature than the assemblage Mrg + Qtz + Cal (compare with Fig. 8.4), and reaction (12) is then replaced by the following reaction:

$$2\ Pg + 2\ Qtz + Cal = 2\ Ab + Mrg + H_2O + CO_2 \tag{17}$$

For a given plagioclase composition, contours of reactions (5) and (17) are separated by a few degrees only. As an example, for a plagioclase composition of An_{70} and for fluid compositions of $X_{CO_2} = 0.2$–0.5, equilibrium (17) is located 5°C below equilibrium (5), but for graphical reason this is not shown in Fig. 8.5.

The production of K-feldspar and plagioclase from Ms + Qtz + Cal for plagioclase compositions > An30 takes place at higher temperatures than the breakdown of Pg + Qtz + Cal (Fig. 8.5).

Prograde metamorphism of Al-rich marls will be discussed separately for low and very low X_{CO_2}.

8.3.2.1 Prograde Metamorphism at Low X_{CO_2}

Let us now consider the prograde metamorphism of a rock consisting of Ms + Pg + Prl + Qtz + Cal and an initial fluid composition of $X_{CO_2} = 0.1$. The beginning of the metamorphic evolution will be identical to that described for the CAS-HC system in Fig. 8.3, i.e. formation of margarite according to reaction (1), Table 8.2. The resulting mineral assemblage with Ms + Pg + Mrg + Qtz + Cal is characteristic for the lower greenschist facies, and such assemblages with three coexisting white micas have been described, e.g. by Frey (1978). Whether a small amount of sodic plagioclase is already formed from Pg + Qtz + Cal at this grade of metamorphism is not documented in the literature. Between 450 and 470°C, the first plagioclase of composition An_{30} is produced by reaction (5). Within the stability field of margarite + plagioclase + quartz + calcite (grey–coloured area in Fig. 8.5), phase relationships are complex and depend, besides PTX conditions, on the modal rock composition considered, including the amount of fluid. Margarite and paragonite will react, together with quartz and calcite, to form plagioclase according to reactions (5) and (12), respectively. On the other hand, reaction (17) will produce some additional margarite. Concomitantly, the fluid composition, X_{CO_2}, will slightly increase. If during prograde metamorphism the upper limit of the grey–coloured area in Fig. 8.5 is reached, the net result will be the consumption of all margarite. For most marly compositions, also all paragonite will be used up at this stage. However, for rocks rich in Al and Na, remaining Pg + Qtz + Cal will react to An-rich plagioclase above the grey–coloured area in Fig. 8.5, between

515 and 540°C. From this it follows that the following mineral assemblages will be encountered in the middle and upper greenschist facies: Ms + Mrg + Pl + Qtz + Cal ± Pg and Ms + Pg + Pl + Qtz + Cal (for Al- and Na-rich compositions only). After the final breakdown of margarite and paragonite in the presence of quartz and calcite, at temperatures up to 590°C, the mineral assemblage in Al-rich metamarls will be Ms + Pl + Kfs + Qtz + Cal. Finally, at even higher temperatures, after the final breakdown of muscovite in the presence of quartz and calcite, the assemblage Pl + Kfs + Qtz + Cal will remain.

According to the contouring of reactions (5), (12), (15) and (17) in Fig. 8.5, at a given temperature and pressure, these reactions will simultaneously produce plagioclase of different An-content. Provided that these differences in plagioclase composition are not eliminated by diffusion processes, it is to be expected that Al-rich marly rocks of the greenschist and amphibolite facies will display a range of plagioclase An-content within a single thin section. Such a situation has actually been observed in nature (e.g. Frank 1983). However, such varying plagioclase composition may be also due to miscibility gaps within the plagioclase feldspar series.

8.3.2.2 Prograde Metamorphism at Very Low X_{CO_2}

Phase relationships for the system KNCAS-HC involving the phases albite, anorthite, clinozoisite, grossular, K-feldspar, margarite, muscovite, paragonite, pyrophyllite with excess quartz and calcite are shown in Fig. 8.6. Calculation of this petrogenetic grid was performed using pure end–member mineral compositions, except for clinozoisite, where an activity of 0.64 was computed based on mineral composition data of Frank (1983). These clinozoisites contain about 6 wt% of Fe_2O_3. Note that Fig. 8.6 is similar to the water-rich portion of Fig. 8.4, but with an enlarged stability field for clinozoisite [reaction (6) is now at X_{CO_2} of ca. 0.13 instead of 0.035 in Fig. 8.4] and with addition of reaction (7).

Figure 8.7 shows phase relationships as in Fig. 8.6, complemented by contours for anorthite-involving reactions as discussed earlier for Fig. 8.5. The stability field of the assemblage Pl + Mrg + Qtz + Cal is emphasized by grey colour and is limited by reaction (5): its high- and low-temperature boundaries are given by the An_{100} and An_{30} contours, respectively. As already discussed above for Fig. 8.5, the assemblage Pg + Qtz + Cal in the presence of intermediate plagioclase composition breaks down by reaction (17), and not by reaction (12). This is indicated in Fig. 8.7 for the An_{70} contours of margarite- and paragonite-involving reactions, which meet at an invariant point at about 470°C and $X_{CO_2} = 0.11$. The configuration around this invariant point is shown in the inset of Fig. 8.7.

The prograde metamorphism of a rock consisting of Ms + Pg + Prl + Qtz + Cal and an initial fluid composition of $X_{CO_2} = 0.02$ will be considered next. The prograde path in PTX_{CO_2} space is shown in Fig. 8.7 by a curve running from a to k and the changing mineral modal composition is displayed in Fig. 8.8. Calculations were performed with the computer program THERIAK (De

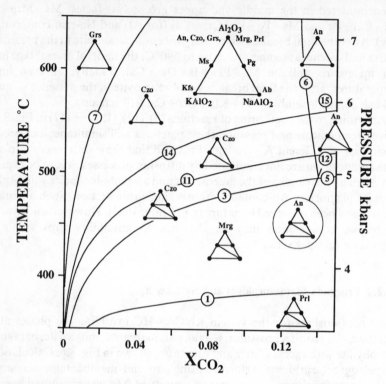

Fig. 8.6. PTX phase relationships for the KNCAS-HC system with excess quartz and calcite at $0 < X_{CO_2} < 0.15$. Pure end–member mineral compositions are used, except for clinozoisite ($a_{Czo} = 0.64$). The *vertical axis* represents an orogenic geotherm characteristic for kyanite-type terrains. The *inset* shows the chemography projected from quartz, calcite, H_2O and CO_2 onto the $KAlO_2$-$NaAlO_2$-Al_2O_3 plane. Compatibility diagrams are given for each divariant field

Capitani and Brown 1987). Note that these calculations are valid only for a specific bulk composition. The prograde path can be divided into several steps as follows.

a-b) (Fig. 8.7) The starting mineral assemblage $Ms + Pg + Prl + Qtz + Cal$ is heated up at a constant X_{CO_2} of 0.020.

b-c) (Fig. 8.7) Margarite is formed according to reaction (1), Table 8.2., between 359 and 364°C and X_{CO_2} increases to 0.025.

c-d) (Fig. 8.7) At point c all pyrophyllite is used up. The rock is now composed of $Ms + Pg + Mrg + Qtz + Cal$ and is heated up at a constant X_{CO_2} of 0.025. This mineral assemblage is characteristic for the lower greenschist facies.

d-e) (Fig. 8.7) Clinozoisite is formed for the first time according to reaction (3) between 425 and 436°C and X_{CO_2} increases to 0.034.

Quartz and calcite in excess

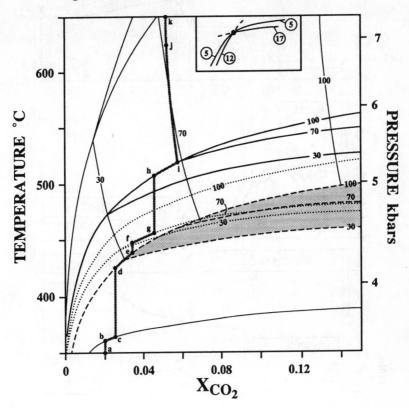

Fig. 8.7. PTX phase relationships for the KNCAS-HC system with excess quartz and calcite at $0 < X_{CO_2} < 0.15$. The vertical axis represents an orogenic geotherm characteristic for kyanite-type terrains. Pure end member mineral compositions are used, except for clinozoisite ($a_{Czo} = 0.64$) and plagioclase. Polybaric divariant equilibria for a particular composition of plagioclase, An_x, are shown as *full lines* (muscovite-involving reactions), *stippled lines* (paragonite-involving reactions) or *dashed lines* (margarite-involving reactions). *Figures* refer to mol% An-content in the plagioclase. The stability field for the assemblage Mrg + Pl (+ Qtz + Cal) is shown in *grey colour*. The prograde path from a to k is emphasized and discussed in the text

e-f) (Fig. 8.7) At point e all margarite is used up and the resulting mineral assemblage Ms + Pg + Czo + Qtz + Cal is heated up at a constant X_{CO_2} of 0.034.

f-g) (Fig. 8.7) Plagioclase of composition An_{33-36} is formed according to reaction (11) between 450 and 458 °C and X_{CO_2} increases to 0.045.

g-h) (Fig. 8.7) At point g all paragonite is used up. The mineral assemblage Ms + Pl + Czo + Qtz + Cal is present between 458 and 507 °C and is characteristic for the upper greenschist facies. In this temperature interval, reactions (6), (11) and (12) are causing minor changes of the mode and mineral

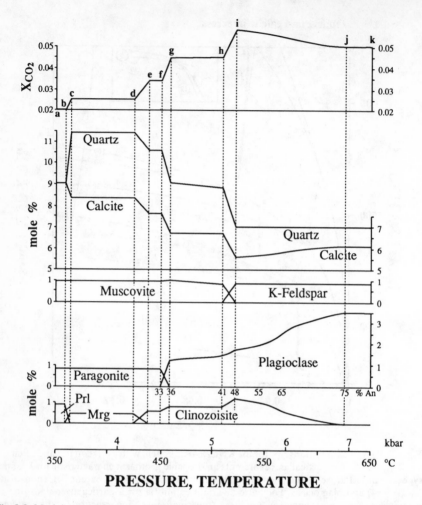

Fig. 8.8. Modal composition (mol%), fluid composition (X_{CO_2}) and plagioclase composition (An%) along the prograde path *a-k* shown in Fig. 8.7

composition (note that these three reactions are linearly dependent). Reactions (11) and (12) are responsible for a decrease of paragonite component in muscovite, and all three reactions are producing additional plagioclase. The very small increase in X_{CO_2} is not visible in Figs. 8.7 and 8.8.

h-i) (Fig. 8.7) K-feldspar and additional clinozoisite are formed by reaction (14) between 507 and 520 °C and X_{CO_2} increases to 0.058. Some additional plagioclase is produced as in the previous step.

i-j) (Fig. 8.7) At point i all muscovite is used up. The mineral assemblage Kfs + Pl + Czo + Qtz + Cal is now present and is characteristic for the lower and middle amphibolite facies. Additional plagioclase is produced by reaction (6), the An content of plagioclase is increasing from 48 to 75 mol%, and X_{CO_2} decreases slightly.

j-k) (Fig. 8.7) At point k at 625 °C all clinozoisite is used up and the resulting mineral assemblage Kfs + Pl + Qtz + Cal is heated up at a constant X_{CO_2} of 0.051 without further reaction.

Summary. Phase relationships of relatively Al-rich metamarls with excess quartz and calcite are complex. Nevertheless, several diagnostic assemblages (always with excess Qtz + Cal) are found along a kyanite geotherm. The assemblage Ms + Pg + Prl is characteristic for subgreenschist facies conditions. The assemblage Ms + Pg + Mrg is characteristic for the lower greenschist facies, Ms + Mrg + Pl and Ms + Czo + Pl for the upper greenschist facies. The assemblages Ms + Pl + Kfs and Czo + Pl + Kfs are characteristic for the lower and middle amphibolite facies, Pl + Kfs for the upper amphibolite facies. The first plagioclase has an An content of ca. 30 mol%, and the An content is increasing with metamorphic grade. In short, sheet silicate-rich assemblages containing two or three white micas at low grade are replaced by feldspar-rich assemblages of increasing An content at high grade.

References and Further Reading

Bucher-Nurminen K, Frank E, Frey M (1983) A model for the progressive regional metamorphism of margarite-bearing rocks in the Central Alps. Am J Sci 283A:370–395

Castelli D (1991) Eclogitic metamorphism in carbonate rocks: the example of impure marbles from the Sesia-Lanzo Zone, Italian Western Alps. J Metamorph Geol 9:61–77

Dachs E (1990) Geothermobarometry in metasediments of the southern Grossvenediger area (Tauern Window, Austria). J Metamorph Geol 8:217–230

De Capitani C, Brown TH (1987) The computation of chemical equilibrium in complex systems containing non-ideal solutions. Geochim Cosmochim Acta 51:2639–2652

Ferry JM (1976) Metamorphism of calcareous sediments in the Waterville-Vassalboro area, south-central Maine: mineral reactions and graphical analysis. Am J Sci 276:841–882

Ferry JM (1981) Petrology of graphitic sulfide-rich schists from south-central Maine: an example of desulfidation during prograde regional metamorphism. Am Mineral 66:908–930

Ferry JM (1983a) Mineral reactions and element migration during metamorphism of calcareous sediments from the Vassalboro Formation, south-central Maine. Am Mineral 68:334–354

Ferry JM (1983b) Regional metamorphism of the Vassalboro Formation, south-central Maine, USA: a case study of the role of fluid in metamorphic petrogenesis. J Geol Soc Lond 140:551–576

Ferry JM (1984) A biotite isograd in south-central Maine, USA: Mineral reactions, fluid transfer, and heat transfer. J Petrol 25:871–893

Ferry JM (1987) Metamorphic hydrology at 13-km depth and 400–500 °C. Am Mineral 72:39–58

Ferry JM (1988) Contrasting meachnisms of fluid flow through adjacent stratigraphic units during regional metamorphism, south-central Maine, USA. Contrib Mineral Petrol 98:1–12

Ferry JM (1992) Regional metamorphism of the Waits River formation, eastern Vermont: delineation of a new type of giant metamorphic hydrothermal system. J Petrol 33:45–94

Frank E (1983) Alpine metamorphism of calcareous rocks along a cross-section in the Central Alps: occurrence and breakdown of muscovite, margarite and paragonite. Schweiz Mineral Petrogr Mitt 63:37–93

Frey M (1978) Progressive low-grade metamorphism of a black shale formation, Central Swiss Alps, with special reference to pyrophyllite and margarite bearing assemblages. J Petrol 19:93–135

Frey M, Orville PM (1974) Plagioclase in margarite-bearing rocks. Am J Sci 274:31–47

Gordon TM, Greenwood HJ (1971) The stability of grossularite in H_2O-CO_2 mixtures. Am Mineral 56:1674–1688

Griffin WL, Styles MT (1976) A projection for analyses of mineral assemblages in calc-pelitic metamorphic rocks. Norsk Geol Tidsskrift 56:203–209

Hewitt DA (1973) The metamorphism of micaceous limestones from south-central Connecticut. Am J Sci 273A:444–469

Hoschek G (1980) Phase relations of a simplified marly rock system with application to the Western Hohe Tauern (Austria). Contrib Mineral Petrol 73:53–68

Hoschek G (1980) The effect of Fe-Mg substitution on phase relations in marly rocks of the Western Hohe Tauern (Austria). Contrib Mineral Petrol 75:123–128

Hoschek G (1984) Alpine metamorphism of calcareous metasediments in the Western Hohe Tauern, Tyrol: mineral equilibria in COHS fluids. Contrib Mineral Petrol 87:129–137

Labotka TC (1987) The garnet + hornblende isograd in calcic schists from a andalsuite-type regional metamorphic terrain, Panamint Mountains, California. J Petrol 28:323–354

Léger A, Ferry JM (1991) Highly aluminous hornblende from low-pressure metacarbonates and a preliminary thermodynamic model for the Al content of calcic amphibole. Am Mineral 76:1002–1017

Menard T, Spear FS (1993) Metamorphism of calcic pelitic schists, Strafford Dome, Vermont: compositional zoning and reaction history. J Petrol 34:977–1005

Nesbitt BE, Essene EJ (1983) Metamorphic volatile equilibria in a portion of the southern Blue Ridge Province. Am J Sci 283:135–165

Selverstone J, Spear FS, Frank G, Morteani G (1984) High-pressure metamorphism in the SW Tauern Window, Austria: P-T paths from hornblende-kyanite-staurolite schists. J Petrol 25:501–531

Selverstone J, Munoz JL (1987) Fluid heterogeneities and hornblende stability in interlayered graphitic and nongraphitic schists (Tauern Window, Eastern Alps). Contrib Mineral Petrol 96:426–440

Sillanpää J (1986) Mineral chemistry study of progressive metamorphism in calcareous schists from Ankarvattnet, Swedish Caledonides. Lithos 19:141–152

Tanner PWG (1976) Progressive regional metamorphism of thin calcareous bands from the Moinian rocks of N.W. Scotland. J Petrol 17:100–134

Thompson AB (1975) Mineral reactions in a calc-mica schists from Gassetts, Vermont, U.S.A. Contrib Mineral Petrol 53:105–127

Thompson PH (1973) Mineral zones and isograds in "impure" calcareous rocks, an alternative means of evaluating metamorphic grade. Contrib Mineral Petrol 42:63–80

Trommsdorff V (1968) Mineralreaktionen mit Wollastonit in einem Kalksilikatfels der alpinen Disthenzone (Claro, Tessin). Schweiz Mineral Petrogr Mitt 48:655–666

Will TM, Powell R, Holland T, Guiraud M (1990) Calculated greenschist facies mineral equilibria in the system CaO-FeO-MgO-Al_2O_3-SiO_2-CO_2-H_2O. Contrib Mineral Petrol 104:353–368

Yardley BWD (1989) An introduction to metamorphic petrology. Longman, Edinburgh, 248 pp

Zen E-an (1981) A study of progressive regional metamorphism of pelitic schists from the Taconic allochthon of southwestern Massachusetts and its bearing on the geologic history of the area. US Geological Survey Professional Paper 1113, US Government Printing Office, Washington DC, 128 pp

9 Metamorphism of Mafic Rocks

9.1 Mafic Rocks

9.1.1 Geological Occurrence

Metamorphic mafic rocks (e.g. mafic schists and gneisses, amphibolites) are derived from mafic igneous rocks, mainly basalts and andesites, and of lesser importance, gabbros (Chap. 2). Metamorphic assemblages found in mafic rocks are used to define the intensity of metamorphism in the metamorphic facies concept (Chap. 4).

Extrusive mafic igneous rocks, basalts and andesites comprise by far the largest amount of mafic rocks and greatly predominate over their plutonic equivalents (Tables 2.1 and 2.2). Basalts and andesites are widespread in the form of massive lava flows, pillow lavas, hyaloclastic breccias, tuff layers, sills and dykes. Basaltic rocks constitute a major portion of the oceanic crust and most appear to have been subjected to ocean-floor metamorphism immediately after formation at a spreading ridge. When transported to a continental margin, oceanic mafic rocks are again recrystallized at or near convergent plate junctions; the alteration in mineralogy depends on whether the oceanic crust was subducted under a continental plate or was obducted onto continental crust. On the other hand, andesitic rocks and their related calc-alkaline volcanics as well as associated greywackes are the dominant lithologies within island arcs and Pacific-type continental margins. These rocks are subjected to alteration by hydrothermal fluids as evidenced by present-day geothermal activity in many island arc districts or are transformed during burial and orogenic metamorphism.

Metamorphosed mafic rocks are very susceptible to changes in temperature and pressure. This is the reason why most names of individual metamorphic facies are derived from mineral assemblages of this rock group, e.g. greenschists amphibolites, granulites, blueschists, and eclogites (cf. Chaps. 2 and 4). In addition, mafic rocks that are metamorphosed under very weak conditions below the greenschist and blueschist facies often show systematic variations in mineralogy that permits a further subdivision into characteristic metamorphic zones. All these distinct low-grade zones may be given separate metamorphic facies names if one wishes. However, we prefer to use the expression sub-greenschist facies for very-low grade conditions of incipient metamorphism.

9.1.2 Hydration of Igneous Mafic Rocks

Basalts and gabbros have solidus temperatures on the order of 1200°C. Consequently, hydrates are not typical members of the solidus mineralogy of basalts and other mafic rocks. At the onset of metamorphism, mafic igneous rocks and pyroclastics are, therefore, in their least hydrated state as opposed to "wet" sedimentary rocks that start metamorphism in their maximum hydrated state. Because newly formed minerals in metamafics at low temperature are predominantly hydrous phases, access of water is absolutely essential to initiate metamorphism. Otherwise, igneous rocks will persist unchanged in metamorphic terrains. Partial or complete hydration of mafic rocks may occur during ocean-floor metamorphism in connection with hydrothermal activity in island arcs or during orogenic metamorphism where deformation facilitates the influx of water. In metamorphism of mafic igneous rocks it is not self-evident that the condition $p_{tot} = p_{H_2O}$ is continuously maintained during the prograde reaction history. On the other hand, worldwide experience with mafic rocks shows that partial persistence of igneous minerals and microstructures is widespread in subgreenschist facies rocks and it is typically absent from the greenschist facies upward. Primary igneous mesostructures such as magmatic layering and pillow structures may be preserved even in eclogite and granulite facies terrains. Coarse-grained gabbroic rocks have the best chance to conserve primary igneous minerals up to high-grade metamorphic conditions. Gabbro bodies often escape pervasive internal deformation and this, in turn, prevents access of water, hampers recrystallization and hinders hydration of the igneous minerals.

As pointed out in Chapter 3, dehydration reactions are strongly endothermic. Consequently, hydration of basalt is exothermic and releases large amounts of heat. The heat of reaction released by replacing the basalt assemblage Cpx + Pl by a collection of low-T hydrates such as prehnite + chlorite + zeolites could raise the temperature by as much as 100°C (in a heat-conserving system). Another interesting aspect of exothermic reactions is their self-accelerating and feedback nature. Once initiated, they will proceed as long as water is available. The increasing temperature will initially make the reactions go faster, but it will also bring the reaction closer to its equilibrium where it eventually would stop. Also note that the reactions that partially or completely hydrate igneous mafic rocks are metastable reactions and the reactions tend to run far from equilibrium. Consider, for instance, the reaction diopside + anorthite + $H_2O \Rightarrow$ chlorite + prehnite $\pm$ quartz that has been mentioned above. In the presence of water at low-T, this reaction will always run to completion and the product and reactant assemblage will never reach reaction equilibrium. The unstable or metastable nature of the hydration reactions that replace high-T anhydrous igneous assemblages by low-T hydrate assemblages has the consequence that low-grade mafic rocks often show disequilibrium assemblages. Unreacted high-T igneous assemblages may occur together with various generations of low-T assemblages of the subgreenschist facies in intimate spatial association. The hydration process commonly leads to the development of zoned metasomatic structures. Examples are: networks of veins and concentric shells of mineral zones that reflect progressive hydration of basaltic pillow lavas. The highly permeable inter-pillow zones serve as aquifers for hydration water and the hydration process progresses towards the pillow centres. The nature of the product assemblages depends on the pressure and temperature during the hydration process but also on the chemical composition of the hydration fluid. CO_2 in the fluid plays a particularly important role. CO_2-rich aqueous fluids may result in altered basalts with modally abundant carbonate minerals (calcite, ankerite). CO_2-rich fluids also tend to favor the formation of less hydrous assemblages compared with pure H_2O fluids. The cation composition of aqueous fluids that have already reacted with large volumes of basalt is dramatically different from that of sea water. Fluid that moves fast through permeable fractured rocks will have a different composition than fluid that trickle slowly through porous basalt. Fluid-rock interaction typically also leads to extensive redistribution of chemical elements in piles of basaltic pillow lavas. This may still be witnessed in high-grade terrains. In the Saas-Zermatt complex of the central Alps, for example, mesoscopic pillow lava structures are often preserved with pillow cores of 25 kbar eclogite and inter-pillow material of nearly monomineralic glaucophane. Finally, the alteration products of the fluid-basalt interaction process will, of course, reflect these compositional differences of the hydration fluid.

The heterogeneous degree of hydration results also in features such as incipient to extensive development, sporadic distribution, and selective growth of low-grade minerals in vesicles and fractures, and the topotaxic growth of these minerals after igneous plagioclase, clinopyroxene, olivine, hornblende, opaques and volcanic glass. From the viewpoint of igneous petrologists, these minerals are often referred to as "secondary minerals". Depending on the effective bulk composition of local domains of the rock, different associations of secondary minerals may develop in vesicles, veins and from replacement of primary phases even within a single thin section. Furthermore, relict igneous phases such as plagioclase and clinopyroxene are common. From the view point of metamorphic petrologists the low-grade hydrates and carbonates, the "secondary minerals", constitute the stable low-grade assemblage and the starting material for prograde metamorphism. In the conceptually simplest case, no relic igneous minerals remain in the mafic rock and its igneous assemblage has been completely replaced by a stable low-T assemblage in its maximum hydrated state.

9.1.3 Chemical and Mineralogical Composition of Mafic Rocks

The characteristic composition of alkali-olivine basalt is listed in Table 2.1 and of MORB basalt in Table 2.3. Mid-ocean ridge basalt (MORB), that is produced along ocean spreading centres, is by far the most common type of igneous mafic rocks. It may serve as a reference composition for all mafic rocks discussed in this chapter. Mafic rocks are characterized by SiO_2 contents of about 45–60 wt% and are also relatively rich in MgO, FeO, CaO and Al_2O_3. It is general custom in petrology to call igneous rocks with 45–52 wt% of SiO_2 basic, and their metamorphic derivatives are then called metabasic rocks or, in short, metabasites. Metabasalts represent the most commonly encountered group of metabasites. Andesitic rocks, on the other hand, contain higher SiO_2, Al_2O_3, and alkalis, but lower MgO, FeO and CaO than basaltic compositions (e.g. Carmichael 1989); basaltic andesites and andesites belong to the intermediate igneous rocks, defined by SiO_2 contents of 52–63 wt%.

Igneous mafic rocks contain appreciable amounts of at least the following eight oxide components: SiO_2, TiO_2, Al_2O_3, Fe_2O_3, FeO, MgO, CaO, and Na_2O. K_2O, H_2O, and CO_2 may also be present in small amounts. These components are stored in relatively few different minerals. The mineralogical inventory of the mafic protolith comprises the major constituents plagioclase, clinopyroxene, quartz, orthopyroxene, olivine and nepheline in various associations. Plagioclase and clinopyroxene are the prime and most common minerals of most mafic rocks and many gabbros and basalts contain more than 90–95 vol% of these two minerals. Other basalts and gabbros may be composed of Pl + Cpx + Opx + Qtz or Pl + Cpx + Ol + Ne, troctolites contain plagioclase + olivine only, anorthosites more than 90 vol% plagioclase and so on. However, all rocks are always combinations of a few different mineral species. A large variety of minor and accessory minerals including ilmenite, magnetite, spinel, garnet, and even hornblende and biotite can be found in igneous mafic rocks. The latter two hydrates, are often poor in OH groups as a result of oxidation and/or halogen substitution.

The complex chemical bulk rock composition of basalts will be redistributed from the few igneous mineral species into a variety of new minerals during metamorphism. The high calcium content of basalts that is stored in Cpx and Pl results in the formation of numerous separate calcium-bearing metamorphic minerals. This is the basic difference to metapelitic rocks (Chap. 7), where CaO is very low and does not form separate calcic phases but is rather found as a minor component in ordinary Fe-Mg minerals such as garnet. The most important minerals found in metamafic rocks are listed in Table 9.1 as a refresher.

9.1.4 Chemographic Relationships and the ACF Projection

The composition space that must be considered in an analysis of phase relationships in metamafic rocks is rather complex and requires eight or more system components. This follows from above and from the chemical analysis of MORBasalt in Table 2.3. Remember the metapelites (Chap. 7) where much of the phase relationships could be discussed in the six-component "AFM" system. A graphic analysis of phase relationships in metamafic rocks is often accomplished by means of the ACF projection. The principles of the ACF projection were introduced in Chap. 2.5.3 and the general chemography and a sample ACF diagram are shown in Fig. 2.10. The ACF diagram is used in

Table 9.1. Minerals and compositions in metabasaltic rocks

Nesosilicates

Garnet	$(Fe, Mg, Ca)_3(Al, Fe)_2Si_3O_{12}$

Sorosilicates

Kyanite	Al_2SiO_5
Zoisite	$Ca_2Al_3Si_3O_{12}(OH)$
Epidote	$Ca_2FeAl_2Si_3O_{12}(OH)$
Pumpellyite	$Ca_4Mg_1Al_5Si_6O_{23}(OH)_3 \cdot 2\ H_2O$
Vesuvianite	$Ca_{19}Mg_2Al_{11}Si_{18}O_{69}(O\ H)_9$
Lawsonite	$CaAl_2Si_2O_7(OH)_2 \cdot H_2O$
Chloritoid	$Mg_1Al_2Si_1O_6(OH)_2$

Pyroxenes

Diopsid	$CaMgSi_2O_6$
Jadeite	$NaAlSi_2O_6$
Hyperstene	$(Mg, Fe)_2Si_2O_6$
Omphacite	$(Ca, Na)(Mg, Fe, Al)Si_2O_6$

Amphiboles

Tremolite	$Ca_2Mg_5Si_8O_{22}(OH)_2$
Actinolite	$Ca_2(Fe, Mg)_5Si_8O_{22}(OH)_2$
Glaucophane	$Na_2(Fe, Mg)_3(Al)_2Si_8O_{22}(OH)_2$
Barroisite	$(Ca, Na)_2(Fe, Mg, Al)_5Si_8O_{22}(OH)_2$
Tschermakite	$Ca_2(Fe, Mg)_3(Al)_2(Al_2S\ i_6O_{22}(OH)_2$
Hornblende	$(Na, K)Ca_2(Fe, Mg, Al)_5(Si, Al)_8O_{22}(OH, F, Cl)_2$

Sheetsilicates

Muscovite	$KAl_3Si_3O_{10}(OH)_2$
Celadonite	$KMgAlSi_4O_{10}(OH)_2$
Paragonite	$NaAl_3Si_3O_{10}(OH)_2$
Phlogopite	$KMg_3[AlSi_3O_{10}](OH)_2$
Biotite	$K(Mg, Fe, Al, Ti)_3[(Al, Si\ )_3O_{10}](OH, F, Cl)_2$
Clinochlore	$Mg_5Al_2Si_3O_{10}(OH)_8$
Chlorite	$(Fe, Mg)_5Al_2Si_3O_{10}(OH)_8$
Prehnite	$Ca_2Al_2Si_3O_{10}(OH)_2$

Tectosilicates

Quartz	SiO_2
Anorthite	$CaAl_2Si_2O_8$
Albite	$NaAlSi_3O_8$
Analcite	$NaAlSi_2O_6 \cdot H_2O$
Scapolite	$Ca_4(Al_2Si_2O_8)_3(CO_3, SO_4, Cl_2)$

Zeolites

Laumontite	$CaAl_2Si_4O_{12} \cdot 4\ H_2O$
Heulandite	$CaAl_2Si_7O_{18} \cdot 6\ H_2O$
Stilbite	$CaAl_2Si_7O_{18} \cdot 7\ H_2O$
Wairakite	$CaAl_2Si_4O_{12} \cdot 2\ H_2O$

Carbonates

Calcite	$CaCO_3$
Aragonite	$CaCO_3$
Dolomite	$Ca(Mg, Fe)(CO_3)_2$

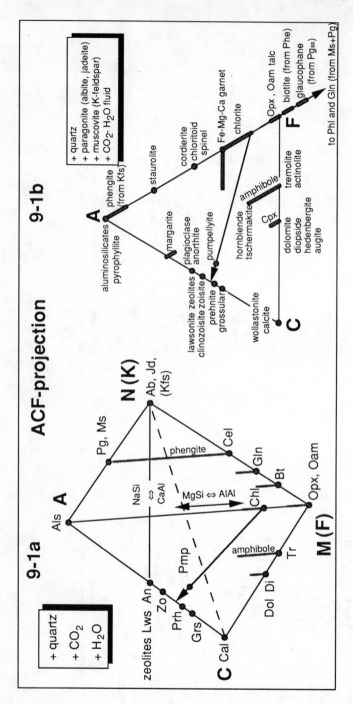

Fig. 9.1. ACF projections

general for display and analysis of phase relationships that involve calcic minerals. Because of its importance for the subsequent treatment of metamafic rocks, some important features of the ACF projection will be briefly reviewed.

The following discussion makes extensive use of Fig. 9.1a that is a pseudo-3D display of the NACF tetrahedron, and Fig. 9.1b, a conventional ACF diagram. The ACF diagram represents, as usual, a mole fraction triangle and displays the three composition variables Al_2O_3, CaO and FeO. Any mineral that is composed of only these three components can be directly displayed on an ACF diagram (e.g. hercynite). Any other mineral composition is projected onto the ACF triangle or is used as a projection point. In the latter case, this mineral must be present in all assemblages in order to deduce meaningful phase relationships. ACF diagrams are projections from SiO_2 and this means that at any given pT condition the stable polymorph of SiO_2 (e.g. quartz) must be present in all assemblages. This restriction imposed by the conventional ACF diagram often causes trouble when dealing with metamorphic silica-poor igneous mafic rocks such as troctolites, nepheline basalts, olivine basalts and olivine gabbros. However, other meaningful diagrams may be designed for such quartz-free metamafics following the suggestions given in Chap. 2.5. ACF diagrams are also projections from H_2O and CO_2 in order to permit the display of hydrates and carbonates. This means that a fluid phase of a specified composition must be present (or the chemical potentials of H_2O and CO_2 must be defined otherwise) under the condition of the diagram. This again may cause trouble, particularly at low metamorphic grades, because the universal presence of a fluid phase is not unavoidably self-evident in anhydrous igneous protolith rocks (see discussion above). Also note that pumpellyite (Pmp) is on the ACF surface in Fig. 9.1a and its composition projects between zoisite and prehnite from the clinochlore composition.

ACF diagrams also represent projections parallel to the $MgFe_{-1}$ exchange vector. This is a rigid and lamentable feature of ACF diagrams. All complex relationships, continuous and discontinuous reactions arising from variations in the Fe-Mg ratio of ferro-magnesian minerals cannot be properly analyzed and understood with the aid of such diagrams. The AF binary on ACF and AFM diagrams are identical. Consequently, many of the relationships discussed in Chapter 7 on metapelites are also valid here. The AF binary is expanded to the AFM triangle in Chapter 7 and, in order to include Ca minerals, to the ACF triangle in this chapter. An ACFM tetrahedron would permit a rather thorough analysis of phase relationships in metamafic rocks, however, it is very inconvenient to work with three-dimensional composition phase diagrams.

The $MgFe_{-1}$ projection on ACF diagrams has the consequence that crossing tie-line relationships among ferro-magnesian silicates (with contrasting X_{Fe}) and Ca minerals do not necessarily represent a discontinuous reaction relationship but rather span a composition volume in the ACFM space. An example (Fig. 9.1b): the four minerals garnet, hornblende, kyanite, and zoisite can be related to the reaction 2 tschermakite + 1 kyanite = 2 garnet + 2 zoisite + 1 quartz + 1 H_2O. This important reaction in high-pressure amphibolites is not discontinuous, as may erroneously be concluded from Fig. 9.1b. The four minerals actually occupy corners of a composition volume in ACFM space and may occur as a stable assemblage over a certain pT interval. The reaction formulated above is thus in reality a continuous reaction. Another effect of the $MgFe_{-1}$ projection is that Fe- and Mg endmembers of a given mineral species project to identical positions (and of course any intermediate X_{Fe} composition of that mineral too, e.g. pyrope and almandine). The consequences of the $MgFe_{-1}$ projection on ACF diagrams must always be kept in mind.

About 3 wt% of Na_2O is present in typical basalts and sodium-bearing minerals are important in metamafics (plagioclase, amphibole, mica, pyroxene, Table 2.3). If one wants to represent sodic phases on ACF diagrams, one may expand the ACF triangle of Fig. 9.1a (that actually is an ACM figure, but this is equivalent, see above) to a tetrahedron with $NaAlO_2$ as an additional corner. In this corner we will find the projected compositions of albite, jadeite and analcite (Qtz, H_2O projection). In a similar fashion, K-feldspar will project to a $KAlO_2$ corner of an analogous K tetrahedron. End member paragonite and muscovite project onto the AN- and AK binary, respectively. All plagioclase compositions project onto the ACN ternary. The compositional variation in plagioclase is mainly related to the $NaSiAl_{-1}Ca_{-1}$ exchange that connects albite and anorthite in Fig. 9.1a. The same exchange component can also be found in, for example, amphiboles, pyroxenes and margarite. End member celadonite, glaucophane and phlogopite (biotite) project onto the MN- and MK binary, respectively. The micas and amphibole show strong pT- and assemblage-dependent compositional variations along the $MgSiAl_{-1}Al_{-1}$ exchange direction. This tschermak exchange is parallel to the AM-(AF) binary. Phengites project onto the AMN-(AFN) ternary and the solid solution series connects Ms and Cel. Sodic phases are often depicted on ACF diagrams by projecting from the $NaAlO_2$ apex of the ACFN tetrahedron onto the ACF triangle. This means then that Ab, Jd, Anl (whatever is stable at the pressure and the temperature of the diagram) is present as an extra mineral or that the potentials of these phase components are fixed. It can be seen in Fig. 9.1a that pure albite cannot be shown on ACF diagrams, but that any Ca-bearing plagioclase will project to the An point. This means that the consequences of the plagioclase composition on phase relationships in metamafic rocks cannot be analyzed on ACF diagrams. Ms and Pa will project to the A apex of ACF diagrams, whereas phengites will occupy the entire AF binary when projected from Kfs (Ab). However, Kfs is very rare in metamafics and consequently Kfs projections are not much used. In analogy to AFM diagrams, one may project from the white micas rather than from the feldspar components, particularly for blueschist facies rocks. In this case, the feldspars do not project onto the ACF plane with the exception of anorthite, phengite does not project onto the ACF plane either, whereas glaucophane projects onto the AF binary with a negative A coordinate, like biotite projected from muscovite (Fig. 9.1a, b). In metamafic rocks, just one of the minerals, K-mica, phengite or biotite, is generally present. This makes it possible to ignore the small amount of potassium that is stored in just one extra minor phase. One may also choose to project from albite (jadeite) and muscovite. The consequences of the countless alternatives can be understood by carefully studying Fig. 9.1a. Also note that not all possible compositional variations of minerals are shown in Fig. 9.1a (e.g. Jd-Di solution). Whatever projection you prefer for your own work, it is important to specify the projections on any of your phase diagrams carefully.

Another complication when dealing with metamafic rocks is the fact that they tend to be much more oxidized than, for example, metapelites. Redox reactions that involve phase components with ferric iron are common and important. Notably, in garnet (andradite), amphibole (e.g. ribeckite, crossite) and pyroxene (e.g. acmite) a considerable amount of the total iron is usually present in the tri-valent state. However, a considerable substitution of Fe_2O_3 for Al_2O_3 can also be found in most of the low-grade Ca-Al hydrosilicates. The presence of epidote in rocks always signifies the presence of Fe^{3+} in the total iron. Magnetite is a widespread oxide phase in metamafics. A separate treatment of Redox reactions will not be given in this book, however. One of the effects of variable Fe^{3+}/Fe^{2+} ratios in minerals is a further increase of the variance of the considered assemblages. Consequently, the assemblages occur over a wider pT range compared to the situation where all iron is present as Fe^{2+}. One may also construct ACF diagrams for a fixed oxygen activity. For instance, the coexistence of hematite and magnetite in a rock at p and T will fix a_{O_2} (as shown in Chap. 3) and the conditions will be rather oxidizing. Thus, a significant amount of the total iron of the rock will be present as Fe^{3+}. The effects on the minerals and mineral compositions can be related, to a large portion, to the $Fe^{3+}Al_{-1}$ exchange. This exchange makes epidote from clinozoisite, magnetite from hercynite and ribeckite from glaucophane (for example). By considering ACF projections as projections parallel to the $Fe^{3+}Al_{-1}$ exchange vector, one may represent important minerals in metamafics such as epidote, grandite garnet, magnetite and crossite. Note that magnetite does not project to the F apex but to the same projection point as hercynite and Mg-Al-spinel.

Carbonates are widespread and abundant in low-grade metamafics. Fe-bearing calcite and ankerite (dolomite) are predominant. Carbonates are also often present in high-grade rocks. Carbonates can be displayed on ACF diagrams as explained above. However, any reactions that involve carbonates are in general mixed volatile reactions and must be analyzed accordingly (see Chaps. 3, 6 and 8). The presence of much CO_2 in an aqueous fluid phase has also effects on pure dehydration reactions that do not involve carbonate minerals (Chap. 3). Compared with pure H_2O fluids, in CO_2-rich fluids hydrates (e.g. chlorite, amphibole) may break down at much lower temperatures also in rocks that do not even contain carbonates.

Basalts also contain TiO_2 in the % range. Titanium is mainly shelved in one major Ti-phase, i.e. titanite, ilmenite or rutile. If two of these minerals are present in the rock, Ti-balanced reactions may be useful for pT estimates. An example: $Ky + 3\,Il + 2\,Qtz = 3\,Rt + Grt$. Ti is also found in considerable amounts in rock forming silicates (biotite, amphiboles, garnet etc.) with unavoidable consequences for solution properties and equilibrium conditions of reactions.

In conclusion, the composition of metamafic rocks is rather complex and the minerals that typically occur in the assemblages show extensive chemical variation along several exchange directions. This makes comprehensive graphical analysis of phase relationships in metamorphic mafic rocks a difficult task. The complex chemical variation of solid-solution minerals can be simplified for graphical analysis by projecting parallel to some of the exchange components. The choice of projection depends entirely on the problem one wants to solve and the kind of rocks one is working with. For many metamafic rocks it turned out that a very advantageous and powerful projection is made from SiO_2, $NaAlO_2$, H_2O and CO_2, parallel to $MgFe_{-1}$ and $Fe^{3+}Al_{-1}$ onto the Al_2O_3-CaO-FeO-mole fraction triangle. This projection is known as ACF projection and will be used below to discuss the phase relationships in metamafics.

The chemical complexity of the mafic rocks makes it difficult to discuss phase relationships by means of chemical subsystems and comprehensive petrogenetic grids and phase diagrams in pT space (as for example in previous chapters). Note, however, the complexity of mafic rocks can be quantitatively analyzed but this must be done for each given suite of rocks individually. The presentation below is thus mainly based on the discussion of the ACF system and MORB compositions with some important reactions discussed separately where necessary.

9.2 Overview on the Metamorphism of Mafic Rocks

The best overview of the metamorphism of mafic rocks can be gained from the metamorphic facies scheme shown in Fig. 4.2. The characteristic assemblages in metabasalts are used for the definition of metamorphic facies and serve as a reference frame for all other rock compositions. The assemblages of metamafic rocks that are diagnostic for the different facies are given in Chapter 4. From Fig. 4.2 it is evident that basalt undergoing prograde metamorphism along a Ky-or Sil-geotherm will first show diagnostic assemblages of the subgreenschist facies, later become a greenschist, then an amphibolite and finally end up as a mafic granulite. High-pressure low-temperature metamorphism (HPLT metamorphism) turns basalt first into blueschists and then into eclogites. Any geologic process that brings basalts to great depth (>50 km) will result in the formation of eclogite. In contact metamorphism, basalts are metamorphosed to mafic hornfelses. Partial melting of H_2O-saturated metamafics starts at a temperature that is significantly higher than in metagranitoids and metapelites.

Prograde metamorphism of mafic rocks produces sequences of mineral zones that can be compared with mineral zones defined by minerals in metapelites. Fig. 9.2 shows sequences of minerals formed in mafic and pelitic rocks by prograde metamorphism from northern Michigan as an example. Metapelites contain muscovite and quartz throughout and the sequence is of the low-pressure type with andalusite and sillimanite as aluminosilicates. The

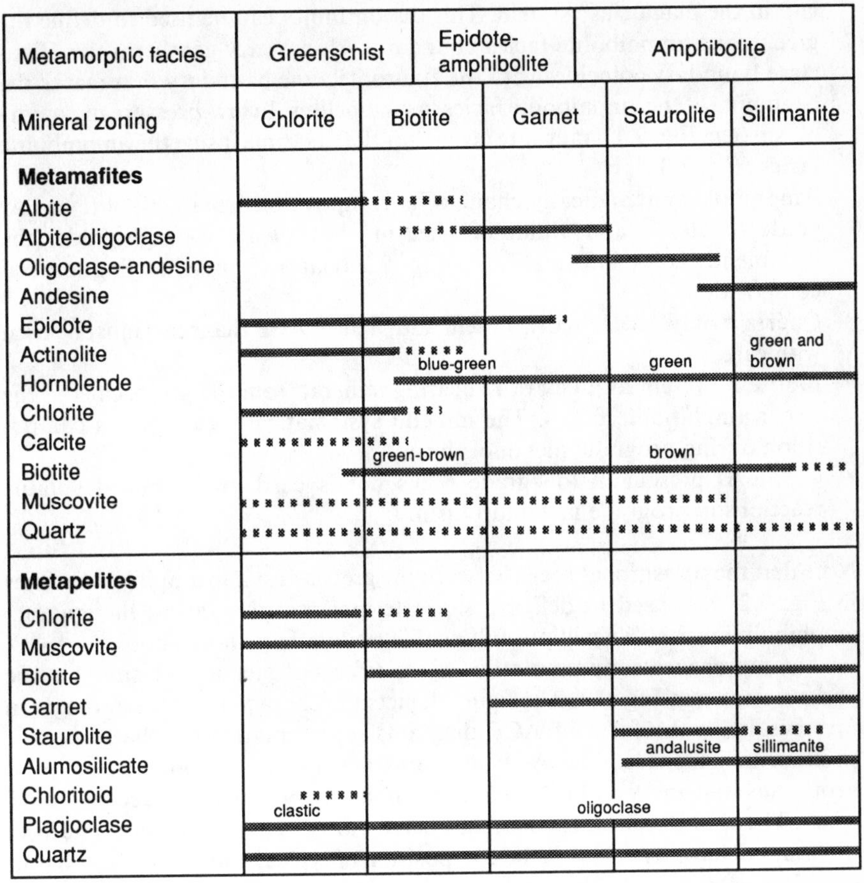

Metamorphic facies	Greenschist		Epidote-amphibolite	Amphibolite	
Mineral zoning	Chlorite	Biotite	Garnet	Staurolite	Sillimanite
Metamafites					
Albite					
Albite-oligoclase					
Oligoclase-andesine					
Andesine					
Epidote					
Actinolite			blue-green		green	green and brown
Hornblende					
Chlorite					
Calcite			green-brown		brown	
Biotite					
Muscovite					
Quartz					
Metapelites					
Chlorite					
Muscovite					
Biotite					
Garnet					
Staurolite				andalusite	sillimanite
Alumosilicate					
Chloritoid		clastic		oligoclase	
Plagioclase					
Quartz					

Fig. 9.2. Progressive mineral changes in northern Michigan. (James 1955)

staurolite zone along the Sil-geotherm is narrow, as expected from Fig. 7.3. The corresponding prograde mineral zonation in metabasites shows a series of features that are very characteristic.

- There are very few different mineral species present in metamafic rocks. In the greenschist facies metamafic rocks contain: albite + chlorite + actinolite + epidote ⇒ greenschist. In the amphibolite facies the minerals are: plagioclase (> An_{17}) + hornblende ± biotite ± epidote ⇒ amphibolite.
- Most minerals occur over many of the mineral zones defined by metapelites.
- The characteristic prograde changes in metabasites pertain to the composition of plagioclase and amphibole.
- Plagioclase systematically changes its composition from albite at low grade to more calcic plagioclase (andesine in the example of Fig. 9.2). The transition from albite to oligoclase is abrupt and marks a sharp mappable boundary in the field. The discontinuous nature is caused by a miscibility

gap in the plagioclase system. This discontinuity can be used to define the greenschist-amphibolite facies boundary. Along a Ky-geotherm the oligoclase boundary coincides with the staurolite zone boundary that marks the beginning of the amphibolite facies in metapelites. In low-pressure metamorphism (e.g. Fig. 9.4) staurolite occurs for the first time inside the amphibolite facies (Fig. 7.3).

- Amphibole systematically changes its composition from actinolite at low grade to alkali- and aluminium-bearing hornblende at high-grade (the hornblende color information in Fig. 9.2 relates to the changing mineral composition).
- Quartz is only occasionally present and, hence, ACF diagrams must be used with care.
- Biotite is present as an extra K-bearing mineral from the greenschist to the upper amphibolite facies. The mineral systematically changes its composition during prograde metamorphism.
- Calcite is present in low-grade rocks but is used up by mixed volatile reactions in prograde metamorphism.

Note that the transitional rocks between the greenschist and amphibolite facies in Fig. 9.2 were used to define a separate metamorphic facies, the epidote-amphibolite facies. We will not follow this custom here in this book.

A comprehensive representation of the effects of metamorphism on mafic rocks is shown in Fig. 9.3. The figure depicts the characteristic assemblages in metabasalts in the form of ACF diagrams representative for the respective position in pT space. The ACF diagrams are arranged along three typical geotherms and the 9°C km^{-1} geotherm limits the geologically accessible pT space. The aluminosilicate phase relationships are given for reference. The inset in the upper left corner shows the composition of typical mid-ocean ridge basalt ($Pl + Cpx \pm Opx \pm Qtz$). Many metabasalts are expected to lie compositionally within the horizontally ruled field. On the ACF diagrams of Fig. 9.3, the assemblages that represent the approximate composition of MORB have been shaded. Metabasalts that have been strongly metasomatized during ocean-floor metamorphism may project to other, often more Ca-rich, compositions in the ACF diagram. In the following presentation, the details of the phase relations in each of the represented ACF chemographies will be discussed separately.

9.3 Subgreenschist Facies Metamorphism

9.3.1 General Aspects and a Field Example

ACF chemographies #1 and #2 (Fig. 9.3) represent mineral assemblages typical for the subgreenschist facies. At very low grade, the characteristic assemblage includes albite + chlorite + carbonate + a variety of zeolites. On the ACF chemography #1 zeolite is typified by laumontite but any of the four zeolites listed

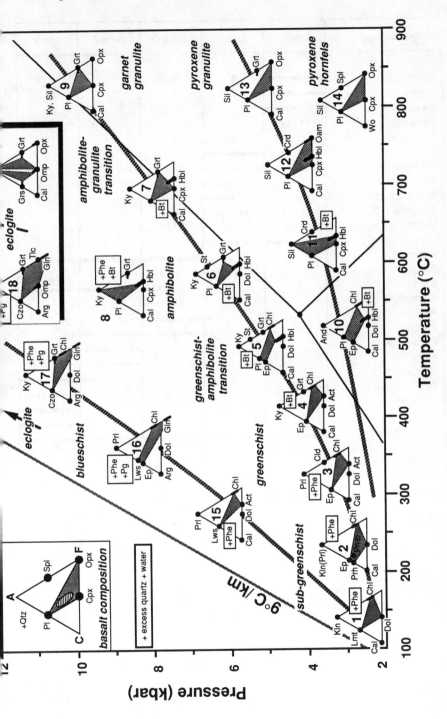

Fig. 9.3. Metamorphism of mafic rocks (meta-basalts) represented by ACF diagrams

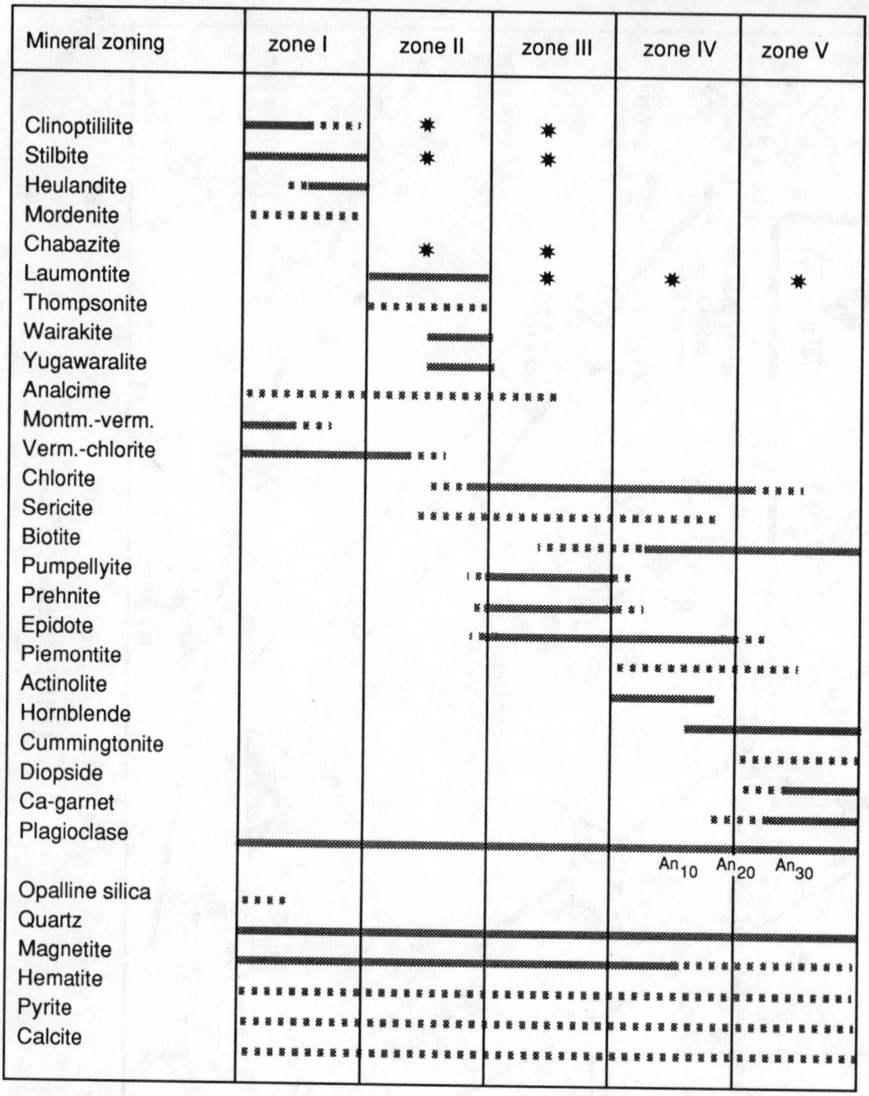

Mineral zoning	zone I	zone II	zone III	zone IV	zone V
Clinoptililite					
Stilbite					
Heulandite					
Mordenite					
Chabazite					
Laumontite					
Thompsonite					
Wairakite					
Yugawaralite					
Analcime					
Montm.-verm.					
Verm.-chlorite					
Chlorite					
Sericite					
Biotite					
Pumpellyite					
Prehnite					
Epidote					
Piemontite					
Actinolite					
Hornblende					
Cummingtonite					
Diopside					
Ca-garnet					
Plagioclase				An$_{10}$ An$_{20}$ An$_{30}$	
Opalline silica					
Quartz					
Magnetite					
Hematite					
Pyrite					
Calcite					

Fig. 9.4. Stability ranges of some metamorphic minerals in Tanzawa Mountains, central Japan (Seki et al. 1969). Mineral zonation in the Tanzawa terrain, Japan. *Zone I* Stilbite (clinoptilolite)-vermiculite; *Zone II* laumontite "mixed-layer" chlorite; *Zone III* pumpellyite-prehnite-chlorite; *Zone IV* actinolite-greenschist; *Zone V* amphibolite. * only in veins

in Table 9.1 would project to the same point. In addition, K-white mica (smectite, illite, sericite, phengite) may be present as well as kaolinite. At slightly higher grade (ACF chemography #2), the zeolites are replaced by the minerals prehnite, pumpellyite and epidote in a number of different combinations. Still albite, chlorite and carbonates are major minerals in prehnite-pumpellyite rocks. Pyrophyllite may be present instead of kaolinite. Zeolite-bearing metamafics are representative for temperatures below about 150–200 °C, metabasalts with prehnite and pumpellyite are most characteristic for the temperature range 150–300 °C.

The zoneography of minerals in metamafic rocks from the Tanzawa mountains in central Japan is shown in Fig. 9.4. The metamorphic terrain has been subdivided into five zones ranging from lower subgreenschist facies to amphibolite facies. The terrain is of the low-pressure sillimanite type. The first two zones are characterized by the presence of zeolites in the assemblages, low-T zeolites in zone I and high-T zeolites in zone II. The zone I zeolites are typically stilbite and heulandite, zone II zeolites are laumontite and wairakite. This is understandable by reference to Table 9.1 where one can see that zone I zeolites clearly contain more crystal water (zeolite water) than zeolites of zone II. The transition of zone I to zone II is therefore related to ordinary dehydration reactions. For the same reason, it is also clear that wairakite is typically the last zeolite in prograde metamorphism. Analcite occurs in both zones sporadically. Mixed-layer clays are typically present in the first two zones. Plagioclase is present in all zones, its composition is that of albite in the first four zones. The protolith is SiO_2-rich, and quartz is present in excess throughout (ACF diagrams). No zeolites survive in zone III, which is characterized by three new metamorphic minerals: pumpellyite, prehnite and epidote. Also chlorite has transformed into a well-defined ferro-magnesian sheet silicate. Sericite is present sporadically. Several zeolite- and chlorite-consuming reactions are responsible for the generation of the new metamorphic minerals Prh, Pmp and Ep. In zone IV typical greenschist facies assemblages replace the zone III mineralogy. In particular, pumpellyite and prehnite disappear and the characteristic greenschist facies mineral association chlorite + epidote + actinolite + albite ± quartz becomes dominant. In zone V the minerals are diagnostic for the amphibolite facies. Calcic plagioclase and hornblende are the dominant minerals but biotite, garnet and CPX may appear as well. In low-pressure terrains such as the Tanzawa Mountains, cummingtonite or other Fe-Mg-amphiboles typically occur together with calcic amphiboles (hornblende).

9.3.2 Metamorphism in the CASH and NCMASH Systems

Quantitative phase relationships in the simple CASH system in low-grade metamorphism are shown in Fig. 9.5. Some of the mineral stability fields are highlighted on that figure. The reaction stoichiometries can be found in Table 9.2. The predicted sequence of Ca-zeolites in progressively metamorphosed metamafic rocks is: stilbite, heulandite, laumontite and wairakite. This sequence is clearly consistent with the sequence of observed mineral zones in the Tanzawa mountains shown in Fig. 9.4. It also follows from Fig. 9.5 that lawsonite favorably forms in terrains that are characterized by high-pressure low-temperature geotherms. In fact, lawsonite is an important mineral in blueschist terrains.

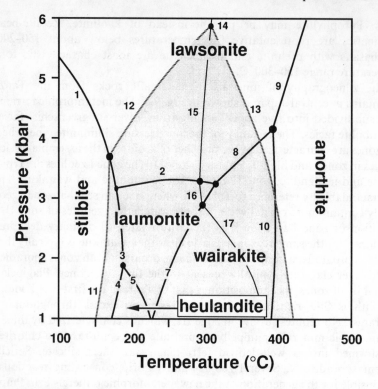

Fig. 9.5. Low-grade metamorphism in the CASH system. Prehnite field *shaded*

Table 9.2. Reaction stoichiometries of reactions shown in the figures of this chapter and reactions discussed in the text

Subgreenschist facies

CASH system (Fig. 9.5)

1	Stb = Lws + 5 Qtz + 5 H_2O
2	Lmt = Lws + 2 Qtz + 2 H_2O
3	Stb = Lmt + 3 Qtz + 3 H_2O
4	Stb = Hul + H_2O
5	Hul = Lmt + 3 Qtz + 2 H_2O
6	Hul = Wa + 3 Qtz + 4 H_2O
7	Lmt = Wa + 2 H_2O
8	Wa = Lws + 2 Qtz
9	Lws = An + 2 H_2O
10	Wa = An + 2 Qtz + 2 H_2O
11	Stb + Grs = 2 Prh + 4 Qtz + 5 H_2O
12	2 Prh = Lws + Grs + Qtz
13	5 Prh = 2 Zo + 2 Grs + 3 Qtz + 4 H_2O
14	5 Lws + Grs = 4 Zo + Qtz + 8 H_2O
15	2 Lws + Prh = 2 Zo + Qtz + 4 H_2O
16	Prh + 2 Lmt = 2 Zo + 5 Qtz + 8 H_2O
17	Prh + 2 Wa = 2 Zo + 5 Qtz + 4 H_2O

Table 9.2 (continued)

NCMASH system (Fig. 9.6)

18	$4\,Hul + Tr = 3\,Prh + Chl + 24\,Qtz + 18\,H_2O$
19	$4\,Lmt + Tr = 3\,Prh + Chl + 12\,Qtz + 10\,H_2O$
20	$20\,Pmp + 3\,Tr + 6\,Qtz = 43\,Prh + 7\,Chl + 2\,H_2O$
21	$86\,Lmt + 17\,Tr = 30\,Pmp + 11\,Chl + 267\,Qtz + 212\,H_2O$
22	$86\,Lws + 17\,Tr = 30\,Pmp + 11\,Chl + 95\,Qtz + 40\,H_2O$
23	$5\,Pmp + 3\,Qtz = 7\,Prh + 3\,Zo + Chl + 5\,H_2O$
24	$6\,Lmt + 17\,Prh + 2\,Chl = 10\,Pmp + 21\,Qtz + 14\,H_2O$
25	$14\,Lmt + 5\,Pmp = 17\,Zo + Chl + 32\,Qtz + 61\,H_2O$
26	$86\,Stb + 17\,Tr = 30\,Pmp + 11\,Chl + 525\,Qtz + 470\,H_2O$

Subgreenschist to greenschist facies transition (Fig. 9.6)

Essential reactions

27	$5\,Prh + Chl + 2\,Qtz = 4\,Zo + Tr + 6\,H_2O$
28	$25\,Pmp + 2\,Chl + 29\,Qtz = 7\,Tr + 43\,Zo + 67\,H_2O$

Additional reactions

29	$4\,Wa + Ab = Pg + 2\,Zo + 10\,Qtz + 6\,H_2O$
30	$4\,Lws + Ab = Pg + 2\,Zo + 2\,Qtz + 6\,H_2O$
31	$14\,Lws + 5\,Pmp = 17\,Zo + Chl + 4\,Qtz + 33\,H_2O$
32	$4\,Lmt + Ab = Pg + 2\,Zo + 10\,Qtz + 14\,H_2O$

Reactions involving carbonates (examples)

33	$3\,Chl + 10\,Cal + 21\,Qtz = 2\,Zo + 3\,Tr + 10\,CO_2 + 8\,H_2O$
34	$15\,Pmp + 9\,Qtz + 4\,CO_2 = 4\,Cal + 25\,Zo + 3\,Tr + 37\,H_2O$

Greenschist facies

CMASH reactions (Fig. 9.7)

35	$2\,Zo + 5\,Prl = 4\,Mrg + 18\,Qtz + 2\,H_2O$
36	$Mrg + 2\,Qtz + 2\,Zo = 5\,An + 2\,H_2O$
37	$Mrg + Qtz = An + And + H_2O$
38	$4\,Mrg + 3\,Qtz = 2\,Zo + 5\,Ky + 3\,H_2O$
39	$2\,Chl + 2\,Zo + 2\,Qtz = 2\,Tr + 5\,Ky + 7\,H_2O$
40	$6\,Zo + 7\,Qtz + Chl = 10\,An + Tr + 6\,H_2O$
41	$2\,An + Chl + 4\,Qtz = Tr + 3\,Sil + 3\,H_2O$

Example of reaction producing tschermak component

42	$7\,Chl + 14\,Qtz + 12\,Zo = 12\,Tr + 25\,TS\,(Al_1Al_1Si_{-1}Mg_{-1}) + 22\,H_2O$

Mica–involving reactions

43	$16\,Tr + 25\,Ms = 25\,Phl + 16\,Zo + 1\,Chl + 77\,Qtz + 4\,H_2O$
44	$4\,Tr + 6\,Chl + 25\,Cel = 25\,Phl + 4\,Zo + 63\,Qtz + 26\,H_2O$
45	$1\,Chl + 4\,Cel = 3\,Phl + 1\,Ms + 7\,Qtz + 4\,H_2O$
46	$2\,Chl + 2\,Zo + 5\,Ab + 4\,Qtz = 2\,Tr + 5\,Pg + 2\,H_2O$

Carbonate–involving reaction (examples)

47	$1\,Ms + 3\,Qtz + 8\,Dol + 4\,H_2O = 1\,Phl + 1\,Chl + 8\,Cal + 8\,CO_2$
48	$9\,Tr + 2\,Cal + 15\,Ms = 15\,Phl + 10\,Zo + 42\,Qtz + 2\,CO_2 + 4\,H_2O$
49	$19\,Cal + 3\,Chl + 11\,CO_2 = 15\,Dol + 2\,Zo + 3\,Qtz + 11\,H_2O$

Table 9.2 (continued)

Greenschist–amphibolite facies transition

50	4 Chl + 18 Zo + 21 Qtz = 5 Ts–amphibole + 26 An + 20 H_2O
51	7 Chl + 13 Tr + 12 Zo + 14 Qtz = 25 Ts–amphibole + 22 H_2O
52	Ab + Tr = Ed + 4 Qtz
53	12 Zo + 15 Chl + 18 Qtz = 8 Grs + 25 Prp + 66 H_2O

Amphibolite facies

54	4 Tr + 3 An = 3 Prp + 11 Di + 7 Qtz + 4 H_2O
55	1 Ts–amphibole + 6 Zo + 3 Qtz = 10 An + 4 Di + 4 H_2O
56	3 Ts–amphibole + 7 Ky = 6 An + 4 Prp + 4 Qtz + 3 H_2O
57	7 Ts–amphibolie + 7 Sil + 4 Qtz = 14 An + 4 Ath + 3 H_2O
58	7 Tr = 3 Ath + 14 Di + 4 Qtz + 4 H_2O

Amphibolite–granulite facies transition

59	Tr = 2 Cpx + 3 Opx + Qtz + H_2O
60	Ts = Cpx + 3 Opx An + H_2O
61	Tr + 7 Grs = 27 Cpx + Prp 6 An + H_2O
62	Tr + Grs = 4 Cpx + Opx An + H_2O

Granulite facies

63	4 En + 1 An = 1 Di + 1 Qtz + 1 Prp

Blueschist facies (Fig. 9.9)

64	Tr + 10 Ab + 2 Chl = 2 Lws + 5 Gln
65	6 Tr + 50 Ab + 9 Chl = 25 Gln + 6 Zo + 7 Qtz + 14 H_2O
66	13 Ab + 3 Chl + 1 Qtz = 5 Gln + 3 Pg + 4 H_2O
67	12 Lws + 1 Gln = 2 Pg + 1 Prp + 6 Zo + 5 Qtz + 20 H_2O
68	4 Pg + 9 Chl + 16 Qtz = 2 Gln + 13 Prp + 38 H_2O
69	10 Pg + 3 Chl + 14 Qtz = 5 Gln + 13 Ky + 17 H_2O

Blueschist-eclogite facies transition (Figs. 9.9, 9.10)

70	Jd + Qtz = Ab
71	1 Gln + 1 Pg = 1 Prp + 3 Jd + 2 Qtz + 2 H_2O

Granulite- and amphibolite-eclogite facies transition

72	CaTS + Qtz = An
73	2 Zo + Ky + Qtz = 4 An + H_2O
74	1 Grs + 2 Ky + 1 Qtz = 3 An
75	4 Tr + 3 An = 3 Prp + 11 Di + 7 Qtz + 4 H_2O
76	4 En + 1 An = 1 Prp + 1 Di + 1 Qtz
77	3 Di + 3 An = 1 Prp + 2 Grs + 3 Qtz
78	1 An + 2 Di = 2 En + 1 Grs + 1 Qtz

Table 9.2 (continued)

Eclogite facies

79	$Pg = Jd + Ky + H_2O$
80	$6\,Zo + 4\,Prp + 7\,Qtz = 13\,Ky + 12\,Di + 3\,H_2O$
81	$Tlc + Ky = 2\,En + Mg\text{–}TS + 2\,Qtz + H_2O$
82	$Grs + Prp + 2\,Qtz = 3\,Di + 2\,Ky$
83	$Grs + 6\,Opx = 3\,Di + Prp$

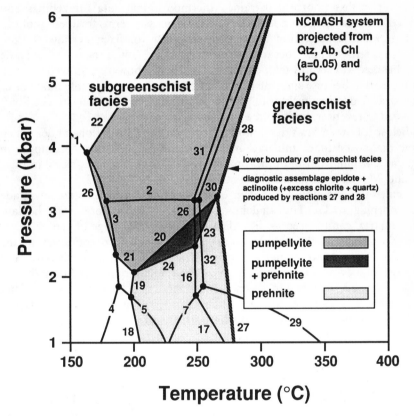

Fig. 9.6. Low-grade metamorphism in the NCMASH system. Prehnite, pumpellyite and prehnite + pumpellyite fields *shaded*

Fig. 9.6 depicts phase relations in the more complex NCMASH system involving the following 14 minerals: actinolite, plagioclase, chlorite, epidote, garnet, heulandite, laumontite, lawsonite, paragonite, prehnite, pumpellyite, quartz, stilbite and wairakite. The relationships in Fig. 9.6 are based on thermodynamic data for Mg end-member phase components in the NCMASH system. The deviations of *average* natural mineral compositions from the model end-member compositions has been accounted for by incorporating appropriate activity terms. However, amphiboles, chlorite, epidote, and pumpellyite from low-grade metabasites have large variations in their chemical composition, depending on mineral assemblage, rock composition, and metamorphic grade. It is therefore important to realize that the petrogenetic grid in Fig. 9.6 is valid only for an average metabasite (meta-MORB). The effect of variable mineral composition on the pT stability fields of various index minerals may be considerable and will be evaluated separately.

Solid solution in zeolite mineral compositions may also be important, especially substitution of Na for Ca in analcite, in heulandite and in wairakite. However, rather poorly constrained thermodynamic data for zeolites and limited chemical data do not justify including activity terms for these zeolites.

Fig. 9.6 is a projection from albite + chlorite (of fixed average composition) + quartz + water, because these phases are ubiquitous in low-grade metabasites. At very low temperatures (T < 200°C), however, this represents a simplification because smectite may be present, either as a discrete phase or as inter stratified layers in chlorite. Thus it is fortunate to note that chlorite is not involved in reactions limiting the stability fields of zeolites.

Below follows a brief discussion of the typical pT conditions under which some diagnostic index minerals may form and subsequently decompose in prograde metamorphism of metamafic rocks. The discussion is based in Figs. 9.5 and 9.6. All reaction stoichiometries are given in Table 9.2.

Analcite is commonly found in quartz-bearing diagenetic and low-grade metamorphic rocks. The analcite-quartz assemblage, along with heulandite + quartz, can be used as an indicator of the initial stage of zeolite facies metamorphism. In the NASH model system, analcite decomposes in the presence of quartz to albite according to the reaction: $Anl + Qtz = Ab + H_2O$ (the reaction is not listed in Table 9.2). The equilibrium conditions of the analcite breakdown reaction very nearly coincides with the three reactions that limit the stilbite field towards higher temperature in Fig. 9.5 [reactions (1), (2), and (3)]. In order to avoid overcrowding of the figure, the analcite reaction has not been shown in Fig. 9.5 explicitly. Natural analcite, however, often shows extensive solid solution with wairakite, thus expanding the stability field of analcite to higher temperature. This has the consequence that the pT fields of analcite and heulandite overlap.

Stilbite, a less common zeolite mineral, dehydrates to heulandite, laumontite or lawsonite with increasing temperature and successively higher pressures in the CASH model system (Fig. 9.5). In nature, however, stilbite is sometimes associated with heulandite and/or laumontite, as, e.g. in geothermal areas in Iceland. Although little is known about the composition of stilbite in low-grade metamorphic rocks, this feature may result from solid solution effects.

Heulandite is one of the most common naturally occurring zeolites besides analcite and laumontite. In Fig. 9.5, its stability field is located at temperatures around 200°C and at pressures below 2 kbar; and limited by the stability fields of three other zeolites, i.e. stilbite towards low-T [reaction (4)], wairakite towards high-T [reaction (6); not shown in Fig. 9.5, reaction is stable below

1 kbar pressure], and laumontite towards high-pT [reaction (5)]. Heulandite and laumontite seem to be equally abundant in ocean-floor thermal, hydrothermal and burial metamorphism. Laumontite, but not heulandite, has been described from subduction zone metamorphism, e.g. from the Franciscan Complex of California and the Sanbagawa Belt of Japan (see Liou et al. 1987, for a summary). This indicates that heulandite is a low-pressure zeolite consistent with the topology shown in Fig. 9.5.

Natural heulandite commonly contains Na or K or both substituted for Ca, and its chemical composition may extend to those of clinoptilolite and alkaliclinoptilolite. This effect, together with other variables (e.g. pore-fluid chemistry), may explain the considerable overlap between the stability regions of heulandite and other zeolites.

Laumontite is widespread in ocean-floor, burial and subduction zone metamorphism. In Fig. 9.5, its field of stable occurrence is limited by five univariant reactions. With increasing temperature, laumontite is formed from heulandite at lower pressure and from stilbite at higher pressure. The upper pressure limit of laumontite with respect to lawsonite is about 3 kbar [reaction (2)], and the upper temperature limit with respect to wairakite [reaction (7)] and epidote [zoisite; reaction (16)] is about 230–260°C. Compositionally, laumontite is near the ideal composition, $CaAl_2Si_4O_{12}$ 4 H_2O, but solid solution in heulandite and wairakite will allow the coexistence of Lmt + Hul and Lmt + Wa.

The presence of laumontite is often taken to indicate low-grade metamorphic conditions, and thus its lower thermal stability limit is of interest. In the CASH model system, this limit is about 180°C. From field evidence, however, it has been inferred that laumontite may have formed at temperatures as low as 50 to 100°C (e.g., Boles and Coombs 1977).

Wairakite is restricted to areas of relatively high geothermal gradients, including geothermal systems and areas with intrusive igneous bodies. In the CASH model system, wairakite is produced from heulandite or laumontite at temperatures between 220 and 260°C. Wairakite has the highest thermal stability of any zeolite and, in the presence of excess albite + quartz, its upper thermal stability is between 260 and 380°C. As mentioned earlier, wairakite shows considerable solid solution with analcite, that may explain the natural occurrence of Wa + Anl and Wa + Lmt (+ Ab + Qtz).

Lawsonite is one of the most definitive of the blueschist suite of minerals.

Pumpellyite is a very common mineral in metamafics and is found in various associations in low-grade metamafic rocks. The shaded pumpellyite field in Fig. 9.6 shows the pT range of the assemblage Pmp + Qtz + Chl + H_2O for average pumpellyite and chlorite composition. The reactions forming and consuming pumpellyite are listed in Table 9.2. In rocks that depart from the average meta-MORB, pumpellyite may occur over a much larger or much smaller pT interval. However, for all rock compositions the centers of the pumpellyite fields cluster around 220° ± 20°C and 4 ± 2 kbar. Most pumpellyite-bearing rocks probably formed close to 200–250°C and at pressures of 2 to 3 kbar.

Prehnite is widespread in low-grade metamafic rocks as a part of the matrix assemblage but also very often in late veins and open fracture spaces. The shaded prehnite fields in Figs. 9.5 and 9.6 clearly suggest that the mineral is diagnostic for low pressure conditions. In natural rocks, the upper pressure limit of prehnite is about 5 kbar but a more typical value is probably around 3 kbar.

The prehnite + pumpellyite assemblage is shown as a dark shaded field in Fig. 9.6. The assemblage may be found in meta-MORB's at 200–250 °C at pressures of 2–3 kbar in the association with quartz and chlorite (of the average composition). Depending on the rock composition (and the resulting average mineral compositions at low grade), the two minerals (Pmp + Prh) may not have a common field of occurrence at all. However, most metamafics may be able to generate the Prh + Pmp assemblage. The assemblage has, however, under any circumstances an extremely narrow pT field where it may be stable. The cumulative conditions for the co-occurrence of Pmp + Prh + Chl + Qtz is about 200–280 °C and 1–4 kbar. In many metamorphic terrains Pmp + Prh can be found in rocks of the subgreenschist facies. This is not surprising because of the natural variation of bulk composition of mafic rocks and the geotherms followed by prograde metamorphism will almost certainly pass through the mentioned cumulative pT field of Pmp + Prh. The higher-grade portion of the subgreenschist facies is therefore characterized by assemblages that involve: Pmp, Prh, Wa, Ep, Chl, Ab, Pg and Qtz.

9.3.3 The Transition to the Greenschist Facies

The transition to the greenschist facies is marked by the first appearance of the diagnostic assemblage actinolite + epidote in the presence of chlorite (of average composition), albite and quartz. The assemblage is produced from prehnite decomposition at low pressures (< 3 kbar) and pumpellyite decomposition at higher pressure (> 3 kbar). Both reactions [reactions (27) and (28)] produce the typical greenschist facies mineralogy at a temperature of about 280° ± 30 °C at pressures below 6 kbar. Along the characteristic Ky- and Sil-geotherms, the first occurrence of actinolite + epidote + chlorite + albite + quartz defines the beginning of the greenschist facies. The assemblage is shown on chemography #3 in Fig. 9.3. Chemography #3 also suggests that the greenschist facies assemblage may also form from dolomite- or calcite-involving reactions [an example is reaction (33)]. As carbonates are present in many low-grade mafic rocks, these mixed volatile reactions are important in removing carbonate minerals from metamafic rocks at an early stage of prograde metamorphism and connecting to the greenschist facies. Typically, these reactions consume chlorite and carbonate and produce epidote + actinolite. The reactions are of the maximum-type mixed volatile reactions (see Chap. 3). Note, however, that pumpellyite- and prehnite-consuming reactions *produce* rather than consume carbonates. Reaction (34) is an example reaction that limits pumpellyite in CO_2-bearing fluids and produces the greenschist facies assemblage. This means that in the presence of CO_2-rich fluids greenschist facies assemblages may appear at lower temperature compared to pure H_2O fluids.

At high pressures (e.g. along a low-T subduction geotherm) the subgreen-schist facies assemblages are replaced by assemblages that are diagnostic for the blueschist facies. These rocks contain glaucophane, lawsonite, paragonite, epidote and other minerals in various assemblages. The blueschist facies assemblages and blueschist metamorphism will be discussed in a separate paragraph below.

9.4 Greenschist Facies Metamorphism

9.4.1 Introduction

Greenschist facies mafic rocks are, as the name suggests, green schists. The green color of greenschists results from the modal dominance of green minerals in the rocks notably chlorite, actinolite and epidote. The most characteristic assemblage found in greenschists is: chlorite + actinolite + epidote + albite ± quartz. The term greenschist is exclusively reserved for a schistose chlorite-rich rock derived from mafic igneous rocks that were metamorphosed under greenschist facies conditions (hence contains the diagnostic assemblage: Chl + Act + Ep + Ab ± Qtz). Note, however, pumpellyite + epidote + chlorite schists of the subgreenschist facies are green schists as well though not in greenschist facies. Serpentinites may be green schists of the greenschist facies but no greenschists.

Chemographies #3 and #4 (Fig. 9.3) represent the typical range of assemblages in the greenschist facies. In addition to the most important minerals named above, some further minerals may be found in greenschists in minor amounts. Their occurrence mostly depends on bulk composition of the protolith and the details of the hydration history prior to greenschist facies metamorphism. Greenschists often lost all relic structures from previous metamorphic and magmatic stages. It is typical and common to find mineralog-ically and structurally perfectly equilibrated greenschists.

9.4.2 Mineralogical Changes Within the Greenschist Facies

9.4.2.1 Reactions in the CMASH System

The difference between chemographies #3 and #4 (Fig. 9.3) is the presence of pyrophyllite and phengite in the lower greenschist facies and of kyanite and biotite in the upper greenschist facies. The phase relationships in the ASH system that covers the changes in the A apex of ACF diagrams has been discussed in Chapter 7 and they are displayed in Fig. 7.2. The reactions in the ASH system also represent the backbone of Fig. 9.7 depicting some phase relationships in the CMASH system (using the same average mineral composi-tions as in Fig. 9.6, unless noticed otherwise). The reaction stoichiometries of

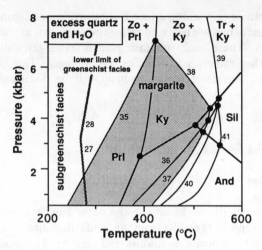

Fig. 9.7. Phase relationships in the greenschist facies. Margarite field *shaded*

reactions shown in Fig. 9.7 are listed in Table 9.2. Note, however, that not all equilibria shown are listed in Table 9.2, particularly the details of the phase relations in the sillimanite field of Fig. 9.7 were omitted because we are concerned with the greenschist facies here.

It is evident from Fig. 9.7, that the first reaction that is relevant for mafic rocks in greenschist facies is reaction (35). It replaces epidote + pyrophyllite by margarite in quartz-bearing rocks. The margarite field that is created by the stable reaction (35) and is terminated by other reactions which will be discussed below is shaded in Fig. 9.7. Margarite is not shown in Fig. 9.3 (ACF figure). Fig. 9.7 suggests that large central portions of the greenschist facies are within the stability field of margarite. However, margarite in greenschists is not very common. The reason for this apparent mismatch is that most metamafic rocks fall compositionally into the shaded Chl-Ep-Act triangle in Fig. 9.3. However, if bulk composition is more aluminous than the epidote-chlorite tie-line, then the stable assemblages may be read from Fig. 9.7. The sequence is (for Qtz-saturated rocks following a Ky-type geotherm): $Prl + Ep \Rightarrow Mrg + Ep \Rightarrow Ky + Ep$.

Fig. 9.7 also suggests that the assemblage epidote + kyanite is diagnostic for relatively high pressures (> about 5 kbar). The Ep + Ky assemblage is produced by reaction (38). Other reactions that terminate the margarite + quartz assemblage in greenschists are: reaction (36) and reaction (37). The assemblage actinolite (amphibole) + kyanite commences with the higher greenschist facies. It is diagnostic for pressures typically above 5–6 kbar. The composition of amphibole that occurs stable with kyanite is, however, far removed from common actinolite and reaction (39), that produces the Act + Ky assemblage, runs close to the amphibolite facies boundary.

Inside the Chl-Ep-Act triangle of Fig. 9.3, there are also reactions that modify the typical assemblage of the greenschist facies. The processes are best explained by using reaction (42) as an example. The reaction describes the production of tschermak component that is taken up by amphibole and chlorite. Both minerals systematically change their Al content across the greenschist facies. The effect can be shown on ACF diagrams by a displacement of the Chl + Ep + Act triangle along the TS vector. Note, however, concurrent rotations of FM relationships cannot be displayed on ACF diagrams. In general, chlorite and actinolite become more aluminous across the greenschist facies. AF relationships are complex and can be evaluated along the lines described in Chapter 7 (metapelites). Redox reactions may change the composition of epidote and other Fe-bearing minerals.

9.4.2.2 Reactions Including Micas

Muscovite (in general K-white mica, sericite, phengite, illite) is the most common K-bearing mineral in low-grade meta-mafic rocks. This is also true for micas in the lower greenschist facies. However, in the middle of the greenschist facies (~400 °C) biotite appears for the first time in metamafics replacing K-white mica. This is shown in Fig. 9.3 (chemographies #3 and #4). The biotite formation can be modelled by reaction (43). It involves all greenschist minerals and it transfers Ms component to Phl component. The equilibrium of reaction (43) is close to 600 °C in the pure KCMASH system and it is rather insensitive on pressure. For real mineral compositions of greenschist facies rocks, biotite formation from reaction (43) takes place around 400 °C ($a_{Phl} = 0.05$; $a_{Ms} = 0.2$; all others see Fig. 9.6). The first biotite that appears in prograde metamorphism of mafic rocks is often green under the microscope. This is an indication of high Fe^{3+} in low-grade biotite and demostrates the importance of Redox reactions in low-grade metamafics (so be aware of that when calculating phase relationships involving biotite).

Biotite production is a continuous process. The biotite producing reaction (44) is similar to reaction (43). However, it consumes celadonite component in K-white mica in order to produce biotite. The reaction will consume celadonite in white mica leaving behind a white mica that is closer to muscovite end-member composition. This is in accordance with field evidence that K-white mica becomes progressively depleted in Cel component and enriched in Ms component during prograde metamorphism along a Ky-geotherm. The effect can also be understood by investigating reaction (45), where celadonite component is consumed in white mica and biotite + muscovite components are formed in prograde metamorphism.

Again, biotite forms in metamafics at about 400–450 °C. K-white mica is typically celadonitic (phengite, sericite) in low-grade rocks it becomes more muscovitic (muscovite) at the upper end of the greenschist facies. However, the small amount of potassium in mafic rocks is usually bound in K-white mica below 400 °C and in biotite at temperatures above 400 °C, and there is not much overlap of the two micas.

Paragonite occurs in greenschists at high pressures. The equilibrium conditions of reaction (46) are extremely sensitive to small compositional variations in the protolith and the resulting metamorphic minerals. The typical actinolite + paragonite assemblage forms, however, at pressures above about 6 kbar, and this pressure is insensitive to temperature.

Garnet may form in mafic schist from reactions very similar to the reactions that form garnet in pelitic schist. The first garnets are manganiferous and contain little grossular component. These garnets appear in the uppermost part of the greenschist facies for the first time, but garnets are not really typical in mafic rocks until the amphibolite facies.

Stilpnomelane is a characteristic mineral in many low-grade mafic schists and, if present, is diagnostic for lower greenschist facies conditions. The brown color of this pleochroic mineral often resembles biotite under the microscope. However, stilpnomelane is diagnostic for the lower greenschist facies (or blueschist facies) and it is replaced by green biotite at around 400 °C. Mafic rocks containing both stilpnomelane and biotite occur over a narrow temperature interval near 400 °C.

Carbonate minerals may still be present in greenschists. Both dolomite (ankerite) and calcite can be present and the carbonates participate in many mixed volatile reactions that involve the characteristic greenschist facies silicates, Chl, Ep, and Act. The reactions may also involve micas such as reaction (47) that replaces K-white mica by biotite in Dol- and Cal-bearing rocks. Another example is reaction (48) where the assemblage actinolite + muscovite + calcite is replaced by epidote + biotite. Reaction (49) involves two carbonates in rocks containing epidote + actinolite. Some of the carbonate-involving reactions are also relevant to pure carbonate rocks (CMS-HC system) or to marly rocks. These have been discussed in Chapters 6 and 8, respectively, and we will not present an extensive discussion of phase equilibria in carbonate-bearing greenschists here. There is, however, a great potential for useful information contained in carbonate-bearing greenschists.

9.4.3 Greenschist-Amphibolite Facies Transition

The greenschist facies assemblage experiences two basic changes as metamorphic temperatures approach about 500 °C:

- albite disappears and it is replaced by plagioclase. The composition of the first Ca-bearing plagioclase is typically An_{17} (oligoclase);
- amphibole becomes capable of taking up increasing amounts of aluminum and alkalis. Actinolite disappears and it is replaced by alkali-bearing aluminous hornblende.

The combined transformation results in the replacement of the Ab + Act pair by the Pl + Hbl pair, that is in other words, replacement of greenschist by amphibolite. In orogenic metamorphism along a Ky-type geotherm, the greenschist-amphibolite facies transition occurs at temperatures of about 500° C (5 kbar).

Anorthite component is produced by a series of continuous reactions. However, albite present in greenschist does not continuously change its composition along the albite-anorthite binary. Under the conditions of the upper greenschist facies the plagioclase solid-solution series is not continuous but rather shows several miscibility gaps. The first Ca-bearing plagioclase that forms in this manner is an oligoclase and its typical composition is An_{17}. Because of the abrupt appearance of oligoclase, due to the miscibility gaps in the plagioclase series, the first appearance of plagioclase with An_{17} can be mapped in the field as an oligoclase-in isograd in mafic schists. This isograd defines the beginning of the amphibolite facies; it separates greenschist facies from amphibolite facies terrains.

The systematic compositional changes in the amphibole solid-solution series are generally more continuous in nature (although miscibility gaps do exist there as well). The amphibole composition changes basically by: (1) taking up tschermak-component produced by a series of continuous reactions, (2) incorporation of edenite component produced by albite-consuming reactions, and (3) the unavoidable FM exchange. Other effects are related to Ti incorporation and REDOX reactions.

The resulting amphibole is a green tschermakitic to pargasitic hornblende that is now present together with plagioclase. The two minerals constitute, by definition, an amphibolite and define the amphibolite facies. Chemography #5 (Fig. 9.3) shows this new situation; hornblende has changed its composition along the TS exchange direction and plagioclase shows up on ACF projections. In the beginning, chlorite and epidote may still by present in amphibolites.

The most important mineralogical changes at the greenschist-amphibolite transition can be related to reaction (50). The reaction consumes epidote and chlorite from the greenschist assemblage and produces anorthite component of plagioclase and tschermak component of amphibole. The reaction will eventually consume all epidote or all chlorite in the rock leaving behind a chlorite-, or epidote-bearing amphibolite. All three FM minerals of the greenschist facies assemblage are consumed by reaction (51). It produces tschermak component and consumes actinolite component of the amphibole. Reaction (52) describes the formation of the edenite component of the amphibolite facies hornblende. The combined effect of the three reactions (51), (52) and (53) is that the chlorite + epidote assemblage gradually disappears, plagioclase becomes increasingly calcic and amphibole systematically changes its composition from actinolite to alkali- and aluminium-bearing green hornblende.

Garnet may also appear at the transition to the amphibolite facies. Its formation is accomplished by similar reactions as in metapelitic rocks (Chap. 7). Garnet grows initially mostly at the expense of chlorite. Chlorite decomposition contributes the bulk of the almandine- and pyrope component of garnet. Epidote-consuming reactions produce much of the grossular and andradite component found in garnets in metamafic rocks. As in metapelitic rocks, low-grade garnets are generally rich in manganese and the strong fractionation of Mn into garnet makes the mineral appear in mafic schists at temperatures as low as 450°C. Reaction (54) consumes chlorite and epidote and produces both grossular and Fe-Mg-garnet component, which results in the typical ternary Ca-Fe-Mg-garnets found in mafic rocks.

9.5 Amphibolite Facies Metamorphism

9.5.1 Introduction

The amphibolite facies is characterized by chemographies #6, #7 and #8 (Fig. 9.3). The minerals plagioclase and hornblende make up the bulk volume of amphibolites. All other minerals that can be present, such as quartz, epidote, muscovite, biotite, garnet and CPX are modally subordinate. Calcite can be found in some amphibolites. Mineralogical changes within the amphibolite facies mostly result from reactions that run continuously over wide pT ranges. The main effect of these continuous reactions is seen in systematic variations in the composition of the two master minerals, plagioclase and hornblende. The continuous reactions also cause epidote-clinozoisite to decrease in modal amount and eventually they disappear. Muscovite, if it survived from the greenschist facies, also gradually disappears. Garnet persists and becomes modally more important upgrade (garnet amphibolites); clinopyroxene appears at higher temperatures in the amphibolite facies.

In principle, given a MORB composition and the composition of plagioclase and hornblende, the pT conditions of equilibration are uniquely defined. However, at present, experimental data do not permit a rigorous treatment of plagioclase-hornblende relationships. In particular, solution properties of amphiboles are poorly known and few end-member phase components are well constrained. The low-temperature phase relationships in the plagioclase system are also not quantitatively known.

9.5.2 Mineralogical Changes Within the Amphibolite Facies

In prograde orogenic metamorphism of metabasalts, the rocks contain hornblende and plagioclase at the beginning of the amphibolite facies (~500°C) as explained in Section 9.4.3 above. In addition, amphibolites may still contain some epidote and/or chlorite that has not been completely consumed by reactions that produced anorthite and tschermak component. Biotite may be present as well. The same continuous reactions that produced the amphibolite facies mineralogy initially, continue to consume chlorite and epidote within the lower part of the amphibolite facies. Eventually, chlorite completely disappears at about 550°C and epidote is not typically found in amphibolites that were metamorphosed to 600°C. Some of the epidote- and chlorite-consuming reactions produce garnet that, in general, becomes modally more important with increasing grade. Chemography #6 is characteristic for the central portion of the amphibolite facies around 600 °C. Here, amphibolites contain plagioclase (typically andesine) and green hornblende ± garnet ± biotite. At still higher temperatures, clinopyroxene appears in amphibolites (chemography #7). CPX of the diopside-hedenbergite series usually appears around 650 °C (along the Ky-geotherm) and a typical reaction that produces it is listed as reaction (54). The reaction simultaneously produces CPX and garnet. However, CPX is also often found in amphibolites lacking garnet. A typical assemblage is Hbl + Pl + CPX + Bt. Reaction (55) produces diopside and anorthite component from amphibole and epidote or zoisite. This extremely important reaction has several significant effects: (1) it continuously consumes epidote or clinozoisite that may still be present in mid-amphibolite facies rocks and eventually eliminates them; (2) the reaction consumes amphibole component, a process typical for the higher amphibolite facies; (3) the reaction produces CPX that appears in higher grade amphibolites; and, last but not least, (4) the reaction produces even more anorthite component that is incorporated in plagioclase in high-grade amphibolites. Plagioclase in high-grade amphibolites becomes progressively more calcic and in the upper amphibolite facies andesine-labradorite compositions are typical for plagioclase in amphibolites (the range is $An_{30} - An_{70}$). Note, however, that phase relations in the plagioclase system are complex and little understood, so if you find a bytownite or anorthite (An_{95} plagioclase) in an upper amphibolite facies terrain, this is no reason to be confused.

The first appearance of clinopyroxene in amphibolites can be used to define the boundary to the upper amphibolite facies. Reactions in the upper amphibolite facies, such as reactions (54) and (55), begin to break down amphibole components and replace them with pyroxene components. This is a continuous process, however, and the first appearance of CPX in mafic rocks is usually not a sharp isograd in the field. Nevertheless, the "CPX-in" transition zone marks the beginning of the upper amphibolite facies and a representative temperature is about 650 °C. In water-saturated environments metamafic rocks show the first structural evidence in the field of local partial melt formation, migmatization and appearance of quartzo-feldspathic seams, patches, veins and similar mobilisate structures in the upper amphibolite facies. The processes remove quartz, plagioclase and biotite from the rocks and transfer them to a melt phase.

At higher pressures than that of the ordinary Ky-geotherm (Fig. 9.3), amphibolites may contain the diagnostic kyanite + hornblende assemblage. The Ky + Hbl pair may have formed by reaction (39) earlier in the course of prograde metamorphism (see Fig. 9.7). The link to the more common plagioclase + garnet assemblage found in amphibolites is given by reaction (56). The reaction replaces the garnet + plagioclase tie-line of chemographies #6 and #7 by the kyanite + hornblende tie-line of chemography #8. The continuous dehydration reaction runs typically between the chemography #6 and the chemography #8 in Fig. 9.3 for typical meta-basalt compositions. Therefore, kyanite-bearing amphibolites are diagnostic for high-pressure amphibolite facies (pressures typically greater than 7 kbar). Remember, however, that the Ky – Hbl – Grt – Pl assemblage is not normally co-planar (as one may erroneously conclude from ACF figures such as 9.3) but rather defines a phase volume that occurs over a relatively wide pT interval.

9.5.3 Low-Pressure Series Amphibolites

At low metamorphic grades, mafic rocks that are metamorphosed along a Ky-geotherm and a Sil-geotherm are very similar in mineralogy. The most significant difference may be found in the composition of chlorite and, especially, K-white mica. The celadonite component of K-white mica is relatively sensitive to pressure variations and it is controlled by processes such as reaction (45). The tschermak component in mica (and in principle also chlorite or biotite) can be used to monitor pressure conditions in low-grade metamorphism.

The transition to the amphibolite facies at low pressures is similar to the one described in Section 9.4.3. In extremely Al-rich rocks, andalusite may appear instead of kyanite (chemography #10). However, andalusite + amphibole is not very common. The phase relationships shown in Fig. 9.7 suggest that the transition to the amphibolite facies occurs at slightly lower temperatures in low-pressure metamorphism. This is reasonable because the reactions that produce amphibolite from greenschist are ordinary continuous dehydration reactions. The lower boundary of the amphibolite facies along a Sil-geotherm is typically at about 450°C (at 3 kbar). Reactions (36), (37), (40) and the andalusite equivalent of reaction (41) pass through the andalusite field in Fig. 9.3. The reactions generally produce anorthite and amphibole components from greenschist facies mineralogy. At 550°C (Figs. 9–3 and 9–7), epidote and chlorite typically disappeared from low-pressure amphibolites and the characteristic assemblage is hornblende + plagioclase ± andalusite ± biotite.

Compared with orogenic metamorphism along a Ky-geotherm, clinopyroxene forms at significantly lower temperatures in low-pressure amphibolites (600°C or lower). Chemography #11 is characteristic for amphibolites that formed in the range of 3–4 kbar. It shows the usual hornblende + plagioclase assemblage together with clinopyroxene and biotite. Sillimanite may be present in Al-rich amphibolites and the Sil + Hbl pair forms a stable, though rare, assemblage. However, garnet is scarce or even absent in low-pressure amphibolites.

Chemography #12 is representative for amphibolites in the upper amphibolite facies at low pressures (Sil-geotherm). The most significant feature is the presence of ferro-magnesian amphiboles in addition to the calcic amphiboles (hornblende). The types of Fe-Mg amphiboles that occur together with hornblende include anthophyllite, gedrite and cummingtonite. In some amphibolites even three different amphibole species may be present such as hornblende, gedrite and anthophyllite. The abbreviation *Oam* for orthoamphibole in Fig. 9.3 includes all Fe-Mg amphiboles and also gedrite. Phase relationships in such multi-amphibole rocks can be very complex. Miscibility gaps in various amphibole series as well as structural changes in amphiboles complicate the picture. However, such low-pressure amphibolites possess a great potential for detailed analysis of relationships among minerals and assemblages and for the reconstruction of the reaction history of low-pressure amphibolites. Fe-Mg-amphiboles may form by a number of different conceivable mechanisms. The first and obvious one is reaction (57) that links chemography #11 to chemography #12. The reaction breaks down the hornblende + sillimanite assemblage and produces anorthite component that increases the An-content of plagioclase and it also produces anthophyllite component that is a major component in any Fe-Mg-amphibole. As metamorphic grade increases, reaction (58) becomes more important in producing anthophyllite component. The reaction consumes calcic amphibole component and replaces it with clinopyroxene and Fe-Mg-amphibole. Reaction equilibrium of reaction (58) in metabasaltic rocks involves phase components of the typical assemblage: Hbl + Oam + CPX + Pl ± Qtz. As a general rule of thumb, the assemblage is characteristic for the temperature range between 650 and 750° C along the Sil geotherm.

9.5.4 Amphibolite-Granulite Facies Transition

Much of the quartz that has been produced in the subgreenschist facies to greenschist facies metamorphism has been consumed by continuous reactions in the amphibolite facies. Many amphibolites are quartz-free (use ACF diagrams with care!). Most of the water bound in hydrous minerals has been released by reactions during prograde metamorphism. Metabasaltic rocks, that have been progressively metamorphosed to 700°C, contain Pl + Hbl + CPX + Grt ± Bt (Ky geotherm) and Pl + Hbl + CPX + Oam ± Bt (Sil-geotherm). Amphibole is the last remaining hydrous phase and further addition of heat to the rock by tectono-thermal processes will eventually destroy the amphibole. The transition from a hydrous amphibolite facies assemblage to a completely anhydrous granulite facies assemblage is gradual and takes place over a temperature interval of at least 200 °C (from about 650–850 °C). The first clear and unequivocal indication that granulite facies conditions are reached is the appearance of orthopyroxene in CPX-bearing quartz-free rocks. Granulite facies is obvious if OPX turns up in quartz-bearing amphibolites. Orthopyroxene is most common in low-pressure mafic granulites. At higher pressures, the typical anhydrous granulite facies assemblage is Pl + CPX + Grt ± Qtz. The reactions that ultimately produce this assemblage are continuous in nature and the temperature interval is very wide. Hbl gradually decreases in modal amount, leaving behind the anhydrous granulite facies assemblage. Note, however, that CPX + Pl + Grt is widespread in the amphibolite facies and that the assemblage as such is by no means diagnostic for granulite facies conditions. If pressure conditions are too high for OPX to form, then the ultimate granulite facies condition is reached when the last crystal of amphibole disappeared from the rock. The reactions (59) to (62) consume the major amphibole components tremolite and tschermakite and produce pyroxene, garnet and anorthite components and, hence, represent the transition of the amphibolite to the granulite facies. Particularly reaction (60) is of great importance. It decomposes hornblende and makes the two-pyroxene assemblage that is diagnostic for the granulite facies in quartz-absent mafic rocks. It produces, in addition, anorthite component that ultimately leads to the calcic plagioclase found in mafic granulites.

What happens to the biotite? Biotite is a subordinate hydrous phase in high grade amphibolites. The small total amount of potassium stored in biotite will be taken up by pyroxene and ternary feldspar or it is used to form a Kfs-Ab-Qtz melt in the migmatite producing process in high-grade metamorphism of mafic rocks. The total amount of water that can be stored in high-grade amphibolites is very small (on the order of 0.4 wt.% H_2O or less). Consequently, mafic rocks are less susceptible to partial melting compared to their metapelitic or metagranitoid counterparts.

9.6 Granulite Facies and Mafic Granulites

Granulite facies metamorphism of mafic rocks is represented by chemographies #9 and #13 in Fig. 9.3. At high pressures the characteristic assemblage is: plagioclase + clinopyroxene (augite) + garnet. At lower pressures the typical granulite facies assemblage is: plagioclase + clinopyroxene (augite) + orthopyroxene (hyperstene). The two assemblages are linked by the important reaction (63). The reaction separates a field with pyroxene granulites at pressures below about 5–7 kbar from a field with garnet granulites at pressures above 5–7 kbar and below the eclogite field at very high pressures (at 800°C). The continuous nature of reaction (63) generates a wide overlap zone with both garnet and OPX present in Pl + CPX rocks. If strictly concerned with meta-MORBs, however, the overlap is smaller and the boundary occurs close to 6–7 kbar at 800°C.

Because high-grade brown hornblende is an extremely stable mineral, very high temperatures are necessary to destroy the last hydrous phase in mafic rocks. Completely anhydrous mafic granulites form normally above 850°C or at even higher temperatures. Dehydration of amphibole, on the other hand, is often aided by interaction of the rocks with a foreign fluid phase that is poor in H_2O. Alternatively, dehydration of amphibole may be triggered by partial melting and removal of H_2O in a melt phase; see also Chapter 7. (metapelitic granulites), where some aspects of granulite facies metamorphism have been discussed. Fig. 9.8 shows some of the relationships that are important in high-grade and granulite facies metamorphism of mafic rocks. Fig. 9.8 is a schematic temperature versus fluid composition diagram at an approximate total pressure of 6 kbar. Along the 750°C isotherm, three different assemblage fields are

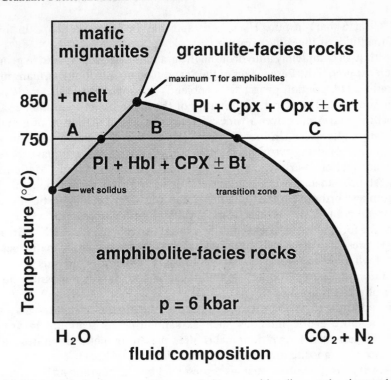

Fig. 9.8. Schematic isobaric temperature versus fluid composition diagram showing amphibo-lite - granulite facies relationships

intersected. If H_2O-rich fluids are present (section A), the mafic rocks undergo partial melting and mafic migmatites are characteristic for such conditions. Intermediate fluid compositions are consistent with amphibolite facies rocks with the typical mineralogy: Pl + Hbl + CPX (section B). Migmatite structures are absent. Fluids low in H_2O coexist with granulite facies rocks at the same temperature and the characteristic assemblage is: Pl + CPX + OPX ± Grt (section C). Note that the transition zone between the granulite and amphibolite facies assemblages (Fig. 9.8) covers in real rocks a wide range of fluid composition. The important message of Fig. 9.8 is, however, that at the same temperature (e.g. 750°C) mafic migmatites, amphibolites and mafic granulites may be found intimately associated in a metamorphic complex or terrain, depending on the composition of the fluid phase present in the volume of rocks under consideration. The situation may be complicated by variations of fluid composition over time at a particular site. Also note that the first melt in mafic rocks forms if pure H_2O-fluids are present (wet solidus in Fig. 9.8). With increasing temperature, migmatites form with fluids that contain decreasing amounts of water. The amphibolite facies assemblage has a temperature maximum in Fig. 9.8. The actual value of the maximum temperature for amphibolite can be correlated with the amphibole-out isograd that is at about

900°C (at 6 kbar) in pure H_2O fluids and may be closer to 850°C in mixed volatile fluids at the same pressure.

Although completely anhydrous mafic granulites are widespread (e.g. in the Jotun nappe and the Bergen arcs of the Scandinavian Caledonides), many mafic granulites still contain prograde hornblende (as opposed to retrograde post-granulite Hbl). The transitional nature of the amphibolite to granulite facies transition that occurs over a wide pT range requires a definition of granulite facies in mafic rocks. In low-pressure terrains this is a simple task: the first appearance of orthopyroxene in mafic rocks (quartz-free or quartz-bearing does not matter) marks the onset of granulite facies conditions. Mafic rocks will, at this stage, always contain clinopyroxene as a major mineral. The assemblage orthopyroxene + clinopyroxene will then be diagnostic for two-pyroxene granulites. The case is more difficult at high pressures where OPX does not form in mafic rocks and the typical assemblage Pl + CPX + Grt may also be present in the amphibolite facies or hornblende persists in mafic rocks to very high temperatures (1000°C at 10 kbar). Obviously, in this case, the definition of an amphibolite – granulite facies boundary is impossible on the basis of mafic rocks alone.

In addition to the minerals discussed so far, mafic granulites may also contain more exotic minerals such as sapphirine or scapolite in various assemblages. Also hercynitic spinel is often present in mafic granulites and its presence is diagnostic for low-pressure conditions (< 4 kbar).

At very low pressures, such as represented by chemography #14, pyroxene hornfelses form by high-temperature contact metamorphism of mafic rocks. Ultimately, the mineralogy of the protolith basalt is expected to form in ultra-high temperature contact metamorphism of mafic rocks. The assemblage Pl + CPX + OPX ± Spl (chemography #14) is then identical to the basalt mineralogy shown in the upper left corner of Fig. 9.3. The upper boundary of granulite facies metamorphism of mafic rocks is given by the dry solidus for basalt (~1100°C).

Geological causes of granulite facies metamorphism at low pressures are commonly attributed to intrusions of mafic (gabbroic/basaltic) or charnock-itic magma into the continental crust (see References and Further Reading; Chap. 7). Large volumes of dry or CO_2-rich hot magma from the mantle or lower crust serves as heat source for extensive high temperature, metamor-phism at shallower crustal levels. The dry or CO_2-rich nature of the magma facilitates dehydration of the crustal rocks. There is little tectonic activity associated with this type of granulite facies metamorphism. After heating to high temperature the rocks cool essentially isobarically at the crustal level at which they reside. Many granulite facies terrains are characterized by isobaric cooling paths. In general, the exhumation of granulite facies terrains requires a later contractional orogenic event that involves stacking and tilting of crustal slices and the formation of fold nappe structures that ultimately permit the erosion surface to intersect with granulites of the middle or lower crust. If water becomes available during slow cooling, the high-grade assemb-lages may be completely erased and granulites may lose their memory of the granulite facies event. Particularly, if OPX disappears from mafic granulites during amphibolitization, it is very hard to tell that the rock ever went through a granulite facies episode.

9.7 Blueschist Facies Metamorphism

9.7.1 Introduction

Blueschists are rocks that contain a significant amount of blue alkali-amphibole with a very high proportion of glaucophane endmember component. Such rocks are, macroscopically, blue in color, note, however, that pure glaucophane is colorless under the microscope. If a rock contains a blue pleochroic amphibole in thin section it means that it is not a pure glaucophane, and appreciable amounts of ribeckite and other sodic ferro-ferric amphibole components are important (crossite). Note also that Cl-rich hornblende may have a blue color under the microscope. However, if mafic rocks of basaltic bulk composition contain sodic amphibole, the rocks are likely to have been metamorphosed under the conditions of the blueschist facies.

The general pT field of the blueschist facies is delineated in Fig. 4.2. The blueschist facies is represented in Fig. 9.3 by the chemographies #16 and #17. The lower pressure version of the blueschist facies (#16) is characterized by glaucophane (sodic amphibole) + lawsonite + chlorite assemblages. In the higher-grade blueschist facies (#17), typical metamafics contain various assemblages among glaucophane (sodic amphibole), zoisite (clinozoisite-epidote), garnet, paragonite, phengite, chlorite, talc, kyanite, rutile, ankerite and other minerals.

A detailed rendering of the blueschist field is shown in Fig. 9.9. It is bounded towards the subgreenschist and greenschist facies by reactions (64), (65) and (66). Its high-pressure boundary is given by reaction (71). Blueschist facies terrains are worldwide associated with subduction zone metamorphism that typically occurs along destructive plate margins where oceanic crust (basalt) is recycled to the mantle under continental lithosphere. Two typical geotherms related to subduction tectonics are depicted in Fig. 9.9. The slow subduction geotherm only marginally passes through the blueschist field and eclogite assemblages are produced at pressures as low as 13 to 14 kbar. If subduction is fast, prominent blueschist facies terrains may develop and eclogite formation does not occur until pressures of 18 to 20 kbar are reached (corresponding to subduction depths of 50–60 km). Because there are natural limits to attainable subduction velocities in global tectonic processes, the pT conditions in the upper left wedge in Fig. 9.9 are not accessible conditions for rock metamorphism.

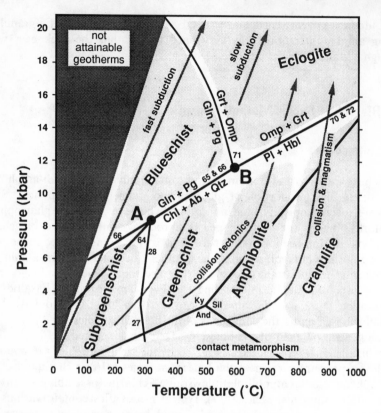

Fig. 9.9. High-pressure metamorphism overview

9.7.2 Reactions and Assemblages

As discussed earlier, reactions (27) and (28) (Fig. 9.9) mark the boundary
between the subgreenschist facies and the greenschist facies. The boundary
between the subgreenschist facies and the blueschist facies is given by the
reaction (64). This reaction replaces the low-grade assemblage actinolite + chlorite + albite by the high-pressure assemblage glaucophane + lawsonite.
The reaction is H_2O-conserving, and as such independent of water pressure or
the composition of the fluid phase. Equilibrium of reaction (64) is shown in Fig.
9.9 for average mineral compositions found in low-grade equivalents of basaltic
rocks.

Point A in Fig. 9.9 marks an important pT range where blueschist,
greenschist and subgreenschist facies assemblages meet. It is at about 8 kbar and
300 °C for rocks of basaltic composition. The pT range is defined by the
approximate intersection of the reactions (22), (28), (30), (31), (64), (66) and
(70). Note, however, that point A is not an invariant point in the Schreinemakers sense but rather a relatively narrow pT range of the locations of various

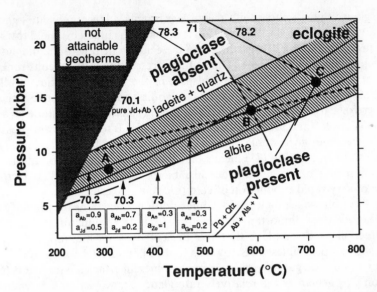

Fig. 9.10. Selected mineral equilibria in high-pressure metabasites. The *dashed line* is the approximate high-pressure boundary for plagioclase in mafic rocks, lined field; plagioclase present-absent transition zone and coexistence of Na-pyroxene, plagioclase and quartz. On the high-T side of (71) and at high-p side of *dashed line*; eclogites. $a_{H2O} = 1$. $a_{Pg} = 0.9$ in reaction (78), other activities in (78) correspond to those of the respective reaction (70), $a_{Prp} = 0.3$ and $a_{Jd} = 0.4$ in reaction (71)

invariant points generated by the intersection of continuous reactions in the NCMASH system.

The blueschist facies is separated from the greenschist facies by reactions (65) and (66). The stable equilibrium of these reactions is shown in Fig. 9.9. Reaction (65) consumes the low-grade assemblage chlorite + actinolite + albite and replaces it by the most typical blueschist facies assemblage glauco-phane + epidote (zoisite). The equilibrium of reaction (65) connects the points A and B in Fig. 9.9, respectively. It almost coincides with another very important reaction, reaction (66). The reaction replaces the greenschist facies assemblage chlorite + albite by the characteristic blueschist assemblage glaucophane + paragonite. The combined effect of reactions (65) and (66) is that the chlorite + albite + actinolite assemblage, that is diagnostic for low-grade meta-mafic rocks, disappears, and the blueschist assemblage glaucophane + epi-dote + paragonite is formed. If, for any reason, some albite should still be present in the rocks after all chlorite has been used up, that albite will be destroyed by reaction (70). The reaction boundary of reaction (66) roughly coincides with the diagonally ruled boundary that separates a plagioclase present from a plagioclase absent region (Fig. 9.10). As in the eclogite facies, blueschists do not contain stable albite (plagioclase). In most low-grade mafic rocks, however, chlorite is much more abundant than is albite. Consequently, reactions (65) and (66) will normally destroy all albite.

The glaucophane + paragonite assemblage is replaced by omphacite + garnet as a result of reaction (71). The reaction terminates the blueschist field towards higher pressures, where it is grading into the eclogite field with the diagnostic omphacite + garnet assemblage. The sodium present in mafic igneous rocks is stored at low grade in the albite feldspar, in the blueschist facies in alkali amphibole (glaucophane, crossite) and paragonite, and in the eclogite facies in sodic pyroxene (omphacite, jadeite). Reaction (71) has a negative slope on a pT diagram (Figs. 9.9 and 9.10) and terminates at point B towards lower pressure where it becomes metastable relative to feldspar involving reactions. The negative slope of reaction (71) means that the most favourable conditions for the development of extensive blueschist facies terrains are met in tectonic settings of very rapid subduction of cold (old) oceanic crust.

Inside the blueschist facies field, some additional reactions may modify the assemblages. Most important is reaction (67). It terminates the lawson-ite + glaucophane assemblage that is represented by chemography #16 and garnet shows up in blueschists towards higher pressure (chemography #17). Another important garnet producer in the blueschist facies is reaction (68). Reaction (69) generates the relatively rare glaucophane + kyanite assemblage that is diagnostic for very low-T high-p glaucophane schists.

The continuously running reactions in the blueschist facies field result in a systematic, also gradual, change in the mineralogical and modal composition of meta-basaltic rocks during prograde metamorphism. Let us follow a metabasal-tic rock with a subgreenschist facies assemblage that undergoes subduction at a rate intermediate to the slow and fast subduction geotherms in Fig. 9.9. Such a geotherm will cover a temperature interval in the blueschist facies field from 250–500°C and the pressure ranges from 7 kbar to 16 kbar. At pressures between 6–8 kbar, lawsonite + alkali amphibole will begin to replace the low-grade chlorite + actinolite + albite assemblage. The alkali amphibole is initially fairly Fe-rich, containing high proportions of ribeckite component. As pressure further increases, albite gradually disappears and amphibole becomes increas-ingly sodic and depleted in actinolite-component, the sodium gradually replaces calcium on the M4 position of the amphibole structure by an exchange mechanism that can be formulated as the glaucophane exchange component: $Na^{M4}Al\,Ca^{M4}_{-1}Mg_{-1}$. The exchange leaves the A site of the amphibole structure untouched. The continuous reactions produced the lawsonite + crossite + chlorite + paragonite ± phengite assemblage that is most characteristic for pressures around 10 kbar. At still higher pressures, lawsonite gradually disappears, alkali amphibole becomes increasingly rich in glaucophane compo-nent, zoisite and epidote become important members of the assemblages, and chlorite continuously decreases in modal amount. The characteristic assem-blage for this intermediate blueschist facies stage is: glaucophane + epi-dote + chlorite + paragonite ± phengite. Typical pressures for that assemblage are in the range of 12–14 kbar (along the geotherm we are presently looking at). With increasing pressure, garnet becomes a member of the assemblage. Chlorite further decreases in modal amount giving rise to increasing amounts of garnet. High-pressure blueschists most typically contain: glaucophane + epidote (clinozoisite, zoisite) + garnet + paragonite ± phengite. Other minerals that might be present in high-p low-T blueschists include Mg-chloritoid, talc and kyanite. For a discussion of systematic compositional changes in the only important potassic mineral in blueschists, that is phengite, with increasing pressure you may consult Section 7.8 in the chapter on metamorphism of pelites. Also note that blueschists often contain carbonate minerals that give rise to mixed volatile reactions that are not discussed here. Carbonates found in blueschists include calcite and its high-pressure polymorph aragonite, ankerite, dolomite, breuner-ite and magnesite. The origin of carbonates in blueschists is related to either ocean-floor metamorphism of basalt and metasomatism associated with it or they are derived from carbonate sediment deposited concurrently with the extrusion of pillow basalt lavas on the ocean floor.

The blueschist facies assemblages are gradually replaced with the typical eclogite facies assemblage omphacite + garnet at pressures greater than about 14 to 16 kbar. Because high-pressure blueschists typically contain garnet by now, it is the appearance of omphacite (sodic pyroxene) that marks the transition to the eclogite facies. The boundary is rather gradual and there is a wide pressure range where glaucophane + paragonite + epidote + garnet + omphacite may occur in a stable association. Strictly speaking, the assemblage is diagnostic for both the blueschist and the eclogite facies. Nevertheless, along an intermediate high-p low-T geotherm, typical eclogite assemblages are formed at about 16–18 kbar and a corresponding temperature of about 500°C.

9.8 Eclogite Facies Metamorphism

9.8.1 Eclogites

Eclogites are metamorphic mafic rocks that contain the stable mineral pair **garnet and omphacite** in significant modal amounts. Eclogites are free of plagioclase. Eclogites are very dense rocks with a density even greater than some ultramafic mantle rocks ($\varrho_{eclogite} > 3300$ kg m^{-3}). Because of the high density of eclogites, their origin can be readily related to very high pressure conditions during formation. Omphacite + garnet is the diagnostic assemblage in metabasaltic rocks that recrystallized above about 12–14 kbar pressure outside the stability field of plagioclase. Typical eclogites form at pressures of 18 to 22 and more kilobars. Many of the eclogites found in collisional mountain belts such as the Alps or the Scandinavian Caledonides still display clear compositional and structural evidence for being derived from basaltic lavas. In the Zermatt region of the Central Alps, for example, preserved MORB compositions and pillow lava structures may be found in eclogites that formed at pressures greater than 20 kbar corresponding to a subduction depth of 60 km. This, in turn, reveals compelling evidence that surface rocks (basaltic lava) can be transported by active tectonic processes to depths of 60 km below the surface (or in the case of coesite-bearing eclogites to 100 or more km).

The pT field of the eclogite facies shown in Fig. 9.9 can be accessed in prograde metamorphism from all three neighbouring facies fields, blueschist, amphibolite and granulite facies, depending on the tectonic setting. In subduction tectonics, eclogites originate from blueschists, as outlined above. These types of subduction-related eclogites often register very high pressures of formation.

In a setting of continent-continent collision, the continental crust often attains twice its normal thickness and the deeper parts of the double crust are exposed to pressures in the range of 12 to 24 kbar. Any mafic rocks of suitable composition in the deeper part of the double crust could be transformed to eclogite (Fig. 9.9). An example of such a setting is the collision of the Eurasian and the Indian plate and the resulting double-crust formation of the Tibetan plateau. A Ky-type geotherm, typical of collision tectonics, is shown in Fig. 9.9. It can be seen that eclogites in such a setting will be created from amphibolites. Eclogites that form in such a setting of crustal stacking and nappe formation do not record extremely high pressures of formation. Pressures in the range of 14 to 18 kbar are most characteristic.

If collision is accompanied by substantial magmatic heat transfer to the crust, eclogites may form from granulites and the "hot" geotherm shown in Fig. 9.9 may be more appropriate in this situation. Mantle-generated basaltic magmas often sample pieces of rocks along the conduits during ascent to the surface. The rock fragments typically found as xenolith in lavas of basalt volcanoes include ultramafic rocks from the mantle, granulites and eclogites. The association is characteristic of abnormally hot geotherms in conjunction with collision and magmatism, initial stages of extension or magmatism alone. In such settings eclogites may have been created from mafic granulites or directly crystallized from basaltic magma at great depth. The pressure range of high-temperature eclogites is very large; however, typical pressures are in the same range as for the eclogites formed from amphibolites. Basaltic magmas may, on the other hand, also collect some eclogite xenoliths from great depth in the mantle and transport them to the surface. The ultimate origin of these deep eclogite samples is ambiguous, however. They also may represent pieces of recycled former oceanic crust.

From the comments above, it follows that eclogites may form in a wide range of tectonic settings and the eclogite facies comprises the widest region of temperature and pressure conditions of any of the metamorphic facies fields. The temperature ranges from about 400 to 1000 °C. It may be reasonable to distinguish three general types of eclogites depending on geological setting and temperature of formation:

1. Low-temperature eclogites are related to subduction tectonics and form from blueschists (high-pressure low-temperature eclogites = **LT-eclogites**).
2. Intermediate temperature eclogites form in continent collision settings from amphibolites (medium-temperature eclogites = **MT-eclogites**).
3. High-temperature eclogites form in settings of collision or extension where the geotherm is abnormally "hot" due to magmatic heat transfer from the

mantle (e.g. basalt magma underplating) from mafic granulites or crystallize directly as eclogite from mafic magma (high-temperature eclogites = HT-eclogites).

The three different genetic types of eclogite are also characterized by typical mineral associations because of dramatically different temperatures of formation and because of widely varying H_2O pressures associated with their formation. While the LT eclogites that formed in subduction zones often contain modally large amounts of hydrate minerals, HT eclogites often show "dry" assemblages.

In dealing with high-pressure rocks, important issues also include the kind of mechanism by which they return to the surface, what kind of detailed pT path they follow on their way to the surface, and, most importantly, what kind of modifications the rocks experience as they are returned to the surface. For example, if an eclogite formed by a subduction mechanism and equilibrated at 650°C and 20 kbar, it is crucial for the fate of the eclogite assemblages whether the rocks reach the surface via the blueschist field (cooling and decompression occur simultaneously), via the amphibolite and greenschist fields (first decompression then cooling), or via the granulite field (first heating and decompression then cooling). Most eclogites display some mineralogical impact of reactions along the return path to the surface. The access of H_2O along that path is, of course, essential whether or not the high-T assemblages are conserved or extensive retrogression occurs. On the other hand, as will be shown below, LT eclogites often contain abundant hydrate minerals at peak pressure. Such rocks will undergo extensive dehydration reactions that will eventually eliminate the eclogite assemblage if decompression occurs without concurrent cooling.

9.8.2 Reactions and Assemblages

Reactions that connect the blueschist facies to the **low-temperature eclogite facies** have been discussed in Section 9.7.2. The most typical assemblages in LT eclogites involve the following minerals (bold = compulsory minerals: ± = optional minerals): **garnet** + **omphacite** ± zoisite ± chloritoid ± phengite ± paragonite ± glaucophane ± quartz ± kyanite ± talc ± rutile ± dolomite.

Omphacite is a sodic high-pressure clinopyroxene that is composed of the major phase components jadeite, acmite, diopside and hedenbergite (a solid solution of: $NaAlSi_2O_6$-$NaFe^{3+}Si_2O_6$-$CaMgSi_2O_6$-$CaFeSi_2O_6$). The M2 site of the pyroxene structure is filled with about 50% Na and 50% Ca in typical omphacite. LT eclogites often contain pure jadeite together with omphacite. Omphacite contains commonly minor amounts of chromium giving the mineral a grass-green color. This is particularly the case in meta-gabbros. Note that Cr-diopside in garnet-peridotite also shows a characteristic grass-green color. However, because ultramafic rocks contain very little sodium, the jadeite component in Cpx of high pressure ultramafic rocks is generally low.

Garnet in eclogite is a solid solution of the major phase components almandine, pyrope and grossular (a solid solution of: $Fe_3Al_2Si_3O_{12}$-$Mg_3Al_2Si_3O_{12}$-$Ca_3Al_2Si_3O_{12}$). However, due to its extreme refractory nature, garnet in eclogite can be inherited from earlier stages of metamorphism or be relics from the protolith (e.g. garnet-granulite).

Chloritoid in LT eclogites is very magnesium-rich and X_{Mg} may be as high as 0.5. Chloritoid and paragonite are diagnostic minerals in LT eclogites. The two minerals are removed by various reactions from eclogites at higher temperatures and are typically absent in MT- and HT eclogites.

Phengite and other sheet silicates in LT eclogites derived from gabbros also often contain small amounts of chromium giving them a green color (Cr-phengite = fuchsite).

9.8.2.1 Amphibolite and Granulite to Eclogite Facies Transition

Amphibolites and granulites contain plagioclase feldspar that is rich in anorthite component. Reaching the eclogite facies from the amphibolite facies or the granulite facies, the reactions that destroy feldspar at high pressures must also decompose the anorthite component of plagioclase.

Reaction (72) produces Ca-tschermak component that is taken up by omphacite. The equilibrium of reaction (72) is nearly independent of temperature and it can be used for reliable pressure estimates. Reaction (73) replaces anorthite by zoisite and kyanite. This reaction is the most important anorthite consuming reaction in high pressure metamorphism of both MT- and HT eclogites. The reaction is a hydration reaction that consumes H_2O as well as anorthite component. In the absence of water, anorthite will not decompose to zoisite. However, if water is not available as solvent then reaction kinetics is so slow and transport distances are so small that also albite and other low-pressure phases or phase components will survive the high pressures. For example, in an olivine gabbro from the Alps (Allalin gabbro, see below) the primary magmatic minerals survive pressures in excess of 20 kbar at 650–700 °C without forming eclogite except where water had access to the rocks. Another example is from the Bergen Arcs (Norwegian Caledonides), a crystalline nappe complex that experienced p ~ 20 kbar and T ~ 700 °C during Caledonian metamorphism. It was found that the mineralogy of Precambrian mafic granulite is perfectly preserved over large areas except along an anastomosing network of shears where the access of water permitted the formation of the stable rock under the Caledonian metamorphic conditions, that is an MT eclogite.

The equilibrium conditions of reaction (73) are shown in Fig. 9.10. It runs roughly parallel to the albite breakdown reaction (70). Reaction (74) decomposes anorthite component and produces grossular component in eclogite garnet together with kyanite and quartz (Fig. 9.10). The slope of the reaction is similar to those of reactions (70) and (73).

Reaction (70) has been used in Chapter 3 to illustrate some general principles of chemical reactions in rocks. In high pressure mafic rocks it is the most important of all reactions. It limits a plagioclase present from a plagioclase absent region of the pT space (Fig. 9.10). If all three minerals are present (quartz, plagioclase and sodic pyroxene), the equilibrium conditions of reaction (70) can be used for pressure estimates (if the temperature can be estimated from a thermometer, e.g., Fe-Mg distribution between garnet and omphacite). If quartz is not present, a maximum pressure can be estimated from the pyroxene-feldspar pair. If feldspar is not present (the normal case in eclogites), the omphacite + quartz assemblage can be used to estimate a minimum pressure for eclogite formation by assuming unit activity for the phase component albite in the equilibrium constant expression for reaction (70). Other important reactions of the amphibolite (or granulite) to eclogite transition zone must consume calcic amphibole or pyroxene. An example is reaction (75) that destroys tremolite and anorthite component (hornblende + plagioclase = amphibolite) and produces garnet + pyroxene. Reactions (76), (77) and (78) are examples of anorthite-consuming reactions that are relevant for the transition from the granulite to the HT eclogite facies.

As a result of the reactions that produce **medium-temperature eclogites (from amphibolites)** the typical mineralogy found in meta-MORB is: **garnet + omphacite** ± zoisite (clinozoisite) ± phengite ± kyanite ± amphibole (Na-Ca amphibole) ± quartz (at very high pressure coesite) ± rutile. **High-temperature eclogites (from granulites) contain typically: garnet + omphacite** ± kyanite ± orthopyroxene ± amphibole ± quartz (at very high pressure coesite) ± rutile.

9.8.2.2 Reactions in Eclogites

Prograde metamorphic mineral reactions that affect eclogites are basically ordinary dehydration reactions. The reactions transform assemblages involving hydrous minerals of LT eclogites and substitute them with less hydrous assemblages of MT eclogites and ultimately HT eclogites. Examples are the paragonite breakdown reaction (79), the zoisite breakdown (80), and the replacement of talc + kyanite by orthopyroxene [reaction (81)]. Other reactions gradually remove chlorite from quartz-free MT eclogites, some of them producing OPX. Reaction (80) replaces zoisite and garnet and produces kyanite + CPX. This continuous reaction relates MT- to HT eclogites. The reaction ultimately leads to the change from chemography #18 to #19 shown in Fig. 9.3.

Other important reactions in HT eclogites are (82) and (83). The last reaction is important in the relatively rare OPX-bearing eclogites that, for example, occur in the Western Gneiss Region of the Scandinavian Caledonides. Phengite is the only K-bearing phase in LT eclogites. At higher temperatures the mica continuously decomposes under production of K-amphibole component. The K-bearing Na-Ca-amphiboles found in high-grade eclogites are very stable and remain the prime K-carrier in HT eclogites.

9.8.3 Eclogite Facies in Non-Basaltic Mafic Rocks

Mafic or basic rocks cover a very wide range in bulk composition. Typical eclogites with garnet and omphacite as the dominant minerals usually develop from basaltic protoliths. The above discussion of eclogites refers to MORB-type mafic rocks. However, many gabbros are compositionally dissimilar to MORB. In particular, gabbros are often (but not always, e.g. ferro-gabbros) much more Mg-rich compared with MORB. It can be expected that high-pressure equivalents of troctolites and olivine-gabbros may develop mineral assemblages that are quite different from normal eclogite facies garnet + omphacite rocks. In fact, Mg-rich olivine gabbro metamorphosed at 25 kbar and 650°C may not contain garnet + omphacite at all as a peak assemblage. The low X_{Fe} of the protolith prevents the formation of garnet. An example of LT eclogite metamorphism of an olivine gabbro from the Alps will be presented in Section 9.8.2.1. Such mafic rocks have been metamorphosed under eclogite facies conditions, however, they are not eclogites. We also recommend to avoid confusing rock names like pelitic eclogites or pelitic blueschists. The proper expressions for such kind of rocks are metapelites in blueschist facies or pelitic gneisses in eclogite facies (see Sect. 7.7).

Anorthosites, also basic rocks, cannot be converted into eclogites by high pressure metamorphism because the rocks contain more than 90–95% plagioclase. Plagioclase of the composition An_{50} will be transformed into a jadeite + zoisite + kyanite + quartz assemblage in the LT- and MT eclogite facies.

9.8.3.1 Blueschist and Eclogite Facies Metamorphism of an Olivine Gabbro, a Case History from the Central Alps

The 2-km gabbro lens forms the Allalinhorn between Zermatt and Saas, it is referred to as the Allalin gabbro in the literature (Bearth 1967; Chinner and Dixon 1973; Meyer 1983). The gabbro belongs to a typical ophiolite complex that once formed the oceanic lithosphere of the Mesozoic Thetys ocean. The ophiolites make up a part of the South Pennine nappes of the Alps. During the early Alpine orogeny, the ophiolites were subducted to great depth and metamorphosed under LT eclogite facies conditions. Pillow basalts were turned first into ordinary LT eclogites, later retrogressed to blueschists, and finally overprinted by the Tertiary metamorphism that has reached the greenschist facies in the Zermatt area. Abundant eclogite garnet represents the only relic mineral from the eclogite stage in locally strongly retrogressed eclogites. The garnet-bearing greenschists of the Zermatt-Saas ophiolites demonstrate the very refractory nature of garnet.

The Allalin gabbro is an Mg-rich olivine gabbro with irregular layers of troctolite. The succession of minerals that formed in the gabbro through its entire pT history is given in Fig. 9.11. The data shown in Fig. 9.11 are based on Meyer (1983). The pT diagram in the upper left corner of Fig. 9.11 shows the path followed by the gabbro from the magmatic crystallization, via the LT

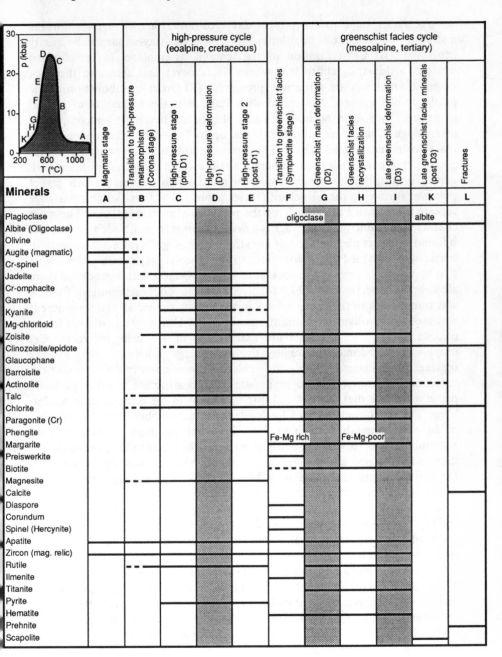

Fig. 9.11. Minerals in the eclogite facies Allalin gabbro, Swiss Alps. (Meyer 1983)

eclogite facies stage to the present-day erosion surface. The metamorphic history of the gabbro can be subdivided into various stages that will be briefly explained. (A) Crystallization of the igneous assemblage. In undeformed sections within the gabbro lens where water never had access to the rock magmatic minerals are often well preserved. (B) Onset of subduction of the partly cooled gabbro. Associated with stage B is the formation of typical coronites. They form because olivine + plagioclase become an incompatible assemblage [Chap. 5, reaction (15)]. The up to 4-cm-large olivine megacrysts react with matrix plagioclase. The reaction products, orthopyroxene and garnet, form concentric monomineralic shells around the primary olivine with garnet forming the outer shell toward the matrix. This remains the only garnet that ever formed in the LT eclogite facies gabbro. The Opx-Grt coronas were repeatedly modified afterwards by the progress of metamorphism. The most characteristic mineral of this stage is jadeite. It forms from albite component. It is found in matrix plagioclase only as extremely small grains along fractures and cracks that provided access for water to the igneous plagioclase grains. Note that there is no quartz in the rock at any stage. Quartz that is produced from albite breakdown [reaction (70)] dissolved in the aqueous metamorphic fluid. It was transported to the reaction sites in the nearby corona structures where it was used up by olivine-consuming reactions. (C) During the first stage of the eclogite facies, the characteristic assemblage of LT eclogites form; Cr-omphacite pseudomorphs augite, the assemblage jadeite + zoisite + kyanite replaces plagioclase; kyanite + talc + chloritoid pseudomorph olivine and the coronas. The coarse-grained magmatic gabbro structure is often perfectly preserved. Note that Mg-rich chlorite is present in the eclogite-stage assemblage. The 650 °C and 25 kbar LT eclogite facies metagabbro contains a number of hydrate minerals (Chl, Cld, Tlc, Zo) and the high-pressure mineral assemblage owes its very existence to the access of water to the gabbro during the subduction process. (D) Access of water is made possible by deformation D1 (brittle fractures healed with Mg-chloritoid and euhedral omphacite). Deformation D1 induces, locally, a complete recrystallization of the rock but usually leaves the magmatic gabbro structure intact (micro-fractures and cracks). (E) The high-pressure stage 2 is characterized by the formation of pure glaucophane, Cr-paragonite and phengite. The assemblage is characteristic of the blueschist facies. Note, however, that the blueschist facies here represents a retrogressive overprint on an LT eclogite facies assemblage. Its extent depends, again, on the availability of water and it is irregularly developed. (F) A transitional stage is associated with the return to shallower levels of the crust. It is characterized by fine-grained reaction products of local hydration such as margarite, preiswerkite and the Na-Ca-amphibole barroisite; but also spinel, corundum and diaspore are formed at this stage. The barroisitic amphibole links the blueschist facies glaucophane to the greenschist facies actinolite. Barroisite is typical and widespread in the retro-eclogites of the Zermatt area. The F-stage assemblages are often found in symplectites that pseudomorph and replace eclogite stage minerals. Note also that plagioclase (oligoclase) returned to the rock. (G, H and I) The Tertiary greenschist facies overprint is accompanied by two distinct phases of deformation. The impact of these stages on the Allalin gabbro varies from place to place and ranges from virtually nothing to complete retrogression. The greenschist facies deformation and hydration is often localized along shear zones and deformation is ductile in contrast to the eclogite-stage deformation that has been brittle. The typical assemblage of the greenschist stage is albite + zoisite (epidote) + actinolite + chlorite + biotite. The assemblage is characteristic for conditions around 4–5 kbar and 400–450 °C; K, L) The minerals of stages of the subgreenschist facies are mainly found in veins (albite veins, calcite, prehnite in veins).

The Allalin gabbro also contains sulfides, oxides and other accessory minerals that formed and were stable at various stages of the pT history of the rock. This example of a complex reaction history and successive development of minerals and mineral assemblages in the Allalin gabbro shows that:

- Deformation is essential in metamorphic processes. It allows for water to enter dry igneous rocks. In the absence of water, igneous and metamorphic assemblages will survive under even extreme pT conditions (e.g. magmatic Pl + Aug + Ol at 25 kbar and 650 °C).
- Mafic rocks that undergo eclogite facies metamorphism do not necessarily become eclogites. The bulk composition of many gabbros may not be favorable for the formation of garnet.
- LT eclogite facies mafic rocks that form by subduction of oceanic lithosphere, are characterized by the presence of a wide variety of hydrous minerals at peak-pressure conditions of the eclogite facies (Zo, Chl, Tlc, Cld).
- LT eclogite facies rocks may pass through the blueschist facies on their return path to the surface. The blueschist facies overprint is then retrograde in nature and adds another selection of hydrous minerals to the high-pressure low-temperature rock (Pg, Phe, Gln).
- All stages (A through L) of the complex pT path followed by the Allalin gabbro and the reaction history involving a large number of minerals can be preserved in a single rock body of not more than about 2 km in extent. The predominant assemblage that is found in a single piece of rock or within a small area of a given outcrop depends on how much water entered the rock at a given pT stage.
- The availability of water is in turn related to the local extent of deformation. Given a certain stage, if much water is available, the formation of the maximum hydrated assemblage for that stage is possible. Water is consumed by hydration reactions that tend to dry out the rock. Water must be continuously supplied in order to keep the reactions running that produce the maximum hydrated assemblage at that stage. The access of water to the originally dry rock is provided by micro-fractures and other brittle deformation during the high-pressure metamorphism and by pervasive ductile deformation during greenschist facies metamorphism.

References

Abbott RN Jr (1982) A petrogenetic grid for medium and high grade metabasites. Am Mineral 67:865–876

Andersen T, Austrheim H, Burke EAJ, Elvevold S (1993) N_2 and CO_2 in deep crustal fluids: evidence from the Caledonides of Norway. Chem Geol 108:113–132

Banno S (1986) The high pressure metamorphic belts in Japan: a review. Geol Soc Am Mem 164:365–374

Bearth P (1963) Chloritoid und Paragonit aus der Ophiolith-Zone von Zermatt-Saas Fee. Schweiz Mineral Petrogr Mitt 43:269–286

Bearth P (1967) Die Ophiolithe der Zone von Zermatt-Saas Fee. Beiträge zur geologischen Karte der Schweiz, v NF 132, Schweizerische Geologische Kommission, Basel. 130 pp

Bégin NJ (1992) Contrasting mineral isograd sequences in metabasites of the Cape Smith Belt, northern Québec, Canada: three new bathograds for mafic rocks. J Metamorph Geol 10; 5:685–704

Behrmann JH, Ratschbacher L (1989) Archimedes revisited: a structural test of eclogite emplacement models in the Austrian Alps. Terra Nova 1, 3:242–252

Bohlen SB, Liotta JJ (1986) A barometer for garnet amphibolites and garnet granulites. J Petrol 27:1025–1034

Boles JR, Coombs DS (1975) Mineral reactions in zeolitic Triassic tuff, Hokonui Hills, New Zealand. Geol Soc Am Bull 86:163–173

Brown EH (1974) Comparison of the mineralogy and phase relations of blueschists from the North Cascades, Washington and greenschists from Otago, New Zealand. Geol Soc Am Bull 85:333–344

Brown EH (1977) The crossite content of Ca-amphibole as a guide to pressure of metamorphism. J Petrol 18:53–72

Bryhni I, Green DH, Heier KS (1970) On the occurrence of eclogite in western Norway. Contrib Mineral Petrol 26:12–19

Carmichael RS (1989) Practical handbook of physical properties of rocks and minerals. CRC Press, Boca Raton

Chinner GA, Dixon JE (1973) Some high-pressure parageneses of the Allalin gabbro, Valais, Switzerland. J Petrol 14:185–202

Coleman RG, Lee DE, Beatty LB, Brannock WW (1965) Eclogites and eclogites: their differences and similarities. Geol Soc Am Bull 76:483–508

Coombs DS, Ellis AJ, Fyfe WS, Taylor AM (1959) The zeolite facies, with comments on the interpretation of hydrothermal syntheses. Geochim Cosmochim Acta 17:53–107

Cooper AF (1972) Progressive metamorphism of metabasic rocks from the Haast Schist group of southern New Zealand. J Petrol 13:457–492

Crawford ML (1966) Composition of plagioclase and associated minerals in some schists from Vermont, USA and South Westland, New Zealand, with inferences about the peristerite solvus. Contrib Mineral Petrol 13:269–294

Dal Piaz GV, Ernst WG (1978) Areal geology and petrology of eclogites and associated metabasites of the Piemonte Ophiolithe Nappe, Breuil-St.Jacques area, Italian Western Alps. Tectonophysics 5:199–126

Donato MM (1989) Metamorphism of an ophiolitic tectonic melange, northern California Klamath Mountains, USA. J Metamorph Geol 7; 5:515–528

Ernst WG (1972) Occurence and mineralogical evolution of blueschist belts with time. Am J Sci 272:657–668

Eskola P (1921) On the eclogites of Norway. Skr Vidensk Selsk Christiania, Mat.-nat. Kl. I, 8:1–118.

Evans BW (1990) Phase relations of epidote-blueschists. Lithos 25:3–23

Evans BW, Brown EH (1986) Blueschists and eclogites. Geol Soc Am Mem vol 164 423 pp

Ferry JM (1984) Phase composition as a measure of reaction progress and an experimental model for the high-temperature metamorphism of mafic igneous rocks. Am Mineral 69:677–691

Forbes RB, Evans BW, Thurston SP (1984) Regional progressive high-pressure metamorphism, Seward peninsula, Alaska. J Metamorph Geol 2:43–54

Frey M, De Capitani C, Liou JG (1991) A new petrogenetic grid for low- grade metabasites. J Metamorph Geol 9:497–509

Frimmel HE, Hartnady CJH (1992) Blue amphiboles and their significance for the history of the Pan-African Gariep belt, Namibia. J Metamorph Geol 10; 5:651–670

Griffin WL (1987) "On the eclogites of Norway" – 65 years later. Mineral Mag 51:333–343

Griffin WL, Austrheim H, Brastad K, Bryhni I, Krill A, MØrk, M B E, Qvale H, TØrudbakken B (1983) High-pressure metamorphism in the Scandinavian Caledonides. In: Gee DG, Sturt BA (eds) The Caledonide orogen – Scandinavia and related areas. Wiley, Chichester

Grove TL, Ferry JM, Spear FS (1983) Phase transitions and decomposition relations in calcic plagioclase. Am Mineral 68:41–59

Grove TL, Ferry JM, Spear FS (1986) Phase transitions in calcic plagioclase: a correction and further discussion. Am Mineral 71; 7–8:1049–1050

Harte B, Graham CM (1975) The graphical analysis of greenschist to amphibolite facies mineral assemblages in metabasites. J Petrol 16:347–370

Helms TS, McSween HYJ, Labotka TC, Jarosewich E (1987) Petrology of a Georgia blue ridge amphibolite unit with hornblende + gedrite + kyanite + staurolite. Am Mineral 72:1086–1096

Hirajima T, Hiroi Y, Ohta Y (1984) Lawsonite and pumpellyite from the Vestgotabreen Formation in Spitsbergen. Nor Geol Tidsskr 4:267–273

Hirajima T, Banno S, Hiroi Y, Ohta Y (1988) Phase petrology of eclogites and related rocks from the Motalafjella high-pressure metamorphic complex in Spitsbergen (Arctic Ocean) and its significance. Lithos 22:75–97

Holland TJB (1979a) Experimental determination of the reaction paragonite = jadeite + kyanite + H_2O, and internally consistent thermodynamic data for part of the system Na_2O-Al_2O_3-SiO_2-H_2O, with applications to eclogites and blueschists. Contrib Mineral Petrol 68:293–301

Holland TJB (1979b) High water activities in the generation of high pressure kyanite eclogites of the Tauern window, Austria. J Geol 87:1–27

Holland TJB, Richardson SW (1979) Amphibole zonation in metabasites as a guide to the evolution of metamorphic conditions. Contrib Mineral Petrology 70:143–148

Hosotani H, Banno S (1986) Amphibole composition as an indicator of subtle grade variation in epidote-glaucophane schists. J Metamorph Geol 4:23–36

Humphris SE, Thompson G (1978) Hydrothermal alteration of oceanic basalts by seawater. Geochim Cosmochim Acta 42:107–125

James HL (1955) Zones of regional metamorphism in the Precambrian of northern Michigan. Geol Soc Am Bull 66:1455–1488

Johnson CD, Carlson WD (1990) The origin of olivine-plagioclase coronas in metagabbros from the Adirondack Mountains, New York. J Metamorph Geol 8:697–717

Kawachi DE (1975) Pumpellyite-actinolite and contiguous facies metamorphism in the Upper Wakatipu district, southern New Zealand. N Z J Geol Geophys 17:169–208

Kienast JR, Lombardo B, Biino G, Pinardon JL (1991) Petrology of very-high-pressure eclogitic rocks from the Brossacasco-Isasca Complex, Dora-Maira Massif, Italian Western Alps. J Metamorph Geol 9:19–34

Lardeaux JM, Spalla MI (1991) From granulites to eclogites in the Sesia zone (Italian Western Alps): a record of the opening and closure of the Piedmont ocean. J Metamorph Geol 9:35–60

Leake BE (1964) The chemical distinction between ortho- and para-amphibolite. J Petrol 5:238–254

Liou JG, Maruyama S, Cho M (1987) Very low-grade metamorphism of volcanic and volcaniclastic rocks – mineral assemblages and mineral facies. In: Frey M (ed) Low temperature metamorphism. Blackie, Glasgow, pp 59–113.

Lucchetti G, Cabella R, Cortesogno L (1990) Pumpellyites and coexisting minerals in different low-grade metamorphic facies of Liguria, Italy. J Metamorph Geol 8:539–550

Matthews A, Schliestedt M (1984) Evolution of the blueschist and greenschist facies rocks of Sifnos, Cyclades Greece. A stable isotope study of subduction-related metamorphism. Contrib Mineral Petrol 88:150–168

Meyer J (1983a) The development of the high-pressure metamorphism in the Allalin metagabbro (Switzerland). Terra Cognita 3; p 187

Meyer J (1983b) Mineralogie und Petrologie des Allalingabbros. Doc Dissertation, University of Basel, Basel, 329 pp

Miyashiro A (1979) Metamorphism and metamorphic belts. Allen and Unwin, Winchester, 492 pp

Mongkoltip P, Ashworth JR (1986) Amphibolitization of metagabbros in the Scottish Highlands. J Metamorph Geol 4:261–283

Newton RC (1986) Metamorphic temperatures and pressures of Group B and C eclogites. Geol Soc Am Mem 164:17–30

O'Brien PJ (1993) Partially retrograded eclogites of the Münchberg Massif, Germany: records of a multistage Variscan uplift history in the Bohemian Massif. J Metamorph Geol 11; 2:241–260

Patrick BE, Day HW (1989) Controls on the first appearance of jadeitic pyroxene, northern Diablo Range, California. J Metamorph Geol 7:629–639

Platt JP (1987) The uplift of high-pressure-low-temperature metamorphic rocks.: Philos Trans R Soc Lond A321:87–102

Platt JP (1993) Exhumation of high-pressure rocks: A review of concepts and processes. Terra Nova 5:119–133

Pognante U (1991) Petrological constraints on the eclogite- and blueschist-facies metamorphism and p-T-t paths in the Western Alps. J Metamorph Geol 9:5–18

Pognante U, Kienast J-R (1987) Blueschist and eclogite transformations in Fe-Ti gabbros: a case from the western Alps ophiolites. J Petrol 28:271–292

Powell WG, Carmichael DM, Hodgson CJ (1993) Thermobarometry in a sub-greenschist to greenschist transition in metabasites of the Abitibi greenstone belt, Superior Province, Canada. J Metamorph Geol 11; 1:165–178

Ridley J (1984) Evidence of a temperature-dependent "blueschist" to "eclogite" transformation in high-pressure metamorphism of metabasic rocks. J Petrol 25:852–870

Russ-Nabelek C (1989) Isochemical contact metamorphism of mafic schist, Laramie Anorthosite Complex, Wyoming: Amphibole compositions and reactions. Am Mineral 74:530–548

Seki Y, Oki Y, Matsuda T, Mikami K, Okumura K (1969) Metamorphism in the Tanzawa Mountains, central Japan. J Jpn Assoc Mineral Petrol Econ Geol 61:1–29, 50–75

Selverstone J, Spear FS (1985) Metamorphic P-T paths from pelitic schists and greenstones from the south-west Tauern Window, eastern Alps. J Metamorph Geol 3:439–465

Smelik EA, Veblen DR (1989) A five-amphibole assemblage from blueschists in northern Vermont. Am Mineral 74:960–964

Smith DC, Lappin MA (1989) Coesite in the Straumen kyanite-eclogite pod, Norway. Terra Nova 1:47–56

Soto JI (1993) PTMAFIC: Software for thermobarometry and activity calculations with mafic and ultramafic assemblages. Am Mineral 78:840–844

Spear FS (1980) NaSi ↔ CaAl exchange equilibrium between plagioclase and amphibole. An empirical model. Contrib Mineral Petrol 72:33–41

Spear FS (1981) An experimental study of hornblende stability and compositional variability in amphibolite. Am J Sci 281:697–734

Springer RK, Day HW, Beiersdorfer RE (1992) Prenite-pumpellyite to greenschist facies transition, Smartville Complex, near Auburn, California. J Metamorph Geol 10; 2:147–170

Starkey RJ, Frost BR (1990) Low-grade metamorphism of the Karmutsen Volcanics, Vancouver Island, British Columbia. J Petrol 31:167–195

Thieblemont D, Triboulet C, Godard G (1988) Mineralogy, petrology and P-T-t path of Ca-Na amphibole assemblages, Saint-Martin des Noyers formation, Vendee, France. J Metamorph Geol 6:697–716

Thompson AB (1970) Laumonite equilibria and the zeolite facies. Am J Sci 269:267–275

Thompson AB (1971a) Analcite-albite equilibria at low temperatures. Am J Sci 271:79–92

Thompson AB (1971b) P_{CO2} in low grade metamorphism: zeolite, carbonate, clay mineral, prehnite relations in the system $CaO-Al_2O_3-CO_2-H_2O$. Contrib Mineral Petrol 33:145–161

Thompson JB Jr (1991) Modal space: applications to ultramafic and mafic rocks. Can Mineral 29; 4:615–632

Tomasson J, Kristmansdottir H (1972) High temperature alteration minerals and the thermal brines, Reykjanes, Iceland. Contrib Mineral Petrol 36:123–134

Triboulet C, Thiéblemont D, Audren C (1992) The (Na-Ca) amphibole- albite- chlorite- epidote- quartz geothermobarometer in the system S- A- F- M- C- N- H_2O. 2. Applications to metabasic rocks in different metamorphic settings. J Metamorph Geol 10; 4:557–566

Warburton J (1986) The ophiolite-bearing schistes lustrés nappe in alpine Corsica: a model for the emplacement of ophiolites that have suffered HP/LT metamorphism. Geol Soc Am Mem 164:313–331

Wenk E, Keller F (1969) Isograde in Amphibolitserien der Zentralalpen. Schweiz Mineral Petrogr Mitt 49:157–198

Wenk E, Schwander H, Wenk HR (1991) Microprobe analyses of plagioclases from metamorphic carbonate rocks of Central Alps. Eur J Mineral 3:181–191

Will TM, Powell R, Holland T, Guiraud M (1990) Calculated greenschist facies mineral equilibria in the CaO- FeO- MgO- Al_2O_3- SiO_2- CO_2- H_2O. Contrib Mineral Petrol 104:353–368

Zen E (961) The zeolite facies: an interpretation. Am J Sci 259:401–409

10 Metamorphism of Granitoids

10.1 Introduction

Granitoid rocks comprise granites, alkali-feldspar granites, granodiorites and tonalites, and constitute a large portion of the continental crust. Because the main constituents – alkali-feldspar, plagioclase, quartz, biotite, muscovite, hornblende – are found over a wide range of P-T conditions, this rock group is not a very useful indicator of metamorphic grade and is therefore largely neglected in textbooks on metamorphic petrology. Unlike wet sedimentary rocks, granitoid rocks will enter the metamorphic realm in a predominantly dry state. In order to start metamorphic reactions, some hydration is necessary. The access of a water-rich fluid will be facilitated by tectonic activity. Also, in the absence of penetrative deformation, granitoid rocks retain remarkably well their original igneous structures.

In the following a few selected mineralogical features are mentioned which are useful to determine metamorphic grade, but no attempt is made to treat the progressive metamorphism of granitoids.

10.2 Presence of Prehnite and Pumpellyite

Although prehnite and pumpellyite are commonly known from very low-grade mafic igneous rocks and greywackes, these Ca-Al silicates are sometimes also found in quartzo-feldspathic plutonic rocks (see Tulloch 1979, for references; see also Al Dahan 1989). Both minerals typically occur within biotite, secondary chlorite and hornblende, often forming lenticular or bulbous bodies in them. In the case of the biotite-prehnite association, most writers conclude that the biotite cleavage simply forms a suitable structural site for prehnite crystallizing from a fluid, but prehnite actually replacing its biotite host has also been observed, according to the possible reaction: biotite + anorthite (component in plagioclase) + H_2O = prehnite + chlorite + K-feldspar + titanite + muscovite (Tulloch 1979). The presence of prehnite and pumpellyite in granitoid rocks was earlier attributed to the activity of deuteric solutions, but alteration under prehnite-pumpellyite facies conditions is now commonly proposed if signs of low-temperature metamorphism are also observed in other nearby lithologies.

10.3 Quartz Textures

Voll (1976) studied quartz textures in Variscan granites of the Central Alps which underwent progressive Alpine orogenic metamorphism. The first recrystallized quartz grains are observed at about 300°C. Proceding upgrade, the volume of recrystallized grains steadily increases from 7×10^{-6} mm^3 at ca. 300°C to 0.5 mm^3 at ca. 550°C. Samples collected from a uniform granite and separated by only 300 m often showed distinguishable median grain sizes by measuring 300 recrystallized quartz grains per sample. From these observations it may be possible to map three-dimensional boundary curves of equal grain size distribution. Note that this increase in grain size can reliably be observed only from pure quartz grain aggregates.

10.4 Presence of Stilpnomelane

Stilpnomelane is a Fe-rich hydrous silicate with low Al and K contents which may be mistaken for biotite. The names ferrostilpnomelane and ferristilpnomelane are used respectively for Fe^{2+}- and Fe^{3+}-rich varieties, the latter being formed by secondary oxidation of the former. In metagranitoids, stilpnomelane generally has a sheaf-like habit and is preferentially found within microcline or is replacing igneous biotite and hornblende. Stilpnomelane-forming reactions in low-grade metagranitoids are poorly understood, but biotite and/or chlorite are considered to be the most important reactants. Stilpnomelane occurs as a mineral of low-grade metamorphism along with chlorite, phengite, epidote, quartz, microcline and albite, but at slightly higher grade also together with green biotite. A possible stilpnomelane-consuming reaction in low-grade metagranitoids is

$$Stp + Phe = Bt + Chl + Qtz + H_2O.$$

This reaction has been provisionally investigated by Nitsch (1970). In hydrothermal experiments oxygen fugacity was controlled by the hematite-magnetite buffer, and apparent equilibrium was obtained at 4 kbar/445°C and at 7 kbar/460°C. Summing up, stilpnomelane is a diagnostic index mineral for metagranitoid rocks formed under greenschist and blueschist facies conditions.

10.5 The Microcline/Sanidine Transformation Isograd

A discontinuity in the structural state of K-feldspar allows the mapping of a microcline/sanidine transformation isograd in granitoid rocks of the greenschist facies (Bambauer and Bernotat 1982; Bernotat and Bambauer 1982; Bernotat and Morteani 1982; Bambauer 1984).

To facilitate the understanding of the following sections, a few terms on the mineralogy of K-feldspars are mentioned here. In the stable high-temperature form sanidine, Al and Si are distributed onto two tetrahedral (T) sites, T_1 and T_2, with Al slightly preferring the T_1 site. In the low-temperature form microcline, there are four T sites, T_1O, T_1m, T_2O, T_2m, with Al concentrating onto the T_1O site. Because there exist intermediate structural states, the following K-feldspar modifications are distinguished with increasing temperature: low microcline, high microcline, low sanidine and high sanidine. For low and high microcline, the Al, Si distributions are given as Al site occupancies t_1O, t_1m, t_2O, t_2m. The Al, Si ordering in T_1 tetrahedral sites is determined either by measuring the optic axial angle, $2V_x$, or by X-ray methods.

The most detailed study on the microcline/sanidine transformation isograd has been performed in the Central Swiss Alps. This area will, therefore, serve as a case study here. The structural state of K-feldspar occurring in granitoid rocks distinguishes three zones, arranged according to increasing Alpine metamorphic grade:

Zone 1 comprises unchanged, variable pre-Alpine structural states ranging from low to high microcline, showing a range of $2V_x \sim 55$–85. Metamorphic grade is restricted to the lowest greenschist facies.

Zone 2 comprises dintinctly cross-hatched low microcline ($2V_x \sim 75$–$88°$, t_1O–$t_1m \sim 1.0$, 1–4 mol% Ab). Metamorphic grade covers lower greenschist facies conditions.

Zone 3 comprises frequent orthoclase (monoclinic bulk optics predominating, triclinic or monoclinic by X-rays) displaying a variety of high microcline structural states ($2V_x \sim 53$–$75°$, t_1O – $t_1m = 0.0$–0.4, 2.5–6.5 mol% Ab) together with variable amounts of cross-hatched low microcline (t_1O – $t_1m \sim 1.0$). Metamorphic grade ranges from high greenshist facies to high amphibolite facies conditions.

The microcline/sanidine transformation isograd separating zone 2 from zone 3 is interpreted as a relic of the diffusive transformation sanidine $\Rightarrow$ microcline at the climax of the late Alpine metamorphism in the following way. K-feldspars from zone 2 were annealed long enough at low temperature during Alpine metamorphism to form low microcline. In zone 3, temperatures surpassed T_{diff} upon slow heating and K-feldspars were transformed into the high-temperature phase sanidine. During the following cooling period, temperatures dropped below the temperature of the phase transformation. Note that during the retrograde sanidine $\Rightarrow$ microcline transition, triclinic domains must form within a monoclinic host, resulting in a high kinetic barrier. Because of rapid cooling rates, the diffusion was slowed down and the K-feldspar remained as intermediate to high microcline. The low microcline observed at the beginning of zone 3 (Fig. 10.1) is interpreted as a relic of zone 2.

The temperature of the diffuse phase transformation between microcline and sanidine has not been determined experimentally because of the very slow kinetics of Al, Si ordering. From several lines of evidence, discussed in detail by Bambauer and Bernotat (1982), this temperature was estimated to be $\sim 450°C$

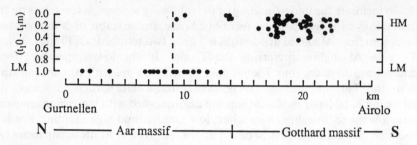

Fig. 10.1. The microcline/sanidine transformation isograd along a traverse of the Central Swiss Alps as determined by the degree of Al, Si order. The K-feldspar discontinuity is indicated by the first appearance of high microcline. *HM* High microcline; *LM* low microline. (After Bambauer and Bernotat 1982, Fig. 7)

for an approximate composition $Or_{95-90}Ab_{5-10}$, with a negligible pressure dependence.

In the Central Swiss Alps, the microcline/sanidine transformation isograd was mapped over a distance of more than 140 km. It fits well into the general pattern of metamorphic zonation of this area, and is found between the stilpnomelane-out and staurolite-in mineral zone boundaries. Furthermore, provisional results indicate that the transformation isograd surface is inclined ~15° towards the north.

10.6 Granitoid Rocks in the Eclogite Facies and the Blueschist Facies

A rare example of granite, granodiorite and tonalite metamorphosed under eclogite facies conditions is known from the Sesia Zone of the Western Alps (Compagnoni and Maffeo 1973; Oberhänsli et al. 1985). These quartzo-feldspathic rocks are remnants of the Variscan basement and were transformed during the earliest phase of Alpine metamorphism in mid-Cretaceous time. The least deformed rocks retain well-preserved igneous textures with relict K-feldspar and biotite. During a first stage, these remarkable rocks have been converted to the assemblage: quartz-jadeite-zoisite-garnet-(3T) phengite-K-feldspar. Jadeite, quartz and zoisite have completely replaced plagioclase, whilst biotite has been partly transformed into coronitic garnet, phengite and rutile. During the later decompression stages, jadeite was replaced by ompha-cite and chloromelanite, followed by glaucophane. The P-T conditions for the climax of metamorphic recrystallization are estimated as > 14 kbar and 500–600°C.

Orthogneisses of granite, granodiorite and tonalite composition from the Seward Peninsula, Alaska, underwent progressive orogenic blueschist facies metamorphism in the Jurassic, followed by partial re-equilibration during decompression under greenschist facies conditions (Evans and Patrick 1987). The mineral assemblage in granitic orthogneiss is quartz-albite-microcline-(3T)phengite-biotite; almandine-rich garnet occurs in metatonalite in the absence of microcline. No typical high-pressure minerals such as jadeite, omphacite or glaucophane, or obvious pseudomorphs thereof, were found in the orthogneisses. Diagnostic blueschist facies paragenesis occur, however, in adjacent metabasites and metapelites, and P-T conditions were approximately 9–10 kbar, 420 °C (Thurston 1985).

The Variscan granitic basement of Corsica was partly involved in early Alpine metamorphism during the middle to late Cretaceous (Gibbons and Horak 1984). The metamorphism of this basement, induced by the overthrusting of a blueschist facies nappe, was confined to a major, ca. 1-km-thick ductile shear zone within which deformation increases upwards towards the overlying nappe. Within a mylonitized biotite-hornblende granodiorite, the mafic minerals are replaced by crossite + low-Ti biotite + phengite + titanite ± epidote, recording metamorphic conditions transitional between blueschist and greenschist facies. P-T estimates for the metamorphism at the base of the shear zone are given as 390–490 °C at 6–9 kbar, but these figures are not well constrained.

References and Further Reading

Al Dahan AA (1989) The paragenesis of pumpellyite in granitic rocks from the Siljan area, central Sweden. Neues Jahrb Mineral Monatsh 367–383

Bambauer HU (1984) Das Einfallen der Mikroklin/Sanidin-Isogradenfläche in den Schweizer Zentral-Alpen. Schweiz Mineral Petrogr Mitt 64:288–289

Bambauer HU, Bernotat WH (1982) The microcline/sanidine transformation isograd in metamorphic regions. I. Composition and structural state of alkali feldspars from granitoid rocks of two N-S traverses across the Aar massif and Gotthard "massif", Swiss Alps. Schweiz Mineral Petrogr Mitt 62:185–230

Bearth P (1958) Über einen Wechsel der Mineralfazies in der Wurzelzone des Penninikums. Schweiz Mineral Petrogr Mitt 38:363–373

Bernotat WH, Bambauer HU (1982) The microcline/sanidine transformation isograd in metamorphic regions. II. The region of Lepontine metamorphism, central Swiss Alps. Schweiz Mineral Petrogr Mitt 62:231–244

Bernotat WH, Morteani G (1982) The microcline/sanidine transformation isograd in metamorphic regions: Western Tauern window and Merano-Mules-Anterselva complex (eastern Alps) Am Mineral 67:43–53

Biino GG, Compagnoni R (1992) Very-high pressure metamorphism of the Brossasco coronite metagranite, southern Dora Maira Massif, Western Alps. Schweiz Mineral Petrogr Mitt 72:347–363

Compagnoni R, Maffeo B (1973) Jadeite-bearing metagranites l.s. and related rocks in the Mount Mucrone area (Sesia-Lanzo zone, western Italian Alps). Schweiz Mineral Petrogr Mitt 53:355–378

Droop GTR (1982) A clinopyroxene paragenesis of albite-epidote-amphibolite facies in metasyenites from the South-East Tauern Window, Austria. J Petrol 23:163–185

Evans BW, Patrick BE (1987) Phengite-3T in high-pressure metamorphosed granitic orthogneisses, Seward Peninsula, Alaska. Can Mineral 25:141–158

Ferry JM (1979) Reaction mechanism, physical conditions, and mass transfer during hydrothermal alteration of mica and feldspar in granitic rocks from south-central Maine, USA. Contrib Mineral Petrol 68:125-139

Frey M, Hunziker JC, Jäger E, Stern WB (1983) Regional distribution of white K-mica polymorphs and their phengite content in the Central Alps. Contrib Mineral Petrol 83:185-197

Gibbons W, Horak J (1984) Alpine metamorphism of Hercynian hornblende granodiorite beneath the blueschist facies schistes lustrés nappe of NE Corsica. J Metamorph Geol 2:95-113

Le Goff E, Ballèvre M (1990) Geothermobarometry in albite-garnet orthogeneisses: A case study from the Gran Paradiso nappe (Western Alps). Lithos 25:261-280

Morteani G, Raase P (1974) Metamorphic plagioclase crystallization and zones of equal anorthite content in epidote-bearing, amphibole-free rocks of the western Tauernfenster, eastern Alps. Lithos 7:101-111

Nitsch KH (1970) Experimentelle Bestimmung der oberen Stabilitätsgrenze von Stilpnomelan. Fortschr Mineral 47, Beih 1:48-49

Oberhänsli R, Hunziker JC, Martinotti G, Stern WB (1985) Geochemistry, geochronology and petrology of Monte Mucrone: an example of eo-alpine eclogitization of Permian granitoids in the Sesia-Lanzo zone, western Alps, Italy. Chem Geol Isotope Geosci Sect 52:165-184

Steck A (1976) Albit-Oligoklas-Mineralgesellschaften der Peristeritlücke aus alpinmetamorphen Granitgneisen des Gotthardmassivs. Schweiz Mineral Petrogr Mitt 56:269-292

Steck A, Burri G (1971) Chemismus und Paragenesen von Granaten aus Granitgneisen der Grünschiefer- und Amphibolitfazies der Zentralalpen. Schweiz Mineral Petrogr Mitt 51:534-538

Thurston SP (1985) Structure, petrology, and metamorphic history of the Nome Group blueschist terrane, Salmon Lake area, Seward Peninsula, Alaska. Geol Soc Am Bull 96:600-617

Tulloch AJ (1979) Secondary Ca-Al silicates as low-grade alteration products of granitoid biotite. Contrib Mineral Petrol 69:105-117

Voll G (1976) Recrystallisation of quartz, biotite and feldspars from Erstfeld to the Leventina nappe, Swiss Alps, and its geological significance. Schweiz Mineral Petrogr Mitt 56:641-647

Wimmenauer W, Stenger R (1989) Acid and intermediate HP metamorphic rocks in the Schwarzwald (Federal Republic of Germany). Tectonophysics 157:109-116

Appendix: Symbols for Rock-Forming Minerals

(Extension of Kretz 1983)

Ab	albite	Chn	chondrodite
Acm	acmite	Chr	chromite
Act	actinolite	Chu	clinohumite
Adr	andradite	Cld	chloritoid
Agt	aegirine–augite	Cls	celestite
Ak	åkermanite	Cp	carpholite
Alm	almandine	Cpx	Ca clinopyroxene
Aln	allanite	Crd	cordierite
Als	aluminosilicate	Crn	carnegieite
Am	amphibole	Crn	corundum
An	anorthite	Crs	cristroballite
And	andalusite	Cs	coesite
Anh	anhydrite	Cst	cassiterite
Ank	ankerite	Ctl	chrysotile
Anl	analcite	Cum	cummingtonite
Ann	annite	Cv	covellite
Ant	anatase	Czo	clinozoisite
Ap	apatite	Dg	diginite
Apo	apophyllite	Di	diopside
Apy	arsenopyrite	Dia	diamond
Arf	arfvedsonite	Dol	dolomite
Arg	aragonite	Drv	dravite
Atg	antigorite	Dsp	diaspore
Ath	anthophyllite	Eck	eckermannite
Aug	augite	Ed	edenite
Ax	axinite	Elb	elbaite
Bhm	boehmite	En	enstatite (ortho)
Bn	bornite	Ep	epidote
Brc	brucite	Fa	fayalite
Brk	brookite	Fac	ferroactinolite
Brl	beryl	Fcp	ferrocarpholite
Brt	barite	Fed	ferroedenite
Bst	bustamite	Flt	fluorite
Bt	biotite	Fo	forsterite
Cal	calcite	Fpa	ferropargasite
Cam	Ca clinoamphibole	Fs	ferrosilite (ortho)
Cbz	chabazite	Fst	fassite
Cc	chalcocite	Fts	ferrotschermakite
Ccl	chrysocolla	Gbs	gibbsite
Ccn	cancrinite	Ged	gedrite
Ccp	chalcopyrite	Gh	gehlenite
Cel	celadonite	Gln	glaucophane
Cen	clinoenstatite	Glt	glauconite
Cfs	clinoferrosilite	Gn	galena
Chl	chlorite	Gp	gypsum

Gr	graphite	Pen	protoenstatite
Grs	grossular	Per	periclase
Grt	garnet	Pg	paragonite
Gru	grunerite	Pgt	pigeonite
Gt	goethite	Phe	phengite
Hbl	hornblende	Phl	phlogopite
Hc	hercynite	Pl	plagioclase
Hd	hedenbergite	Pmp	pumpellyite
Hem	hematite	Pn	pentlandite
Hl	halite	Po	pyrrhotite
Hs	hastingsite	Prg	pargasite
Hu	humite	Prh	prehnite
Hul	heulandite	Prl	pyrophyllite
Hyn	haüyne	Prp	pyrope
Ill	illite	Prv	perovskite
Ilm	ilmenite	Py	pyrite
Jd	jadeite	Qtz	quartz
Jh	johannsenite	Rbk	riebeckite
Kfs	K–feldspar	Rdn	rhodonite
Kln	kaolinite	Rds	rhodochrosite
Kls	kalsilite	Rt	rutile
Krn	kornerupine	Sa	sanidine
Krs	kaersutite	Scp	scapolite
Ktp	kataphorite	Sd	siderite
Ky	kyanite	Sdl	sodalite
Lct	leucite	Ser	sericite
Lm	limonite	Sil	sillimanite
Lmt	laumontite	Sp	sphalerite
Lo	loellingite	Spd	spodumene
Lpd	lepidolite	Spl	spinel
Lws	lawsonite	Spr	sapphirine
Lz	lizardite	Sps	spessartine
Mag	magnetite	Srl	schorl
Mcp	magnesiocarpholite	Srp	serpentine
Mel	melilite	St	staurolite
Mgh	maghemite	Stb	stilbite
Mgs	magnesite	Stp	stilpnomelane
Mkt	magnesiokatophorite	Str	strontianite
Mnt	montmorillonite	Sud	sudoite
Mnz	monazite	Tlc	talc
Mo	molybdenite	Tmp	thompsonite
Mrb	magnesioriebeckite	Toz	topaz
Mrg	margarite	Tr	tremolite
Ms	muscovite	Trd	tridymite
Mtc	monticellite	Tro	troilite
Mul	mullite	Ts	tschermakite
Ne	nepheline	Ttn	titanite
Nrb	norbergite	Tur	tourmaline
Nsn	nosean	Usp	ulvöspinel
Ntr	natrolite	Ves	vesuvianite
Oam	orthoamphibole	Vrm	vermiculite
Ol	olivine	Wa	wairakite
Omp	omphacite	Wo	wollastonite
Opx	orthopyroxene	Wth	witherite
Or	orthoclase	Wus	wüstite
Osm	osumilite	Zo	zoisite
Pct	pectolite	Zrn	zircon

Index

Springer-Verlag
and the Environment

$\mathbf{W}$e at Springer-Verlag firmly believe that an international science publisher has a special obligation to the environment, and our corporate policies consistently reflect this conviction.

$\mathbf{W}$e also expect our business partners – paper mills, printers, packaging manufacturers, etc. – to commit themselves to using environmentally friendly materials and production processes.

$\mathbf{T}$he paper in this book is made from low- or no-chlorine pulp and is acid free, in conformance with international standards for paper permanency.

707 09

Printing: Saladruck, Berlin
Binding: Buchbinderei Lüderitz & Bauer, Berlin